Tumor-Induced Immune Suppression

Dmitry I. Gabrilovich · Arthur A. Hurwitz
Editors

Tumor-Induced Immune Suppression

Mechanisms and Therapeutic Reversal

Editors
Dmitry I. Gabrilovich
H. Lee Moffitt Cancer Center
University of South Florida
Tampa, FL 33612, USA
dmitry.gabrilovich@moffitt.org

Arthur A. Hurwitz
Laboratory of Molecular Immunoregulation
National Cancer Institute
Frederick, MD 21071, USA
hurwitza@ncifcrf.gov

ISBN: 978-0-387-69117-6   e-ISBN: 978-0-387-69118-3

Library of Congress Control Number: 2007941316

© 2008 Springer Science+Business Media, LLC
All rights reserved. This work may not be translated or copied in whole or in part without the written permission of the publisher (Springer Science+Business Media, LLC., 233 Spring Street, New York, NY10013, USA), except for brief excerpts in connection with reviews or scholarly analysis. Use in connection with any form of information storage and retrieval, electronic adaptation, computer software, or by similar or dissimilar methodology now known or hereafter developed is forbidden.
The use in this publication of trade names, trademarks, service marks, and similar terms, even if they are not identified as such, is not to be taken as an expression of opinion as to whether or not they are subject to proprietary rights.

Printed on acid-free paper

9 8 7 6 5 4 3 2 1

springer.com

Contents

Immune-Suppressive Mechanisms and Cancer: Understanding the Implications, Paradoxes, and Burning Questions 1
Arthur A. Hurwitz and Dmitry I. Gabrilovich

Mechanisms of Tumor-Associated T-Cell Tolerance 7
Adam J. Adler

Contribution of B7-H1/PD-1 Co-inhibitory Pathway to T-Cell Dysfunction in Cancer 29
Sheng Yao and Lieping Chen

Regulatory T Cells in Cancer 41
Silvia Piconese and Mario P. Colombo

Cancer-Induced Signaling Defects in Antitumor T Cells 69
Alan B. Frey

Immunobiology of Dendritic Cells in Cancer 101
Michael R. Shurin and Gurkamal S. Chatta

Macrophages and Tumor Development 131
Suzanne Ostrand-Rosenberg and Pratima Sinha

Myeloid-Derived Suppressor Cells in Cancer 157
Paolo Serafini and Vincenzo Bronte

Signaling Pathways in Antigen-Presenting Cells Involved in the Induction of Antigen-Specific T-Cell Tolerance 197
Ildefonso Vicente-Suarez, Alejandro Villagra, and Eduardo M. Sotomayor

Arginine Availability Regulates T-Cell Function in Cancer 219
Paulo C. Rodríguez and Augusto C. Ochoa

Protein–Glycan Interactions in the Regulation of Immune Cell Function in Cancer: Lessons from the Study of Galectins-1 and -3 235
Gabriel A. Rabinovich and Fu-Tong Liu

Role of Reactive Oxygen Species in T-Cell Defects in Cancer 259
Alex Corzo, Srinivas Nagaraj, and Dmitry I. Gabrilovich

Tumor Stroma and the Antitumor Immune Response 281
Bin Zhang, Donald A. Rowley, and Hans Schreiber

Subject Index 295

Contributors

Adam J. Adler
Center for Immunotherapy of Cancer and Infectious Diseases and Department of Immunology, University of Connecticut Health Center, Farmington, CT 06030-1601, aadler@up.uchc.edu

Vincenzo Bronte
Istituto Oncologico Veneto, Via Gattamelata 64, 35128 Padua, Italy, enzo.bronte@unipd.it

Gurkamal S. Chatta
Department of Medicine, University of Pittsburgh Medical Center and University of Pittsburgh Cancer Institute, Pittsburgh, PA

Lieping Chen
Johns Hopkins Medicine, 209 David H. Koch Cancer Research Building, Baltimore, MD 21231, lchen42@jhmi.edu

Mario P. Colombo
Immunotherapy and Gene Therapy Unit, Department of Experimental Oncology, Fondazione IRCCS Istituto Nazionale dei Tumori, Milan, Italy, mario.colombo@istitutotumori.mi.it

Alex Corzo
H. Lee Moffitt Cancer Center and Research Institute and University of South Florida, Tampa, FL, 33612

Alan B. Frey
Department of Cell Biology- MSB623, New York University, School of Medicine, 550 First Avenue, New York, NY 10016, freya01@med.nyu.edu, alanbfrey@yahoo.com

Dmitry I. Gabrilovich
H. Lee Moffitt Cancer Center, University of South Florida, Tampa, FL 33612, dmitry.gabrilovich@moffitt.org

Arthur A. Hurwitz
Laboratory of Molecular Immunoregulation, National Cancer Institute, Frederick, MD 21071, hurwitza@ncifcrf.gov

Fu-Tong Liu
Department of Dermatology, School of Medicine, University of California, Davis, Sacramento, CA

Srinivas Nagaraj
H. Lee Moffitt Cancer Center and Research Institute and University of South Florida, Tampa, FL, 33612

Augusto C. Ochoa
Stanley S. Scott Cancer Center and Department of Pediatrics, Louisiana State University Health Sciences Center, 533 Bolivar Street, New Orleans, LA 70112

Suzanne Ostrand-Rosenberg
Department of Biological Sciences, University of Maryland, 1000 Hilltop Circle, Baltimore, MD 21250, srosenbe@umbc.edu

Silvia Piconese
Immunotherapy and Gene Therapy Unit, Department of Experimental Oncology, Fondazione IRCCS Istituto Nazionale dei, Tumori, Milan, Italy

Gabriel A. Rabinovich
Instituto de Biología y Medicina Experimental, Consejo, Nacional de Investigaciones Científicas y Técnicas de Argentina, Vuelta de Obligado 2490, Buenos Aires, C1428ADN, Argentina, gabyrabi@ciudad.com.ar

Paulo C. Rodríguez
Stanley S. Scott Cancer Center, Louisiana State University Health Sciences Center, 533 Bolivar Street, New Orleans, LA 70112

Donald A. Rowley
The Department of Pathology and the Committee on Immunology, University of Chicago, Chicago, IL

Hans Schreiber
Department of Pathology and Committee on Immunology, The University of Chicago, 5841 S. Maryland, MC3083, Chicago, IL 60637, hszz@uchicago.edu

Paolo Serafini
Department of Microbiology & Immunology, Dodson Interdisciplinary Immunotherapy Institute, University of Miami, School of Medicine, Miami, FL

Michael R. Shurin
Clinical Immunopathology, 5725 CHP-MT, 200 Lothrop Street, Pittsburgh, PA 15213, shurinmr@upmc.edu

Contributors

Pratima Sinha
Department of Biological Sciences, University of Maryland Baltimore County, Baltimore, MD 21250

Eduardo M. Sotomayor
Division of Immunology and Division of Malignant Hematology, Department of Interdisciplinary Oncology, H. Lee Moffitt Cancer Center & Research Institute at the University of South Florida, Tampa, FL 33612, sotomed@moffitt.usf.edu

Ildefonso Vicente-Suarez
Division of Immunology and Division of Malignant Hematology, Department of Interdisciplinary Oncology, H. Lee Moffitt Cancer Center & Research Institute at the University of South Florida, Tampa, FL 33612

Alejandro Villagra
Division of Immunology and Division of Malignant Hematology, Department of Interdisciplinary Oncology, H. Lee Moffitt Cancer Center & Research Institute at the University of South Florida, Tampa, FL 33612

Sheng Yao
Department of Dermatology, Oncology and the Sidney Kimmel Comprehensive Cancer Center, Johns Hopkins University School of Medicine, Baltimore, MD 21231

Bin Zhang
The Department of Pathology and the Committee on Immunology, University of Chicago, Chicago, IL

Immune-Suppressive Mechanisms and Cancer: Understanding the Implications, Paradoxes, and Burning Questions

Arthur A. Hurwitz and Dmitry I. Gabrilovich

Since Paul Ehrlich's 1909 prediction that the immune system is capable of suppressing the growth of tumors, a large volume of evidence produced by the work of many investigators has demonstrated the existence of a natural immune protection against cancer. As a tumor develops, it acquires novel epitopes as a result of mutations in self-proteins, frame shifts, or protein splicing identified in some tumor cells. Many tumors acquire an anaplastic or de-differentiated histologic phenotype, losing tissue differentiation antigens and acquiring expression of embryonic or "cancer-testis" antigens. In addition, changes in glycosylation or levels of expression may also change the antigenic repertoire of tumor cells. Finally, virally transformed cells may harbor strongly immunogenic viral antigens.

As a whole, these changes in antigenicity of tumor cells may permit the adaptive immune system to recognize a tumor as "foreign", despite the fact that tumors arise from "normal" self-tissues, against which tolerance is maintained. All these data justify the concept of immunosurveillance of tumors, which proposes that as mutations that lead to transformation occur, the immune system can detect these changes as "foreign" and eliminate the "invader". Recently, this concept has evolved into the concept of "immunoediting", which postulates that as a tumor develops, the immune system can shape the repertoire of a tumor's inherent immunogenicity.

It is now clear that tumors can be recognized and eliminated by the host immune system. However, this idea raises two main questions that have confronted researchers and physicians for many years: why the immune system does not always prevent tumor progression, and how to manipulate the immune system to achieve tumor eradication. The last 20 years have brought a clear realization that one of the major mechanisms of tumor escape that limits the clinical success of cancer immunotherapy is the inadequate function of the host immune system in the context of a developing tumor. During recent years, there has been an explosion of information about the potential immunosuppressive strategies employed by tumor cells.

A.A. Hurwitz
Laboratory of Molecular Immunoregulation, National Cancer Institute, Frederick, Maryland, USA

Intensive studies from many different groups have resulted in the discovery of numerous cellular mechanisms of immune suppression in cancer. With the identification of T-cell priming pathways, it became clear that aberrant T-cell activation can lead to non-responsiveness or anergy. T-cell receptor ligation in the absence of costimulatory signals is generally recognized as a potent way to anergize T cells. Thus, tumor cells that express MHC but lack costimulatory ligands, or immature APCs that cross-present tumor antigens, may be capable of tolerizing tumor-reactive T cells. Developing tumors can also induce production of a variety of suppressive cells. They include regulatory T cells, B cells, myeloid-derived suppressive cells, and different types of macrophages and dendritic cells. These cells suppress T-cell responses via suppressive surface molecules like CTLA-4, PDL-1, PDL-2, galectins, etc., production of inhibitory cytokines like IL-10, TGF-β, VEGF, etc., depletion of T cells of tryptophan and arginine, release of reactive oxygen species and nitric oxide and many others (many of which are discussed in this monograph).

Like suppressor cells, tumors can also express factors that create a suppressive environment. Tumors can express catabolic enzymes like indoleamine dioxygenase or arginase. Tumors have also been demonstrated to express ligands to inhibitory receptors on T cells and expression of these ligands has an inverse correlation to survival, suggesting that tumors use these receptors to evade immune recognition.

Discovery of this multitude of different immune-suppressive factors helped to develop new experimental and clinical methods to improve the immune response in cancer and the effect of cancer vaccines. However, these discoveries also raise several fundamental questions that need to be addressed in order to understand fully the biology of antitumor immunity and effective approaches to its use in therapeutic settings.

1. *Specific vs. non-specific suppression in cancer.* Most of the suppressive mechanisms that have been demonstrated in cancer and described in this monograph do not require the presence of tumor-specific antigens for their negative effect on T cells. The paradox is that despite the apparent presence of a large number of potent immune-suppressive factors, neither tumor-bearing mice nor cancer patients are profoundly immune compromised. Even at a relatively advanced stage of cancer, the host immune system retains the ability to respond to stimulation with viral and bacterial antigens or lectins. At the same time, tumor-specific immune response is repressed. The question arises that if those multiple suppressive mechanisms are truly operational, why is more profound immune deficiency not observed in tumor-bearing hosts? This paradox is currently not resolved. Currently, it appears that the understanding of the mechanisms of tumor escape requires identification of the precise role of tumor-specific immune tolerance vis-à-vis non-specific immune suppression. It is possible that the role of multiple immunosuppressive mechanisms in cancer is exaggerated due to the nature of experimental models employed. However, another explanation is much more likely. It relates to the phenomenon of compartmentalization of immune suppression in cancer. There is certainly a need for development of more experimental models that more closely reflect the "real" situation present in cancer patients. Such models might allow more accurate

characterization of multiple immunosuppressive mechanisms active at the same time.

2. *Spatial characteristics of immune suppression in cancer.* Immune suppression in cancer is not a universal process. It has become increasingly clear that the nature of immune suppression in peripheral lymphoid organs and inside the tumor site is different. Available data may suggest that in peripheral lymphoid organs, tumor-specific T-cell tolerance is more likely to be responsible for tumor escape than non-specific immune suppression. T cells retain their ability to respond to other stimuli. In contrast, tumor microenvironment creates a milieu that inhibits any type of immune reactivity and this immune suppression is not antigen-specific. Multiple studies demonstrated that tumor-infiltrating T lymphocytes are profoundly suppressed. Their function could be recovered only if they are cultured ex vivo in the presence of appropriate cytokines and effective stimulation. However, it is still unclear whether T cells are rendered non-responsive inside a tumor or if they migrate to the tumor site, having already been tolerized in peripheral lymphoid organs. This question is especially important for the attempts to use adoptive transfer of previously activated, antigen-specific T cells. Although adoptive immunotherapy holds promise, the local immunosuppressive environment of the tumor may hamper those attempts. It is very important to establish whether immune suppression at the tumor site is indeed able to block the antitumor effect of adoptively transferred T cells and to determine therapeutic approaches to tilt the balance toward effector T cells. There are no clear answers to these questions. However, the overview of current data presented in this monograph may help to develop them in the future.

3. *Strategies to target negative regulatory pathways.* The fact that tumors develop and progress is a good indication that immune surveillance of cancer is not completely efficient. Successes of cancer vaccines at this time are not impressive. The failure of antitumor immune responses is presumably the consequence of the environment of a large network of tumor-associated immune-suppressive factors. This makes targeting of this network very attractive for the goal of improvement of overall antitumor reactivity.

How best to target immune-suppressive regulatory pathways remains unclear. Negative regulatory mechanisms discussed in detail in this monograph are also essential in preventing excessive immune responses to foreign antigens and autoimmune abnormalities. It is logical that the elimination of these factors will result in the activation of the immune system. The question is whether this activation alone will be sufficient. The potential problem is that the removal of negative "brakes" would result in an accumulation of T cells reactive to any available antigens. Most of the viral and bacterial antigens are much stronger immunogens than the self-antigen present in tumors. The proportion of tumor-specific T cells among this pool of reactive T cells could be quite small. They can still be easily detected since investigators are specifically looking for these cells. However, whether they are sufficient to prevent tumor progression is not apparent.

In addition, antitumor effects will most likely be associated with autoimmune abnormalities. The more effective the antitumor response generated by a potent therapy, the more severe the potential side effects that could be developed. Often, successful anti-melanoma responses are associated with autoimmune vitiligo, where the immune system destroys melanocytes as well as melanoma cells. However, some therapies give rise to more system autoimmune sequellae. It was reported that some of those side effects could be alleviated by corticosteroids (Ribas et al., 2005). However, it is not clear how this may affect the clinical efficacy of the treatment. Current clinical studies will undoubtedly help to address these questions. However, accumulated data presented in this monograph strongly argue in favor of a direct combination of immunostimulatory therapy with targeting immune-suppressive pathways. Partial removal of suppressive mechanisms in the presence of tumor-specific T cells may dramatically enhance their antitumor effect. A number of clinical trials testing this hypothesis have been initiated in recent years. The results of these trials will undoubtedly help to shape future therapeutic strategies.

4. *Combination of immunotherapy and other therapeutic modalities in cancer as a future of cancer treatment.* Another approach to cancer therapy has emerged in recent years. It employs conventional chemotherapy in direct combination with immunotherapy. This approach seems to be counterintuitive since it is well established that potent cancer chemotherapy blunts the immune responses. However, this perception was recently challenged by unexpected results from several clinical trials demonstrating substantial clinical benefits when immunotherapy was immediately followed by chemotherapy (Antonia et al., 2006; Arlen et al., 2006; Gribben et al., 2005; Wheeler et al., 2004). These data, in combination with the results of pre-clinical studies (Emens and Jaffee, 2005), suggest a synergistic effect of immunotherapy and chemotherapy. One of the potential mechanisms of this synergistic effect could be the elimination of immune-suppressive factors by chemotherapy. Chemotherapy is known to be able to deplete regulatory T cells, myeloid-derived suppressor cells, as well as tumor-associated macrophages. Eventually, CTL responses are also ablated by chemotherapy. However, apparently the effect of chemotherapy on tumor microenvironment precedes the effect on CTL, which may explain the clinical benefits of this approach. In addition, chemotherapy may disrupt tumor stroma, which would improve CTL penetration into tumor parenchyma. As discussed in this monograph, it is also possible that chemotherapy can help load stromal cells with tumor-associated antigens and thus help to facilitate antitumor immune responses. This field is at an early phase of development now and more studies are needed to clarify the mechanisms of this phenomenon.

The data accumulated in recent years provide strong indication that targeting immune-suppressive mechanisms in combination with induction of antitumor immune responses may profoundly enhance the effect of cancer immunotherapy. We have become more sophisticated in our understanding of the mechanisms of immune suppression in cancer and in developing new approaches to targeting those mechanisms. This monograph presents the "state of the art" in our understanding of the mechanisms of suppression of tumor immunity. By presenting a comprehensive understanding of how these suppressive mechanisms reduce the ability to elicit potent tumor immunity, we hope to stimulate the study of more powerful and presumably synergistic approaches to treating cancer.

References

Antonia, S. J., Mirza, N., Fricke, I., Chiappori, A., Thompson, P., Williams, N., Bepler, G., Simon, G., Janssen, W., Lee, J. H., Menander, K., Chada, S., and Gabrilovich, D. I. (2006). Combination of p53 cancer vaccine with chemotherapy in patients with extensive stage small cell lung cancer. *Clin Cancer Res* 12:878–887.

Arlen, P. M., Gulley, J. L., Parker, C., Skarupa, L., Pazdur, M., Panicali, D., Beetham, P., Tsang, K. Y., Grosenbach, D. W., Feldman, J., Steinberg, S. M., Jones, E., Chen, C., Marte, J., Schlom, J., and Dahut, W. (2006). A randomized phase II study of concurrent docetaxel plus vaccine versus vaccine alone in metastatic androgen-independent prostate cancer. *Clin Cancer Res* 12:1260–1269.

Emens, L. A., and Jaffee, E. M. (2005). Leveraging the activity of tumor vaccines with cytotoxic chemotherapy. *Cancer Res* 65:8059–8064.

Gribben, J. G., Ryan, D. P., Boyajian, R., Urban, R. G., Hedley, M. L., Beach, K., Nealon, P., Matulonis, U., Campos, S., Gilligan, T. D., Richardson, P. G., Marshall, B., Neuberg, D., and Nadler, L. M. (2005). Unexpected association between induction of immunity to the universal tumor antigen CYP1B1 and response to next therapy. *Clin Cancer Res* 11:4430–4436.

Ribas, A., Camacho, L. H., Lopez-Berestein, G., Pavlov, D., Bulanhagui, C. A., Millham, R., Comin-Anduix, B., Reuben, J. M., Seja, E., Parker, C. A., Sharma, A., Glaspy, J. A., and Gomez-Navarro, J. (2005). Antitumor activity in melanoma and anti-self responses in a phase I trial with the anti-cytotoxic T lymphocyte-associated antigen 4 monoclonal antibody CP-675,206. *J Clin Oncol* 23:8968–8977.

Wheeler, C. J., Das, A., Liu, G., Yu, J. S., and Black, K. L. (2004). Clinical responsiveness of glioblastoma multiforme to chemotherapy after vaccination. *Clin Cancer Res* 10:5316–5326.

Mechanisms of Tumor-Associated T-Cell Tolerance

Adam J. Adler

1 Introduction

The great challenge in the treatment of cancer has been to develop modalities that destroy tumor cells without damaging healthy tissues. In fact, modalities such as chemotherapeutics that are standardly used to treat a wide variety of cancers work on the principle that tumor cells are slightly more sensitive to their cytotoxic effects than are healthy cells, and thus treatment regimens are administered that may or may not fully eradicate the cancer (depending upon the outgrowth of drug-resistant tumor cells) but generally inflict significant side effects on the patient. In this regard, there has been a long-standing interest in programming the adaptive immune system to mediate anti-tumor immunity through the targeting of antigens expressed specifically by tumors. This effort has been accelerated during recent years by advances in the ability to prime robust cytotoxic T-lymphocyte responses. Nevertheless, results from recent clinical trials testing a variety of T cell-based immunotherapeutic approaches have only demonstrated partial successes (Rosenberg et al., 2004; Srivastava, 2006). This is likely to be at least partially due to the ability of tumors to dampen cognate T-cell responses.

Ironically, the first evidence demonstrating that tumors can suppress cognate T-cell responses came from the same studies establishing that tumors can elicit T-cell responses. Thus, mice harboring established carcinogen-induced transplantable tumors can reject a second transplant of the same tumor, and T cells harvested from mice with established tumors can confer protection against tumor growth when transferred into naive syngeneic mice that are simultaneously challenged with the same tumor. This phenomenon of concomitant immunity (reviewed in Gorelik, 1983) thus indicated that while tumors can possess immunogenic properties that allow them to prime cognate T-cell responses, they can simultaneously suppress the function of these effector T cells when they enter the tumor

---

A.J. Adler
Center for Immunotherapy of Cancer and Infectious Diseases and Department of Immunology, University of Connecticut Health Center, Farmington, CT 06030-1601, USA
e-mail: aadler@up.uchc.edu

microenvironment. Although initial murine studies suggested that concomitant immunity was more likely to occur with high-dose carcinogen-induced tumors compared to spontaneously arising tumors (Gorelik, 1983), the subsequent observation that T cells with tumor specificity commonly infiltrate certain human tumors such as melanoma (Topalian et al., 1989) suggested that naturally arising tumors can also elicit cognate T-cell responses while simultaneously inhibiting T-cell effector functions in the tumor microenvironment. Understanding how the tumor microenvironment is able to locally suppress the function of tumor-infiltrating tumor-reactive effector T cells has been the subject of intense study and will be reviewed in detail in several of the accompanying chapters.

The immunogenic properties of certain tumors may be related to their potential to generate inflammation when they invade surrounding tissue or metastasize (Pardoll, 2003). Conversely, other tumors might not elicit inflammation either because they are able to grow and spread without causing tissue damage (e.g., hematopoietic tumors) or because they express activities that minimize inflammation when they do cause tissue destruction (Wang et al., 2004). Overall, the potential of tumors to grow while eliciting minimal inflammation would be consistent with the potential to induce immunological tolerance (Pardoll, 2003). For the purpose of this discussion, *tolerance will be defined as an impaired ability of antigen-specific T cells to respond to antigenic challenge at the systemic level*, as opposed to the above-mentioned immunosuppressive effects that impair T-cell effector function locally in the tumor microenvironment. Evidence from numerous models indicates that T-cell tolerance to tumor-associated antigens can occur, and that this tolerance can negatively impact tumor immunity.

Ultimately, the development of effective T cell-based strategies to treat cancers that have a propensity to induce T-cell tolerance will likely require a component to prevent or reverse tolerance to tumor-associated antigens, which will be facilitated through a detailed understanding of the cellular and molecular mechanisms that regulate tolerance.

2 Tumors Can Tolerize Cognate T cells

Since many human and mouse tumor antigens are expressed on both tumors and the normal tissues from which they derive (i.e., differentiation antigens, e.g., tyrosinase (Wolfel et al., 1994), TRP2 (Wang et al., 1996) and Pmel-17/gp100 (Cox et al., 1994)), it is likely that the pathways which tolerize the T-cell repertoire to tissue-specific self-antigens in order to avoid autoimmunity also negatively impact the ability of these same T-cell specificities to mediate tumor immunity. To model the impact of pre-existing T-cell tolerance to differentiation antigens on tumor vaccine efficacy, Hu et al. developed a transgenic mouse model in which the Friend murine leukemia virus envelope protein (env) was expressed under the control of a lymphoid-specific promoter. Env-specific T cells were tolerant in these animals as demonstrated by their failure to expand following vaccination with an env-expressing recombinant vaccinia virus, and this tolerance was associated with a failure of the vaccine to protect against subsequent challenge with an env-expressing transplantable erythroleukemia (Hu et al., 1993). While this result illustrates that pre-existing tolerance to tumor-associated differentiation antigens can severely dampen tumor vaccine efficacy, tolerance is probably not always absolute. For instance, when a transgenic tumor differentiation antigen is expressed on normal tissues in a more restricted fashion, naive CD8 cells expressing T-cell receptors (TCRs) with low avidity for the tumor differentiation epitope escape tolerization and can be primed through vaccination to mediate tumor immunity (Morgan et al., 1998). The possibility that tumor differentiation antigen-specific T cells that can be primed may tend to express low-avidity TCRs might represent one facet explaining why tumor vaccines are sometimes only able to elicit partially effective tumor immunity.

The finding that T-cell tolerance to tumor-associated differentiation antigens exists and can negatively impact the efficacy of tumor vaccines targeting these antigens is not particularly surprising given that tolerance induction through both central and peripheral mechanisms will have presumably been operative long before the initiation of tumorigenesis. It might therefore seem reasonable that tolerance would be less apparent for tumor-specific antigens such as those deriving from oncogenic viruses or mutated self-antigens given that they would in all probability not be present in the thymus to facilitate negative selection of cognate developing T cells and would not be accessible to the peripheral tolerance-inducing machinery prior to tumorigenesis. Nevertheless, numerous studies have indicated that T-cell tolerance can develop rapidly toward tumor-specific antigens. When Bogen and colleagues transplanted a plasmacytoma into transgenic mice expressing a TCR specific for a class II-restricted peptide that derives from the hypervariable region of the idiotypic immunoglobulin expressed by that plasmacytoma, the idiotype-specific CD4 cells underwent deletion (Bogen, 1996). Given that bolus injection of soluble foreign antigens induces immunological tolerance (in contrast to particulate antigen or antigen admixed with adjuvant that induces immunity) (Chiller et al., 1971; Dresser, 1962), the potent tolerogenic nature of the tumor-specific antigen (i.e., idiotypic immunoglobulin) may have been related to its secretion into the blood stream at very high levels, a situation that would probably not be the case for most other tumor-specific antigens that are either expressed at lower levels or that remain cell-associated. To assess whether T-cell tolerance can develop toward less abundant non-secreted tumor-specific antigens, Levitsky and colleagues developed a model in which naive TCR-transgenic CD4 cells specific for the model antigen influenza hemagglutinin (HA) are adoptively transferred into mice bearing a transplantable B-cell lymphoma that expresses a low level of HA. Over several weeks, these naive HA-specific CD4 cells progressively lost the ability to both proliferate and secrete cytokines in response to subsequent in vitro or in vivo antigenic challenge (Stavely-O'Carroll et al., 1998).

Subsequent studies from various groups have confirmed that both CD4 and CD8 cell tolerance can develop toward antigens expressed on transplantable as well as spontaneously arising tumors (Doan et al., 2000; Drake et al., 2005; Lyman et al.,

2004; Schell et al., 2000; Shrikant et al., 1999). Tolerance does not develop in all tumor systems (Hanson et al., 2000; Nguyen et al., 2002; Ochsenbein et al., 2001; Spiotto et al., 2002), underscoring the notion that different types of tumors vary in their capacity to induce tolerance. As discussed in the introduction, those tumors that elicit cognate effector (rather than tolerogenic) T-cell responses must elaborate immunosuppressive mechanisms to inhibit the tumoricidal activity of the tumor-reactive effector T cells that have infiltrated into the tumor microenvironment. Given the dynamic nature of tumorigenesis (Lengauer et al., 1998), it might be possible that the capacity of a given tumor to either prime or tolerize cognate T cells might change during disease progression. Indirect support for this possibility stems from the observation that melanoma patients can exhibit clonally expanded populations of non-functional tumor-associated antigen-specific CD8 cells (Lee et al., 1999), consistent with a scenario in which these tumor-reactive T cells are initially primed to undergo expansion but subsequently inactivated.

3 Mechanisms of Peripheral Self-Antigen- and Tumor-Associated Antigen-Induced T-Cell Tolerance

Since tolerization of tumor antigen-specific T cells can restrict the repertoire of T-cell specificities that can be primed through vaccination, manipulations that can either block the development of and/or restore the function of tolerant tumor-reactive T cells could enhance tumor vaccine efficacy. In this regard, understanding the cellular and molecular pathways that mediate tolerance will be critical.

For tumor-associated differentiation antigens that are also expressed on normal tissues, T-cell tolerance should be mediated through the central and peripheral pathways that normally operate to prevent autoimmunity. Thus, the majority of self-reactive T cells undergo negative selection during development in the thymus, where immature T cells expressing high-avidity TCRs that recognize MHC-self-peptide complexes presented by thymic antigen-presenting cells (APCs) undergo apoptosis (Kappler et al., 1987; Kisielow et al., 1988; Sebzda et al., 1994; Surh and Sprent, 1994). Subsequently, mature T cells specific for parenchymal self-antigens that are not presented in the thymus can be subjected to a variety of peripheral tolerance mechanisms such as deletion (Jones et al., 1990), functional inactivation (also referred to as anergy; Schwartz, 2003) or suppression by regulatory T cells (Sakaguchi, 2000; Shevach, 2001).

It was initially thought that central tolerance functioned specifically to delete developing T cells with reactivity to self-antigens that were either ubiquitously expressed or that could gain access to the thymus via the circulation, while peripheral mechanisms performed the task of inactivating mature T cells specific for tissue-restricted self-antigens. More recent evidence, however, suggests a degree of overlap between central and peripheral tolerance. Expression of the transcription factor AIRE in thymic medullary epithelial cells (mTECs) induces low-level expression of a variety of tissue-restricted self-antigens that can mediate the deletion of developing cognate T cells (Anderson et al., 2002). Although AIRE extends the range of thymic tolerance, several lines of evidence strongly implicate that peripheral mechanisms are still essential for preventing autoimmunity. First, not all tissue-restricted self-antigens appear to be expressed in mTECs, and those that are expressed are generally present at low levels (Derbinski et al., 2005), suggesting that there is likely to be a high level of leakiness in this process. In fact, a substantial fraction of self-reactive T cells do escape thymic deletion (Bouneaud et al., 2000), and it is well established in a variety of inbred mouse strains and other species that self-reactive T cells in the periphery of normal individuals can be induced to mediate autoimmunity following vaccination with cognate auto-antigen plus adjuvant (von Budingen et al., 2001). The spontaneous development of autoimmunity in mice that either exhibit defective DC apoptosis (Chen et al., 2006) or lack negative regulators of peripheral T-cell responsiveness such as Foxp3, Cbl-b (Bachmaier et al., 2000), TGF-β (Gorelik and Flavell, 2000) and CTLA-4 (Tivol et al., 1995) provides additional evidence that peripheral tolerance is critical for preventing autoimmunity.

Tissue-restricted self-antigens expressed in mTECs include certain tumor-associated antigens (Bos et al., 2005), suggesting that central tolerance does impact tumor immunity. Nevertheless, the understanding and ability to manipulate peripheral tolerance will likely have a greater potential to increase the efficacy of T cell-based therapies to treat cancer. Thus, thymic deletion will have mostly occurred prior to clinical diagnosis and administration of therapy, and T-cell deletion cannot be reversed. In contrast, peripheral tolerance can involve mechanisms such as anergy/hypo-responsiveness that could potentially be reversed in the context of vaccination, and strategies that prevent the tolerization of adoptively transferred tumor-reactive effector T cells in the context of adoptive immunotherapy might also enhance anti-tumor immunity (as will be discussed shortly).

Since tumor-associated differentiation antigens exist as normal self-antigens prior to tumorigenesis, cognate T cells should be subject to normal tolerance mechanisms. Interestingly, mounting evidence suggests that these same mechanisms might also induce tumor-specific T-cell tolerance. The studies by Bogen and colleagues demonstrated that plasmacytomas can secrete sufficient levels of idiotypic antibody into the circulation to reach the thymus and induce the deletion of developing anti-idiotypic T cells (Bogen, 1996; Bogen et al., 1993). Since many other tumor-specific antigens derive from mutated self-proteins, these unique epitopes cannot be encoded in the genome of thymic APCs, and assuming that they are not released into the circulation at high levels, it is unlikely that cognate T cells will undergo thymic deletion. It does appear, however, that tumor-specific antigens can be processed by similar peripheral tolerization pathways as normal parenchymal self-antigens. As a corollary to the system described previously in which naive TCR-transgenic HA-specific CD4 cells become tolerant following adoptive transfer into mice harboring a transplantable tumor expressing HA (i.e., tumor-HA) (Stavely-O'Carroll et al., 1998), an analogous system was developed in which the same HA-specific CD4 cells are adoptively transferred into C3-HA transgenic mice that express HA in a wide variety of normal parenchymal tissues (i.e., self-HA) (Adler et al., 1998,

2000). In both the tumor-HA and self-HA models, the clonotypic CD4 cells initially display a surface marker phenotype indicative of activation, but ultimately develop a non-responsive phenotype similar to anergy (Schwartz, 2003) where they lose the ability to proliferate and secrete IL-2 following secondary exposure to antigen.

In addition to the similarity in the non-responsive phenotype of CD4 cells exposed to tumor-HA vs self-HA, tolerance in both cases was mediated through a similar antigen-processing pathway. Prior to the development of transgenic model systems to study peripheral T-cell tolerance (e.g., Kearney et al., 1994; Rocha and von Boehmer, 1991), in vitro tolerance studies using Th1 clones indicated that anergy is induced when TCR ligation occurs in the absence of costimulation (reviewed in Schwartz, 2003). This observation led to the notion that TCR engagement without costimulation leading to non-responsiveness/anergy might occur in vivo when T cells encounter their cognate antigens presented on either normal parenchyma or tumors (neither of which normally express costimulatory ligands). Additionally, even though B-cell lymphomas do express costimulatory ligands such as B7 (Stavely-O'Carroll et al., 1998), the overall level of costimulatory ligand expression is substantially less compared to dendritic cells (DC) which represent the most potent APC subset (Bannchereau and Steinman, 1998), and normal B cells which also express low levels of costimulatory ligands can induce T-cell tolerance in vivo (Eynon and Parker, 1992; Fuchs and Matzinger, 1992). Thus, it was somewhat surprising when bone marrow chimera studies revealed that CD4 cell tolerance to self-HA was not mediated through direct interaction between the HA-specific CD4 cells and HA-expressing parenchyma, but rather tolerogenic antigen presentation was mediated indirectly via bone marrow-derived APCs that had acquired parenchymal-HA (Adler et al., 1998). This indirect or cross-presentation pathway can also facilitate the peripheral tolerization of self-reactive CD8 cells (Kurts et al., 1997). Subsequent work has suggested that steady-state DC likely represent the predominant cross-tolerizing APC (Belz et al., 2002; Hagymasi et al., 2007; Kurts et al., 2001), although other APC populations also appear to cross-tolerize (Hagymasi et al., 2007). The ability of DC to prime both effector and tolerogenic T-cell responses appears to be regulated by the environment in which the antigen is acquired. Thus, when DC acquire pathogen-derived antigens, the presence of invariant pathogen-derived inflammatory mediators (i.e., pathogen-associated molecular patterns or PAMPs) induce high expression levels of costimulatory molecules and cytokines that endow DC with the ability to prime cognate naive T cells to develop effector and memory functions. In contrast, when DC acquire self-antigens under steady-state conditions, the absence of PAMPs results in a default expression level of sub-optimal costimulation that programs a tolerogenic T-cell differentiation program that can involve the induction of anergy generally followed by deletion (Finkelman et al., 1996; Hawiger et al., 2001; Janeway and medzhitov, 2002; Jenkins et al., 2001; Matzinger, 1994; Medzhitov, 2001).

Returning to the HA-expressing B-cell lymphoma model (Stavely-O'Carroll et al., 1998), given that the tumor appears to exhibit a tolerogenic sub-optimal costimulatory ligand expression profile and also that it metastasizes to lymphoid organs, it seemed reasonable to presume that tumor cells would directly present

HA to naive HA-specific CD4 cells to induce tolerance. Thus, it was notable that cross-presentation proved to be the predominant pathway of tolerance induction (Sotomayor et al., 2001). That peripherally tolerized self-reactive and tumor-reactive T cells can exhibit similar phenotypes that can be induced by the same indirect antigen presentation pathway suggests that the peripheral tolerance machinery that normally operates to prevent autoimmunity might also help tumors to evade immune-neutralization. The similarities between the tumor-HA and self-HA models do not necessarily exclude the possibility that there may be aspects of peripheral tolerance that are unique to tumors, but these similarities do suggest that a more detailed mechanistic understanding of peripheral tolerance to normal self-antigens will be relevant to understanding tolerance to tumor-specific antigens.

With regard to studying peripheral tolerance mechanisms that are common to both tumor and normal self-antigens, transgenic systems designed to examine the latter have certain advantages. For example, different founder lines generated using the same model antigen expression vector can express different levels of the model antigen due to differences in either the genomic location of transgene integration or the number of integrated transgene copies. This allows examination of the effect of antigen dose on T-cell tolerization without introducing other variables such as differences in tumor burden. Additionally, tumor antigen presentation (and hence cognate T-cell recognition and response) in systems where tumors are localized to discrete anatomical locations tends to be concentrated in tumor-draining lymph nodes (Drake et al., 2005; Marzo et al., 1999). While this restricted pattern of tumor antigen presentation is important to examine with regard to understanding T-cell tolerization induced by specific types of tumors, the disadvantage is that relatively few tolerized T cells can be recovered for functional and biochemical analyses. In contrast, transgenic model self-antigen expression systems can be engineered so that the model self-antigen is expressed in multiple tissues, resulting in tolerance induction occurring in multiple lymphoid organs, and hence the potential to recover larger numbers of tolerized T cells for analysis (Long et al., 2006).

Some of the initial model self-antigen TCR-transgenic adoptive transfer studies indicated that in vivo tolerance is more complex than had been predicted from in vitro models. Thus, in vitro TCR ligation of CD4 Th1 clones in the absence of costimulation results in a lack of proliferation as well as a rapid (less than 24 h) induction of anergy that is defined by the inability to produce IL-2 and proliferate in response to subsequent stimulation with antigen plus costimulation (Schwartz, 2003). In contrast, when naive TCR-transgenic clonotypic CD4 or CD8 cells are adoptively transferred into recipients expressing the cognate self-antigen they generally proliferate (as measured either by BrdU incorporation or CFSE dilution) for several days prior to becoming anergic and/or undergoing deletion (Kurts et al., 1997; Pape et al., 1998; Rocha and von Boehmer, 1991). It was subsequently observed that clonotypic T cells encountering cognate tumor-derived antigen can also proliferate prior to becoming tolerant (Anderson et al., 2007; Drake et al., 2005; Shrikant et al., 1999; Zhou et al., 2004). Interestingly, the kinetics of this initial proliferative response elicited by self-antigen that ultimately leads to tolerance can be comparable to that elicited by the same antigen when expressed within a recombinant viral vector that programs Th1 effector differentiation (Adler et al., 2000; Higgins et al., 2002b), indicating that the kinetics of initial proliferation per se does not dictate functional outcome, but rather the context in which the antigen is presented to the T cell may have a more critical role in determining T-cell fate. Because the theoretical expansion in clonotypic T-cell frequencies estimated by the average number of cell divisions far exceeded the actual T-cell expansions, these data also suggested that in vivo anergy may simply represent an intermediate step in the pathway that ultimately leads to deletion (Adler et al., 2000). Further supporting this notion, several studies that have defined deletion as the operative tolerance mechanism have also observed a residual population of T cells that exhibit an anergic phenotype (Rocha and von Boehmer, 1991; Webb et al., 1990).

That naive T cells encountering cognate self-antigen proliferate vigorously prior to becoming tolerant and that toleranT cells can maintain an anergic phenotype prior to deletion seem somewhat counterintuitive insofar as proliferation expends a significant amount of metabolic energy and anergic cells take up space within lymphoid organs. Thus, it is not clear why self-reactive T cells in the periphery do not simply apoptose without initially proliferating, as they do in the thymus. One possibility is that anergic cells might express an important regulatory function, and that proliferation is required for the development of this function. Consistent with this possibility, it has been observed in several peripheral tolerance systems (including when the tolerizing antigen is tumor-derived) that anergic CD4 cells do exhibit regulatory function (Apostolou and von Boehmer, 2004; Jooss et al., 2001; Zhou et al., 2006).

Peripheral T-cell tolerance was initially thought to act mainly on naive rather than effector T cells. Thus, although it had been shown in various autoimmunity models that effector T cells can be tolerized following exposure to large boluses of cognate exogenous soluble auto-antigen (reviewed in Liblau et al., 1997), it had generally been thought that effector T cells would not become tolerant under physiological conditions such as when cognate self-antigen might be expressed at relatively low levels. This notion derived largely from the ability of effector T cells to become activated in vitro without optimal costimulation (Croft et al., 1994; Horgan et al., 1990; Sagerstrom et al., 1993), which might have made them resistant to the effects of steady-state APCs (which induce naive T cells to become tolerant because they express sub-optimal costimulation; Hawiger et al., 2001; Janeway et al., 2002; Jenkins et al., 2001; Matzinger, 1994). It was therefore surprising when it was found that virally primed effector and memory T cells are equally susceptible to peripheral tolerance induction compared to naive counterparts following adoptive transfer into recipients that express cognate self-antigen (Higgins et al., 2002a; Kreuwel et al., 2002). This effector/memory T-cell tolerization pathway might exist to limit the extent of autoimmune damage that ensues during molecular mimicry scenarios (reviewed in Oldstone, 1998) where naive self-reactive T cells that have not yet been tolerized are primed by pathogens that express cross-reactive antigens (Adler, 2005) (Fig. 1). However, this pathway might also have the undesirable effect of inactivating tumor-reactive effector T cells that are either primed through vaccination (Fig. 2) or injected following ex vivo expansion (i.e., adoptive immunotherapy; Yee

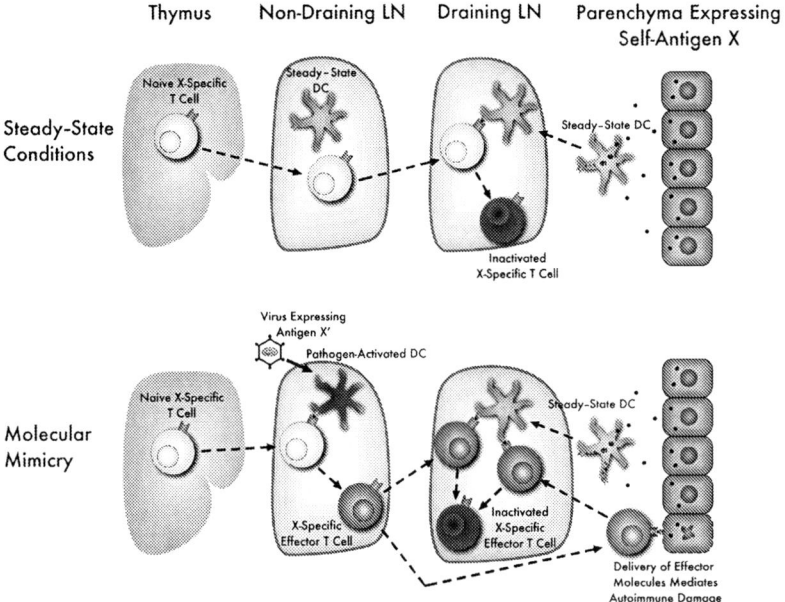

Fig. 1 The normal physiological role of the effector T-cell tolerization pathway might be to limit the extent of autoimmune pathology that ensues during molecular mimicry scenarios. During steady-state conditions naive T cells specific for the self-antigen X leave the thymus and migrate between peripheral lymph nodes (*LN*) until they enter LN draining tissues expressing X, where they are inactivated (i.e., tolerized) following encounter with steady-state tolerogenic DC presenting X. During molecular mimicry, infection with a pathogen expressing antigen X′ that is structurally similar to X leads to activation of DC presenting X′ and subsequent priming of naive X-specific T cells to differentiate into effectors that migrate into X-expressing parenchymal tissues and inflict autoimmune damage. Presentation of X by steady-state DC in the draining LN inactivates the X-specific effectors and thus shortens the duration of the autoimmune effector T-cell response

et al., 1997) and might therefore represent yet another level at which tolerance can negatively impact tumor immunity. Consistent with this possibility, naive prostate tumor-reactive T cells can be primed through vaccination to develop effector functions and partially control tumor growth, but over time effector functions and control of tumor growth diminish (Anderson et al., 2007).

The cellular and molecular mechanisms that regulate peripheral T-cell tolerance in vivo have been studied mostly in systems where naive T cells encounter tolerizing forms of antigen. However, given the relevance of peripheral tolerization of effector and memory T cells to tumor immunity, elucidating the unique aspects associated with these tolerance pathways will also be important. Thus far, it appears that there are similarities as well as interesting differences in the mechanisms by which effector and memory T cells undergo tolerization compared to naive T cells. Similar to naive T cells, both memory CD8 cells (Kreuwel et al., 2002) and Th1 effector CD4 cells (Higgins et al., 2002a) undergo an initial proliferative response prior to becoming tolerant. Additionally, steady-state bone marrow-derived APCs that

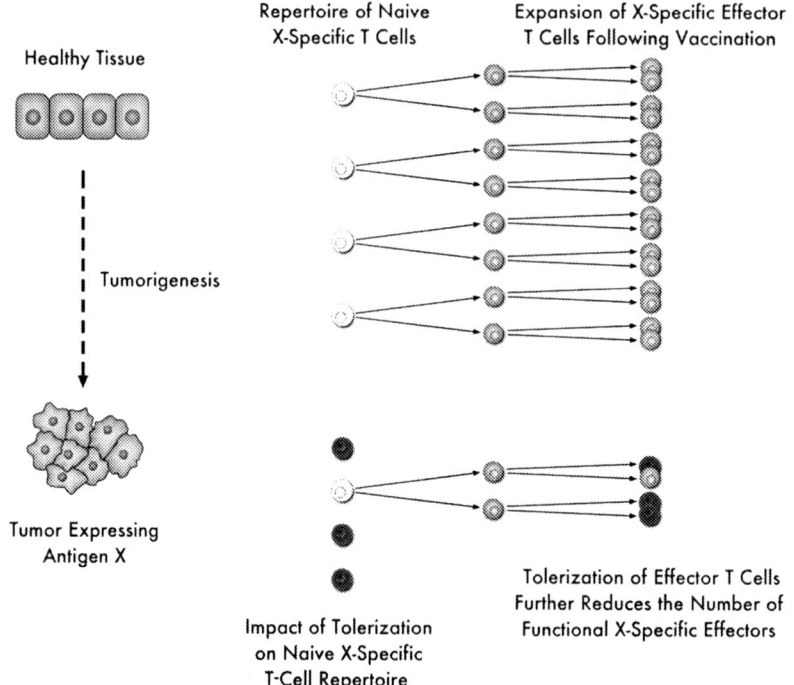

Fig. 2 An undesirable facet of the effector T-cell tolerization pathway is that it might represent an additional level at which tolerance can negatively impact tumor immunity. Tumorigenesis can result in the tolerization of a significant fraction of naive T cells specific for cognate tumor-associated antigens, thus restricting the repertoire of specificities that can respond to vaccination. Tolerization of the expanded tumor-reactive effector T-cell population (that is already reduced in number) could potentially impair tumor immunity even further. Although not shown in this figure, the effector T-cell tolerization pathway might also impede tumor immunity in the context of adoptive immunotherapy, where tumor-reactive effector T cells expanded ex vivo might be inactivated following injection into patients

indirectly present parenchymally derived self-antigen are required for Th1 effector CD4 cell tolerization (Higgins et al., 2002a). Effector T cells are distinguished from their naive progenitors by the expression of effector molecules such as IFN-γ (Th1 effector CD4 cells and effector CD8 cells), IL-4 (Th2 effector CD4 cells) as well as perforins and granzymes (effector CD8 cells) (Glimcher et al., 2004; Murphy and Reiner, 2002). It was therefore of interest to assess whether the regulation of these effector molecules is altered during tolerization. In the case of Th1 effector CD4 cells exposed to self-antigen, their potential to express the effector cytokines IFN-γ and TNF-α becomes impaired as early as 24 h, while the abilities to express IL-2 and to proliferate are lost only after several days (Long et al., 2003). In addition to indicating that the Th1 effector CD4 cell tolerization process is complex, this observation likely has physiological relevance since IFN-γ and TNF-α can both play critical roles in mediating tumor immunity (Hung et al., 1998; Ikeda et al.,

2002; Poehlein et al., 2003; Qin and Blankenstein, 2000). Thus, since neither T-cell proliferation nor IL-2 production is directly tumoricidal, effectors that can produce IL-2 and proliferate but have lost the ability to express IFN-γ and TNF-α would probably not be very effective at destroying tumors.

The TCR-transgenic adoptive transfer experiments demonstrating that effector T cells are highly susceptible to peripheral tolerization were somewhat analogous to adoptive immunotherapy approaches for treating cancer where ex vivo expanded tumor-reactive effector T cells are adoptively transferred into cancer patients (Yee et al., 1997). The relevance of effector T-cell tolerization to adoptive immunotherapy, however, was a bit unclear given that adoptive immunotherapy has demonstrated a degree of clinical efficacy (Dudley et al., 2002; Yee et al., 2002) despite the possibility that in these patients, the targeted tumor-associated antigens might be presented by tolerogenic steady-state APCs. In this regard it is worth noting that these and other adoptive immunotherapy protocols use cytotoxic drugs such as cyclophosphamide (Cytoxan) to condition patients prior to receiving tumor-reactive effector T cells and/or exogenous IL-2 administered thereafter. Cytoxan and IL-2 can also enhance the efficacy of anti-tumor adoptive immunotherapy in mouse models (Greenberg and Cheever, 1984; Hu et al., 1993; North, 1982). The mechanism(s) by which Cytoxan and IL-2 enhance anti-tumor adoptive immunotherapy has not been precisely established, although some studies have suggested that Cytoxan can eliminate tumor-specific regulatory T cells (North, 1982) or elicit the expression of T-cell growth factors (Proietti et al., 1998) or type I interferons (Schiavoni et al., 2000). Given the cytotoxic activity of Cytoxan, it might also enhance the engraftment of adoptively transferred tumor-reactive effector T cells (Greenberg and Cheever, 1984) by creating space (Dummer et al., 2002; Hu et al., 2002). IL-2 has been reported in some systems to enhance the proliferation and survival of effector T cells (Blattman et al., 2003; D'Souza, 2003). Rather than being mutually exclusive, these different potential mechanisms might be synergistic. Along similar lines, Cytoxan plus IL-2 impeded the tolerization of TCR-transgenic clonotypic Th1 effector CD4 cells that were adoptively transferred into cognate self-antigen-expressing recipients (Mihalyo et al., 2004), suggesting that the empirically developed adoptive immunotherapy protocols might be effective in part because they minimize tolerization of the adoptively transferred tumor-reactive effector T cells. It should be noted, however, that in the transgenic mouse model Cytoxan plus IL-2 delayed rather than prevented tolerization; for example, the capacity to express IFN-γ was extended by approximately 4 days (Mihalyo et al., 2004). This result may in part explain why multiple T-cell infusions enhance adoptive immunotherapy protocols, and underscores that the efficacy of adoptive immunotherapy might be further improved by strategies that more effectively preserve T-cell function in the face of tolerizing antigen.

Mitigating T-cell tolerance in the context of T cell-based immunotherapeutic approaches to treat cancer will require a detailed understanding of the intrinsic molecular defects that are associated with T-cell non-responsiveness. Using both in vitro anergy models and TCR-transgenic adoptive transfer systems in which naive T cells are exposed to tolerizing antigen, a variety of cytoplasmic signaling defects that are positioned down-stream of the TCR signaling apparatus and that contribute to impaired IL-2 expression and proliferation have been characterized (reviewed in Mueller, 2004; Schwartz, 2003). Some of these lesions might also play a role in the tolerization of effector T cells, since they also lose the ability to proliferate and express IL-2. Since there are unique functional defects associated with Th1 effector CD4 cell tolerization such as the rapid loss in effector cytokine expression potentials (Long et al., 2003), there are also likely to be unique intrinsic defects that are associated with this tolerance pathway. Recent work has revealed the existence of a yet-to-be identified TCR-proximal signaling defect(s) that contributes to impaired expression of IL-2, IFN-γ and TNF-α, as well as at least two additional defects that selectively impair IFN-γ and TNF-α expression. One of these defects has been identified as the down-modulated expression of the Th1 master regulatory factor T-bet, which contributes to impaired IFN-γ, but not TNF-α, expression (Long et al., 2006). Given the tumoricidal activities of IFN-γ and TNF-α, further identification and characterization of these defects that selectively impair their expression should aid the development of strategies to enhance tumor immunity.

4 The Relationship Between Hormones, T-Cell Tolerance and Tumor Immunity

Certain hormones can influence both tumorigenesis and T-cell function, and therefore understanding how these effects interact will be critical in tailoring appropriate T cell-based therapies. An example of this interplay is the relationship between androgens and prostate cancer (the most common malignancy in American men; Jemal et al., 2005). Androgens are required for the normal growth and differentiation of prostate epithelial cells (the cells that give rise to prostate cancer), and castration (i.e., androgen ablation) induces the apoptotic degeneration of the prostate epithelium (Furuya et al., 1995; Sugimura et al., 1986). Since most prostate tumor cells also require androgens for their growth and survival, androgen ablation has become a standard therapy for advanced prostate cancer (Denmeade and Isaacs, 2002). Unfortunately, disease relapse usually occurs following androgen ablation because a subset of tumor cells develop alterations in either the expression or activity of the androgen receptor that allows activation in the absence of normal androgen levels (Chen et al., 2004; Hakimi et al., 1996; Han et al., 2005; Zhao et al., 2000).

From an immunological perspective, androgen levels are inversely related to disease severity in certain autoimmunity models (Fox, 1992; Roubinian et al., 1978), and androgen ablation can reverse the decline in thymic output associated with aging (Sutherland et al., 2005) as well as enhance peripheral T-cell responsiveness (Roden et al., 2004; Viselli et al., 1995). Since androgen ablation is a standard therapy for advanced prostate cancer, many clinical trials utilizing T cell-based therapies will likely involve patients who have already undergone or who will be scheduled to undergo androgen ablation. Thus, understanding the effects of androgen ablation on the function of prostate-specific T cells will be critical for considering how T cell-based therapies should be administered relative to hormonal therapy.

To study the effects of prostate tumorigenesis and androgen ablation on the function of prostate-specific T cells, Drake et al. [2005] generated Pro-HA transgenic mice in which the prostate epithelial-specific probasin promoter drives the expression of HA antigen that has been modified to be secreted rather than expressed on the cell surface to model secreted prostatic antigens such as PSA. In contrast to the aforementioned C3-HA transgenic mice in which self-HA expressed in multiple parenchymal tissues programs adoptively transferred naive HA-specific CD4 cells to undergo tolerization (Adler et al., 1998; Higgins et al., 2002b), the same HA-specific CD4 cells retain their naive phenotype following adoptive transfer into Pro-HA mice (i.e., they remain "ignorant") (Drake et al., 2005). This lack of antigen recognition in the Pro-HA mice did not appear to be caused solely by a low level of expression (as has been observed in other systems; Kurts et al., 1998), but rather more likely because HA was being secreted into the prostatic lumen rather than the draining lymphatics (Whitmore and Gittes, 1977) where it could potentially be acquired by tolerance-inducing steady-state DC (Adler et al., 1998; Mihalyo et al., 2007). Thus, disruption of the normal prostatic architecture induced by androgen ablation-mediated apoptosis of the prostate epithelium caused adoptively transferred naive HA-specific CD4 cells in the prostate-draining lymph nodes to undergo an abortive proliferative response suggestive of tolerization. Additionally, the development of prostate cancer (induced by crossing the Pro-HA mice to TRAMP transgenic mice that develop spontaneous prostate tumors resulting from SV40 T antigen expression also under the control of the probasin promoter; Greenberg et al., 1995) resulted in a similar abortive proliferative response (Drake et al., 2005) regardless of the stage or rate of disease progression (Mihalyo et al., 2007). Notably, the duration of HA presentation in the draining lymph nodes of healthy androgen ablated mice was relatively short (~3 days) (Drake et al., 2005), perhaps because epithelial degeneration occurs in a synchronous wave and the phagocytic DCs that likely acquire HA from apoptotic epithelia (Liu et al., 2002; Steinman et al., 2000) have a lifespan in the lymph nodes of only a few days (Kamath et al., 2002). The sustained HA presentation associated with prostate cancer, but not the transient presentation caused by androgen ablation in healthy mice, was sufficient to render these prostate-specific T cells systemically tolerant as defined by an impaired ability to respond to subsequent viral immunization (Drake et al., 2005). Notably, androgen ablation of mice with prostate cancer elicited a transient increase in HA presentation in the draining lymph nodes, followed by a diminution (but not complete elimination) of HA presentation. This pattern appeared to parallel the apoptosis and subsequent clearance of the androgen ablation-sensitive sub-population of HA-expressing tumor cells. Most importantly, this diminution in tolerogenic antigen presentation allowed the HA-specific CD4 cells to retain their capacity to respond to vaccination, indicating that while prostate tumorigenesis promotes the tolerization of prostate-specific T cells, androgen ablation mitigates this effect (Fig. 3).

From a clinical standpoint, the observation in the Pro-HA system that androgen ablation reduces the tolerance-inducing capacity of prostate tumors suggests that

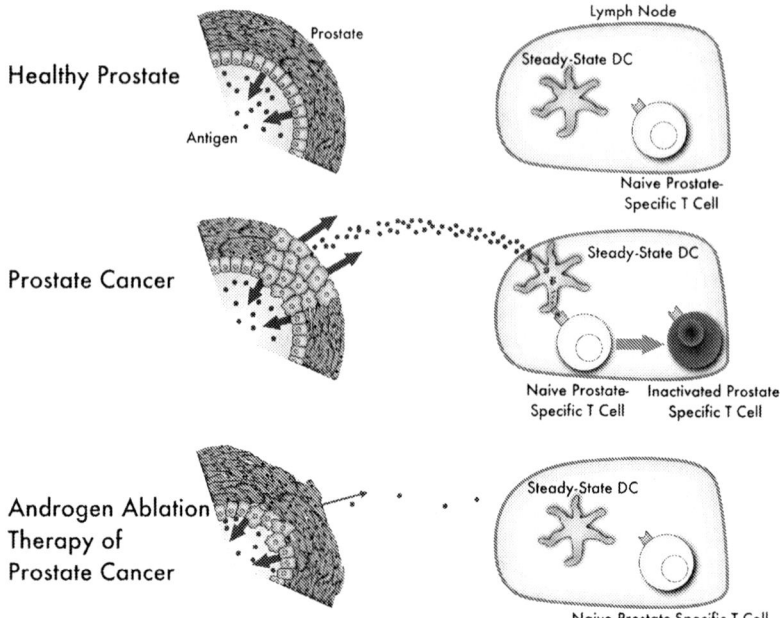

Fig. 3 The influence of prostate tumorigenesis and androgen ablation on the tolerization of prostate-specific T cells. In healthy prostates, prostate epithelial antigens are preferentially secreted in the prostatic lumen, rather than the draining lymphatics, and thus cognate T cells remain in a naive state due to a lack of presentation by steady-state tolerogenic DC. Alterations in the prostatic architecture caused by prostate tumorigenesis allow prostate epithelial/tumor antigen to reach the draining LN and to be presented by tolerogenic DC to inactivate cognate T cells. Androgen ablation induces the apoptosis of a large fraction of prostate epithelia and tumor cells, causing the level of prostate epithelial/tumor antigen to drop below the threshold required for tolerogenic antigen presentation

T cell-based therapies to treat prostate cancer might be the most effective when administered following rather than preceding androgen ablation. Mechanistically, this enhancement could potentially operate at multiple levels. It has been reported in some systems that T-cell anergy can be reversed following removal of the tolerizing antigen (Pape et al., 1998; Ramsdell and Fowlkes, 1992). Thus, androgen ablation might allow anergic prostate-specific T cells to regain the ability to respond to vaccination. Since effector T cells are susceptible to tolerization (Adler, 2005), adoptive immunotherapy targeting prostatic antigens might also have a better opportunity to eliminate the residual androgen ablation-resistant tumor cells after the level of tolerizing antigen has been reduced. Additionally, one of the inherent challenges in developing prostate cancer vaccines is that disease incidence increases with age, and aging is associated with a reduction in thymic output that contributes to a constriction in the repertoire of naive T cells. Since androgen ablation reverses the age-associated reduction in thymic output (Sutherland et al., 2005) as well as transiently augments antigen responsiveness in mature T cells (Roden et al., 2004), in the context of prostate cancer androgen ablation might thus enhance vaccine efficacy by both expanding the repertoire of naive prostate-specific T cells and augmenting the ability of these T cells to respond to vaccination.

Hormones may influence immunity to other types of cancer as well. For example, breast cancer is similar to prostate cancer in many respects that might influence tumor immunity; breast tumors arise from glandular epithelial cells that require estrogens for their growth and differentiation, and tumor cells can often be eliminated through treatment with estrogen receptor antagonists such as tamoxifen, but hormonal therapy-resistant tumor cells often cause disease relapse (Coffey, 2001; Cosman and Lindsay, 1999; Lopez-Otin and Diamandis, 1998). Thus, similar to prostate cancer, the possibility exists that hormonal blockade in the context of breast cancer might enhance the efficacy of T cell-based therapies by reducing the levels of tolerizing antigen.

5 Conclusion

As detailed above, tumors often exploit T-cell peripheral tolerization pathways that normally operate to prevent autoimmunity, to delete or inactivate tumor-reactive T cells. Understanding how these tolerance pathways operate under normal conditions will undoubtedly provide key insights into how tolerance might be mitigated in order to allow tumor vaccines to more effectively prime tumor-reactive effector T-cell responses. It is also becoming apparent that standard treatments for certain cancers can not only impact disease progression, but also influence the functional capacity of tumor-reactive T cells. For instance, chemotherapeutic drugs such as Cytoxan can deplete T cells; however, when administered in the proper sequence they can actually augment certain T cell-based anti-tumor modalities. Additionally, androgen ablation therapy for prostate cancer can induce a state of minimal residual disease that leads to a reduction in the level of tolerizing prostate tumor antigen and hence might restore the ability of cognate T cells to respond to vaccination. In the future it will be important to study in more depth how these other complex processes interact.

Acknowledgments The author is supported by NIH grants AI057441 and CA109339, and thanks Aaron Slaiby for expert assistance in preparing the figures.

References

Adler, A. J. (2005). Peripheral tolerization of effector and memory T cells: implications for autoimmunity and tumor-immunity. *Curr Immunol Rev* 1: 21–28.
Adler, A. J., Huang, C. T., Yochum, G. S., Marsh, D. W., and Pardoll, D. M. (2000). In vivo CD4+ T cell tolerance induction versus priming is independent of the rate and number of cell divisions. *J Immunol* 164:649–655.
Adler, A. J., Marsh, D. W., Yochum, G. S., Guzzo, J. L., Nigam, A., Nelson, W. G., and Pardoll, D. M. (1998). CD4+ T cell tolerance to parenchymal self-antigens requires presentation by bone marrow-derived antigen presenting cells. *J Exp Med* 187:1555–1564.

Anderson, M. J., Shafer-Weaver, K., Greenberg, N. M., and Hurwitz, A. A. (2007). Tolerization of tumor-specific T cells despite efficient initial priming in a primary murine model of prostate cancer. *J Immunol* 178:1268–1276.

Anderson, M. S., Venanzi, E. S., Klein, L., Chen, Z., Berzins, S. P., Turley, S. J., von Boehmer, H., Bronson, R., Dierich, A., Benoist, C., and Mathis, D. (2002). Projection of an immunological self shadow within the thymus by the aire protein. *Science* 298:1395–1401.

Apostolou, I., and von Boehmer, H. (2004). In vivo instruction of suppressor commitment in naive T cells. *J Exp Med* 199:1401–1408.

Bachmaier, K., Krawczyk, C., Kozieradzki, I., Kong, Y. Y., Sasaki, T., Oliveira-dos-Santos, A., Mariathasan, S., Bouchard, D., Wakeham, A., Itie, A., Le, J., Ohashi, P. S., Sarosi, I., Nishina, H., Lipkowitz, S., and Penninger, J. M. (2000). Negative regulation of lymphocyte activation and autoimmunity by the molecular adaptor Cbl-b. *Nature* 403:211–216.

Bannchereau, J., and Steinman, R. M. (1998). Dendritic cells and the control of immunity. *Nature* 392:245–252.

Belz, G. T., Behrens, G. M., Smith, C. M., Miller, J. F., Jones, C., Lejon, K., Fathman, C. G., Mueller, S. N., Shortman, K., Carbone, F. R., and Heath, W. R. (2002). The CD8alpha(+) dendritic cell is responsible for inducing peripheral self-tolerance to tissue-associated antigens. *J Exp Med* 196:1099–1104.

Blattman, J. N., Grayson, J. M., Wherry, E. J., Kaech, S. M., Smith, K. A., and Ahmed, R. (2003). Therapeutic use of IL-2 to enhance antiviral T-cell responses in vivo. *Nat Med* 9: 540–547.

Bogen, B. (1996). Peripheral T cell tolerance as a tumor escape mechanism: deletion of CD4+ T cells specific for a monoclonal immunoglobulin idiotype secreted by a plasmacytoma. *Eur J Immunol* 26:2671–2679.

Bogen, B., Dembic, Z., and Weiss, S. (1993). Clonal deletion of specific thymocytes by an immunoglobulin idiotype. *EMBO J* 12:357–363.

Bos, R., van Duikeren, S., van Hall, T., Kaaijk, P., Taubert, R., Kyewski, B., Klein, L., Melief, C. J., and Offringa, R. (2005). Expression of a natural tumor antigen by thymic epithelial cells impairs the tumor-protective CD4+ T-cell repertoire. *Cancer Res* 65:6443–6449.

Bouneaud, C., Kourilsky, P., and Bousso, P. (2000). Impact of negative selection on the T cell repertoire reactive to a self-peptide: a large fraction of T cell clones escapes clonal deletion. *Immunity* 13:829–840.

Chen, C. D., Welsbie, D. S., Tran, C., Baek, S. H., Chen, R., Vessella, R., Rosenfeld, M. G., and Sawyers, C. L. (2004). Molecular determinants of resistance to antiandrogen therapy. *Nat Med* 10:33–39.

Chen, M., Wang, Y. H., Wang, Y., Huang, L., Sandoval, H., Liu, Y. J., and Wang, J. (2006). Dendritic cell apoptosis in the maintenance of immune tolerance. *Science* 311:1160–1164.

Chiller, J. M., Habicht, G. S., and Weigle, W. O. (1971). Kinetic differences in unresponsiveness of thymus and bone marrow cells. *Science* 171:813–815.

Coffey, D. S. (2001). Similarities of prostate and breast cancer: evolution, diet, and estrogens. *Urology* 57:31–38.

Cosman, F., and Lindsay, R. (1999). Selective estrogen receptor modulators: clinical spectrum. *Endocr Rev* 20:418–434.

Cox, A. L., Skipper, J., Chen, Y., Henderson, R. A., Darrow, T. L., Shabanowitz, J., Engelhard, V. H., Hunt, D. F., and Slingluff, C. L., Jr (1994). Identification of a peptide recognized by five melanoma-specific human cytotoxic T cell lines. *Science* 264:716–719.

Croft, M., Bradley, L. M., and Swain, S. L. (1994). Naive versus memory CD4 T cell response to antigen. Memory cells are less dependent on accessory cell costimulation and can respond to many antigen-presenting cell types including resting B cells. *J Immunol* 152:2675–2685.

Denmeade, S. R., and Isaacs, J. T. (2002). A history of prostate cancer treatment. *Nat Rev Cancer* 2:389–396.

Derbinski, J., Gabler, J., Brors, B., Tierling, S., Jonnakuty, S., Hergenhahn, M., Peltonen, L., Walter, J., and Kyewski, B. (2005). Promiscuous gene expression in thymic epithelial cells is regulated at multiple levels. *J Exp Med* 202:33–45.

Doan, T., Herd, K. A., Lambert, P. F., Fernando, G. J., Street, M. D., and Tindle, R. W. (2000). Peripheral tolerance to human papillomavirus E7 oncoprotein occurs by cross-tolerization, is largely Th-2-independent, and is broken by dendritic cell immunization. *Cancer Res* 60: 2810–2815.

Drake, C. G., Doody, A. D., Mihalyo, M. A., Huang, C. T., Kelleher, E., Ravi, S., Hipkiss, E. L., Flies, D. B., Kennedy, E. P., Long, M., McGary, P. W., Coryell, L., Nelson, W. G., Pardoll, D. M., and Adler, A. J. (2005). Androgen ablation mitigates tolerance to a prostate/prostate cancer-restricted antigen. *Cancer Cell* 7:239–249.

Dresser, D. W. (1962). Specific inhibition of antibody production. II. Paralysis induced in adult mice by small quantities of protein antigen. *Immunology* 5:378–388.

D'Souza, W. N., and Lefrancois, L. (2003). IL-2 is not required for the initiation of CD8 T cell cycling but sustains expansion. *J Immunol* 171:5727–5735.

Dudley, M. E., Wunderlich, J. R., Robbins, P. F., Yang, J. C., Hwu, P., Schwartzentruber, D. J., Topalian, S. L., Sherry, R., Restifo, N. P., Hubicki, A. M., Robinson, M. R., Raffeld, M., Duray, P., Seipp, C. A., Rogers-Freezer, L., Morton, K. E., Mavroukakis, S. A., White, D. E., and Rosenberg, S. A. (2002). Cancer regression and autoimmunity in patients after clonal repopulation with antitumor lymphocytes. *Science* 298:850–854.

Dummer, W., Niethammer, A. G., Baccala, R., Lawson, B. R., Wagner, N., Reisfeld, R. A., and Theofilopoulos, A. N. (2002). T cell homeostatic proliferation elicits effective antitumor autoimmunity. *J Clin Invest* 110:185–192.

Eynon, E. E., and Parker, D. C. (1992). Small B cells as antigen-presenting cells in the induction of tolerance to soluble protein antigens. *J Exp Med* 175:131–138.

Finkelman, F. D., Lees, A., Birnbaum, R., Gause, W. C., and Morris, S. C. (1996). Dendritic cells can present antigen in vivo in a tolerogenic or immunogenic fashion. *J Immunol* 157: 1406–1414.

Fox, H. S. (1992). Androgen treatment prevents diabetes in nonobese diabetic mice. *J Exp Med* 175:1409–1412.

Fuchs, E. J., and Matzinger, P. (1992). B cells turn off virgin but not memory T cells. *Science* 258:1156–1159.

Furuya, Y., Lin, X. S., Walsh, J. C., Nelson, W. G., and Isaacs, J. T. (1995). Androgen ablation-induced programmed death of prostatic glandular cells does not involve recruitment into a defective cell cycle or p53 induction. *Endocrinology* 136:1898–1906.

Glimcher, L. H., Townsend, M. J., Sullivan, B. M., and Lord, G. M. (2004). Recent developments in the transcriptional regulation of cytolytic effector cells. *Nat. Rev. Immunol.* 4:900–911.

Gorelik, E. (1983). Concomitant tumor immunity and the resistance to a second tumor challenge. *Adv Cancer Res* 39:71–120.

Gorelik, L., and Flavell, R. A. (2000). Abrogation of TGFbeta signaling in T cells leads to spontaneous T cell differentiation and autoimmune disease. *Immunity* 12:171–181.

Greenberg, N. M., DeMayo, F., Finegold, M. J., Medina, D., Tilley, W. D., Aspinall, J. O., Cunha, G. R., Donjacour, A. A., Matusik, R. J., and Rosen, J. M. (1995). Prostate cancer in a transgenic mouse. *Proc Natl Acad Sci USA* 92:3439–3443.

Greenberg, P. D., and Cheever, M. A. (1984). Treatment of disseminated leukemia with cyclophosphamide and immune cells: tumor immunity reflects long-term persistence of tumor-specific donor T cells. *J Immunol* 133:3401–3407.

Hagymasi, A. T., Slaiby, A. M., Mihalyo, M. A., Qui, H. Z., Zammit, D. J., Lefrancois, L., and Adler, A. J. (2007). Steady state dendritic cells present parenchymal self-antigen and contribute to, but are not essential for, tolerization of naïve and Th1 effector CD4 cells. *J Immunol* 179:1524–1531.

Hakimi, J. M., Rondinelli, R. H., Schoenberg, M. P., and Barrack, E. R. (1996). Androgen-receptor gene structure and function in prostate cancer. *World J Urol* 14:329–337.

Han, G., Buchanan, G., Ittmann, M., Harris, J. M., Yu, X., Demayo, F. J., Tilley, W., and Greenberg, N. M. (2005). Mutation of the androgen receptor causes oncogenic transformation of the prostate. *Proc Natl Acad Sci USA* 102:1151–1156.

Hanson, H. L., Donermeyer, D. L., Ikeda, H., White, J. M., Shankaran, V., Old, L. J., Shiku, H., Schreiber, R. D., and Allen, P. M. (2000). Eradication of established tumors by CD8+ T cell adoptive immunotherapy. *Immunity* 13:265–276.

Hawiger, D., Inaba, K., Dorsett, Y., Guo, M., Mahnke, K., Rivera, M., Ravetch, J. V., Steinman, R. M., and Nussenzweig, M. C. (2001). Dendritic cells induce peripheral T cell unresponsiveness under steady state conditions in vivo. *J Exp Med* 194:769–779.

Higgins, A. D., Mihalyo, M. A., and Adler, A. J. (2002a). Effector CD4 cells are tolerized upon exposure to parenchymal self-antigen. *J Immunol* 169:3622–3629.

Higgins, A. D., Mihalyo, M. A., McGary, P. W., and Adler, A. J. (2002b). CD4 cell priming and tolerization are differentially programmed by APCs upon initial engagement. *J Immunol* 168:5573–5581.

Horgan, K. J., Van Seventer, G. A., Shimizu, Y., and Shaw, S. (1990). Hyporesponsiveness of "naive" (CD45RA+) human T cells to multiple receptor-mediated stimuli but augmentation of responses by co-stimuli. *Eur J Immunol* 20:1111–1118.

Hu, H. M., Poehlein, C. H., Urba, W. J., and Fox, B. A. (2002). Development of antitumor immune responses in reconstituted lymphopenic hosts. *Cancer Res* 62:3914–3919.

Hu, J., Kindsvogel, W., Busby, S., Bailey, M. C., Shi, Y., and Greenberg, P. D. (1993). An evaluation of the potential to use tumor-associated antigens as targets for antitumor T cell therapy using transgenic mice expressing a retroviral tumor antigen in normal lymphoid tissues. *J Exp Med* 177:1681–1690.

Hung, K., Hayashi, R., Lafond-Walker, A., Lowenstein, C., Pardoll, D., and Levitsky, H. (1998). The central role of CD4(+) T cells in the antitumor immune response. *J Exp Med* 188: 2357–2368.

Ikeda, H., Old, L. J., and Schreiber, R. D. (2002). The roles of IFN gamma in protection against tumor development and cancer immunoediting. *Cytokine Growth Factor Rev* 13:95–109.

Janeway, C. A., Jr, and Medzhitov, R. (2002). Innate immune recognition. *Annu Rev Immunol* 20:197–216.

Jemal, A., Murray, T., Ward, E., Samuels, A., Tiwari, R. C., Ghafoor, A., Feuer, E. J., and Thun, M. J. (2005). Cancer statistics, 2005. *CA Cancer J Clin* 55:10–30.

Jenkins, M. K., Khoruts, A., Ingulli, E., Mueller, D. L., McSorley, S. J., Reinhardt, R. L., Itano, A., and Pape, K. A. (2001). In vivo activation of antigen-specific CD4 T cells. *Annu Rev Immunol* 19:23–45.

Jones, L. A., Chin, L. T., Longo, D. L., and Kruisbeek, A. M. (1990). Peripheral clonal elimination of functional T cells. *Science* 250:1726–1729.

Jooss, K., Gjata, B., Danos, O., von Boehmer, H., and Sarukhan, A. (2001). Regulatory function of in vivo anergized CD4(+) T cells. *Proc Natl Acad Sci USA* 98:8738–8743.

Kamath, A. T., Henri, S., Battye, F., Tough, D. F., and Shortman, K. (2002). Developmental kinetics and lifespan of dendritic cells in mouse lymphoid organs. *Blood* 100:1734–1741.

Kappler, J., Roehm, M., and Marrack, P. (1987). T cell tolerance by clonal elimination in the thymus. *Cell* 49:273–280.

Kearney, E. R., Pape, K. A., Loh, D. Y., and Jenkins, M. K. (1994). Visualization of peptide-specific T cell immunity and peripheral tolerance induction in vivo. *Immunity* 1:327–339.

Kisielow, P., Bluthmann, H., Staerz, U. D., Steinmetz, M., and von Boehmer, H. (1988). Tolerance in T-cell-receptor transgenic mice involves deletion of nonmature CD4+8+ thymocytes. *Nature* 333:742–746.

Kreuwel, H. T., Aung, S., Silao, C., and Sherman, L. A. (2002). Memory CD8(+) T cells undergo peripheral tolerance. *Immunity* 17:73–81.

Kurts, C., Cannarile, M., Klebba, I., and Brocker, T. (2001). Dendritic cells are sufficient to cross-present self-antigens to CD8 T cells in vivo. *J Immunol* 166:1439–1442.

Kurts, C., Kosaka, H., Carbone, F. R., Miller, J. F. A. P., and Heath, W. R. (1997). Class I-restricted cross-presentation of exogenous self-antigens leads to deletion of autoreactive CD8+ T cells. *J Exp Med* 186:239–245.

Kurts, C., Miller, J. F. A. P., Subramaniam, R. M., Carbone, F. R., and Heath, W. R. (1998). Major histocompatibility complex class I-restricted cross-presentation is biased towards high dose antigens and those released during cellular destruction. *J Exp Med* 188:409–414.

Lee, P. P., Yee, C., Savage, P. A., Fong, L., Brockstedt, D., Weber, J. S., Johnson, D., Swetter, S., Thompson, J., Greenberg, P. D., Roederer, M., and Davis, M. M. (1999). Characterization of circulating T cells specific for tumor-associated antigens in melanoma patients. *Nat Med* 5: 677–685.

Lengauer, C., Kinzler, K. W., and Vogelstein, B. (1998). Genetic instabilities in human cancers. *Nature* 396:643–649.

Liblau, R., Tisch, R., Bercovici, N., and McDevitt, H. (1997). Systemic antigen in the treatment of T-cell-mediated autoimmune diseases. *Immunol Today* 18:599–604.

Liu, K., Iyoda, T., Saternus, M., Kimura, Y., Inaba, K., and Steinman, R. M. (2002). Immune tolerance after delivery of dying cells to dendritic cells in situ. *J Exp Med* 196:1091–1097.

Long, M., Higgins, A. D., Mihalyo, M. A., and Adler, A. J. (2003). Effector CD4 cell tolerization is mediated through functional inactivation and involves preferential impairment of TNF-alpha and IFN-gamma expression potentials. *Cell Immunol* 224:114–121.

Long, M., Slaiby, A. M., Hagymasi, A. T., Mihalyo, M. A., Lichtler, A. C., Reiner, S. L., and Adler, A. J. (2006). T-bet down-modulation in tolerized Th1 effector CD4 cells confers a TCR-distal signaling defect that selectively impairs IFN-gamma expression. *J Immunol* 176: 1036–1045.

Lopez-Otin, C., and Diamandis, E. P. (1998). Breast and prostate cancer: an analysis of common epidemiological, genetic, and biochemical features. *Endocr Rev* 19:365–396.

Lyman, M. A., Aung, S., Biggs, J. A., and Sherman, L. A. (2004). A spontaneously arising pancreatic tumor does not promote the differentiation of naive CD8+ T lymphocytes into effector CTL. *J Immunol* 172:6558–6567.

Marzo, A. L., Lake, R. A., Lo, D., Sherman, L., McWilliam, A., Nelson, D., Robinson, B. W., and Scott, B. (1999). Tumor antigens are constitutively presented in the draining lymph nodes. *J Immunol* 162:5838–5845.

Matzinger, P. (1994). Tolerance, danger, and the extended family. *Annu Rev Immunol* 12:991–1045.

Medzhitov, R. (2001). Toll-like receptors and innate immunity. *Nat Rev Immunol* 1: 135–145.

Mihalyo, M. A., Doody, A. D., McAleer, J. P., Nowak, E. C., Long, M., Yang, Y., and Adler, A. J. (2004). In vivo cyclophosphamide and IL-2 treatment impedes self-antigen-induced effector CD4 cell tolerization: implications for adoptive immunotherapy. *J Immunol* 172: 5338–5345.

Mihalyo, M. A., Hagymasi, A. T., Slaiby, A. M., Nevius, E. E., and Adler, A. J. (2007). Dendritic cells program non-immunogenic prostate-specific T cell responses beginning at early stages of prostate tumorigenesis. *Prostate* 67:536–546.

Morgan, D. J., Kreuwel, H. T., Fleck, S., Levitsky, H. I., Pardoll, D. M., and Sherman, L. A. (1998). Activation of low avidity CTL specific for a self epitope results in tumor rejection but not autoimmunity. *J Immunol* 160:643–651.

Mueller, D. L. (2004). E3 ubiquitin ligases as T cell anergy factors. *Nat Immunol* 5:883–890.

Murphy, K. M., and Reiner, S. L. (2002). Decision making in the immune system: the lineage decisions of helper T cells. *Nat Rev Immunol* 2:933–944.

Nguyen, L. T., Elford, A. R., Murakami, K., Garza, K. M., Schoenberger, S. P., Odermatt, B., Speiser, D. E., and Ohashi, P. S. (2002). Tumor growth enhances cross-presentation leading to limited T cell activation without tolerance. *J Exp Med* 195:423–435.

North, R. J. (1982). Cyclophosphamide-facilitated adoptive immunotherapy of an established tumor depends on elimination of tumor-induced suppressor T cells. *J Exp Med* 155: 1063–1074.

Ochsenbein, A. F., Sierro, S., Odermatt, B., Pericin, M., Karrer, U., Hermans, J., Hemmi, S., Hengartner, H., and Zinkernagel, R. M. (2001). Roles of tumour localization, second signals and cross priming in cytotoxic T-cell induction. *Nature* 411:1058–1064.

Oldstone, M. B. (1998). Molecular mimicry and immune-mediated diseases. *FASEB J* 12: 1255–1265.

Pape, K. A., Merica, R., Mondino, A., Khoruts, A., and Jenkins, M. K. (1998). Direct evidence that functionally impaired CD4+ T cells persist in vivo following induction of peripheral tolerance. *J Immunol* 160:4719–4729.

Pardoll, D. (2003). Does the immune system see tumors as foreign or self? *Annu Rev Immunol* 21:807–839.
Poehlein, C. H., Hu, H. M., Yamada, J., Assmann, I., Alvord, W. G., Urba, W. J., and Fox, B. A. (2003). TNF plays an essential role in tumor regression after adoptive transfer of perforin/IFN-gamma double knockout effector T cells. *J Immunol* 170:2004–2013.
Proietti, E., Greco, G., Garrone, B., Baccarini, S., Mauri, C., Venditti, M., Carlei, D., and Belardelli, F. (1998). Importance of cyclophosphamide-induced bystander effect on T cells for a successful tumor eradication in response to adoptive immunotherapy in mice. *J Clin Invest* 101: 429–441.
Qin, Z., and Blankenstein, T. (2000). CD4+ T cell-mediated tumor rejection involves inhibition of angiogenesis that is dependent on IFN gamma receptor expression by nonhematopoietic cells. *Immunity* 12:677–686.
Ramsdell, F., and Fowlkes, B. J. (1992). Maintenance of in vivo tolerance by persistence of antigen. *Science* 257:1130–1134.
Rocha, B., and von Boehmer, H. (1991). Peripheral selection of the T cell repertoire. *Science* 251:1225–1228.
Roden, A. C., Moser, M. T., Tri, S. D., Mercader, M., Kuntz, S. M., Dong, H., Hurwitz, A. A., McKean, D. J., Celis, E., Leibovich, B. C., Allison, J. P., and Kwon, E. D. (2004). Augmentation of T cell levels and responses induced by androgen deprivation. *J Immunol* 173:6098–6108.
Rosenberg, S. A., Yang, J. C., and Restifo, N. P. (2004). Cancer immunotherapy: moving beyond current vaccines. *Nat Med* 10:909–915.
Roubinian, J. R., Talal, N., Greenspan, J. S., Goodman, J. R., and Siiteri, P. K. (1978). Effect of castration and sex hormone treatment on survival, anti-nucleic acid antibodies, and glomerulonephritis in NZB/NZW F1 mice. *J Exp Med* 147:1568–1583.
Sagerstrom, C. G., Kerr, E. M., Allison, J. P., and Davis, M. M. (1993). Activation and differentiation requirements of primary T cells in vitro. *Proc Natl Acad Sci USA* 90:8987–8991.
Sakaguchi, S. (2000). Regulatory T cells: key controllers of immunologic self-tolerance. *Cell* 101:455–458.
Schell, T. D., Knowles, B. B., and Tevethia, S. S. (2000). Sequential loss of cytotoxic T lymphocyte responses to simian virus 40 large T antigen epitopes in T antigen transgenic mice developing osteosarcomas. *Cancer Res* 60:3002–3012.
Schiavoni, G., Mattei, F., Di Pucchio, T., Santini, S. M., Bracci, L., Belardelli, F., and Proietti, E. (2000). Cyclophosphamide induces type I interferon and augments the number of CD44(hi) T lymphocytes in mice: implications for strategies of chemoimmunotherapy of cancer. *Blood* 95:2024–2030.
Schwartz, R. H. (2003). T cell anergy. *Annu Rev Immunol* 21:305–334.
Sebzda, E., Wallace, V. A., Mayer, J., Yeung, R. S., Mak, T. W., and Ohashi, P. S. (1994). Positive and negative thymocyte selection induced by different concentrations of a single peptide. *Science* 263:1615–1618.
Shevach, E. M. (2001). Certified professionals: CD4(+)CD25(+) suppressor T cells. *J Exp Med* 193:F41–F46.
Shrikant, P., Khoruts, A., and Mescher, M. F. (1999). CTLA-4 blockade reverses CD8+ T cell tolerance to tumor by a CD4+ T cell- and IL-2-dependent mechanism. *Immunity* 11:483–493.
Sotomayor, E. M., Borrello, I., Rattis, F. M., Cuenca, A. G., Abrams, J., Staveley-O'Carroll, K., and Levitsky, H. I. (2001). Cross-presentation of tumor antigens by bone marrow-derived antigen-presenting cells is the dominant mechanism in the induction of T-cell tolerance during B-cell lymphoma progression. *Blood* 98:1070–1077.
Spiotto, M. T., Yu, P., Rowley, D. A., Nishimura, M. I., Meredith, S. C., Gajewski, T. F., Fu, Y. X., and Schreiber, H. (2002). Increasing tumor antigen expression overcomes "ignorance" to solid tumors via crosspresentation by bone marrow-derived stromal cells. *Immunity* 17:737–747.
Srivastava, P. K. (2006). Therapeutic cancer vaccines. *Curr Opin Immunol* 18:201–205.
Stavely-O'Carroll, K., Sotomayor, E., Montgomery, J., Borrello, I., Hwang, L., Fein, S., Pardoll, D., and Levitsky, H. (1998). Induction of antigen-specific T cell anergy: an early event in the course of tumor progression. *Proc Natl Acad Sci USA* 95:1178–1183.

Steinman, R. M., Turley, S., Mellman, I., and Inaba, K. (2000). The induction of tolerance by dendritic cells that have captured apoptotic cells. *J Exp Med* 191:411–416.

Sugimura, Y., Cunha, G. R., and Donjacour, A. A. (1986). Morphological and histological study of castration-induced degeneration and androgen-induced regeneration in the mouse prostate. *Biol Reprod* 34:973–983.

Surh, C. D., and Sprent, J. (1994). T-cell apoptosis detected in situ during positive and negative selection in the thymus. *Nature* 372:100–103.

Sutherland, J. S., Goldberg, G. L., Hammett, M. V., Uldrich, A. P., Berzins, S. P., Heng, T. S., Blazar, B. R., Millar, J. L., Malin, M. A., Chidgey, A. P., and Boyd, R. L. (2005). Activation of thymic regeneration in mice and humans following androgen blockade. *J Immunol* 175: 2741–2753.

Tivol, E. A., Borriello, F., Schweitzer, A. N., Lynch, W. P., Bluestone, J. A., and Sharpe, A. H. (1995). Loss of CTLA-4 leads to massive lymphoproliferation and fatal multiorgan tissue destruction, revealing a critical negative regulatory role of CTLA-4. *Immunity* 3:541–547.

Topalian, S. L., Solomon, D., and Rosenberg, S. A. (1989). Tumor-specific cytolysis by lymphocytes infiltrating human melanomas. *J Immunol* 142:3714–3725.

Viselli, S. M., Stanziale, S., Shults, K., Kovacs, W. J., and Olsen, N. J. (1995). Castration alters peripheral immune function in normal male mice. *Immunology* 84:337–342.

von Budingen, H. C., Tanuma, N., Villoslada, P., Ouallet, J. C., Hauser, S. L., and Genain, C. P. (2001). Immune responses against the myelin/oligodendrocyte glycoprotein in experimental autoimmune demyelination. *J Clin Immunol* 21:155–170.

Wang, R. F., Appella, E., Kawakami, Y., Kang, X., and Rosenberg, S. A. (1996). Identification of TRP-2 as a human tumor antigen recognized by cytotoxic T lymphocytes. *J Exp Med* 184:2207–2216.

Wang, T., Niu, G., Kortylewski, M., Burdelya, L., Shain, K., Zhang, S., Bhattacharya, R., Gabrilovich, D., Heller, R., Coppola, D., Dalton, W., Jove, R., Pardoll, D., and Yu, H. (2004). Regulation of the innate and adaptive immune responses by Stat-3 signaling in tumor cells. *Nat Med* 10:48–54.

Webb, S., Morris, C., and Sprent, J. (1990). Extrathymic tolerance of mature T cells: clonal elimination as a consequence of immunity. *Cell* 63:1249–1256.

Whitmore, W. F., and Gittes, R. F. (1977). Studies on the prostate and testis as immunologically privileged sites. *Cancer Treat Rep* 61:217–222.

Wolfel, T., Van Pel, A., Brichard, V., Schneider, J., Seliger, B., Meyer zum Buschenfelde, K. H., and Boon, T. (1994). Two tyrosinase nonapeptides recognized on HLA-A2 melanomas by autologous cytolytic T lymphocytes. *Eur J Immunol* 24:759–764.

Yee, C., Riddell, S. R., and Greenberg, P. D. (1997). Prospects for adoptive T cell therapy. *Curr Opin Immunol* 9:702–708.

Yee, C., Thompson, J. A., Byrd, D., Riddell, S. R., Roche, P., Celis, E., and Greenberg, P. D. (2002). Adoptive T cell therapy using antigen-specific CD8+ T cell clones for the treatment of patients with metastatic melanoma: in vivo persistence, migration, and antitumor effect of transferred T cells. *Proc Natl Acad Sci USA* 99:16168–16173.

Zhao, X. Y., Malloy, P. J., Krishnan, A. V., Swami, S., Navone, N. M., Peehl, D. M., and Feldman, D. (2000). Glucocorticoids can promote androgen-independent growth of prostate cancer cells through a mutated androgen receptor. *Nat Med* 6:703–706.

Zhou, G., Drake, C. G., and Levitsky, H. I. (2006). Amplification of tumor-specific regulatory T cells following therapeutic cancer vaccines. *Blood* 107:628–636.

Zhou, G., Lu, Z., McCadden, J. D., Levitsky, H. I., and Marson, A. L. (2004). Reciprocal changes in tumor antigenicity and antigen-specific T cell function during tumor progression. *J Exp Med* 200:1581–1592.

Contribution of B7-H1/PD-1 Co-inhibitory Pathway to T-Cell Dysfunction in Cancer

Sheng Yao and Lieping Chen

1 Introduction

Ample evidence indicates cancer patients often mount adaptive immune response against cancer, and tumor-specific CD8 effector T cells could infiltrate into tumor lesions. However, these immune responses are often incapable of controlling cancer growth. It is becoming clear in recent years that tumor cells utilize various elaborate tactics to passively reduce their immunogenicity to avoid detection by immune system and to actively induce immune tolerance by surface expression or secreting inhibitory and pro-apoptotic molecules to evade immune system.

In the last few years, expression of several B7 family co-signaling molecules, including B7-H1 (Thompson et al., 2006) and B7-H4 (Krambeck et al., 2006), has been linked to accelerated tumor progression and poor prognosis. Furthermore, B7-H1-PD-1 pathway has been vigorously investigated and found to be associated with T-cell dysfunction in chronic infections, including human immunodeficiency virus (HIV) (Day et al., 2006; Trautmann et al., 2006), hepatitis C virus (HCV) (Urbani et al., 2006) and hepatitis B virus (HBV) (Boni et al., 2007). In this chapter, we will focus on the discussion of the mechanisms of B7-H1 and PD-1 pathway in negative regulation of T-cell function and potential therapeutic manipulations targeting B7-H1 in the treatment of cancer and chronic viral infections.

2 T-Cell Co-signaling Pathways

Current theory of T-cell activation could be delineated into a simple two-signal model (Mueller et al., 1989). Interaction of T-cell receptor (TCR) with MHC–peptide complex provides the primary signal; ligation of co-signaling receptors and ligands serves as the second signal, which could be positive or negative and ultimately determines the outcome of a T-cell response (Fig. 1). Without TCR–MHC

L. Chen
Johns Hopkins Medicine, 209 David H. Koch Cancer Research Building, Baltimore, MD 21231, USA
e-mail: lchen42@jhmi.edu

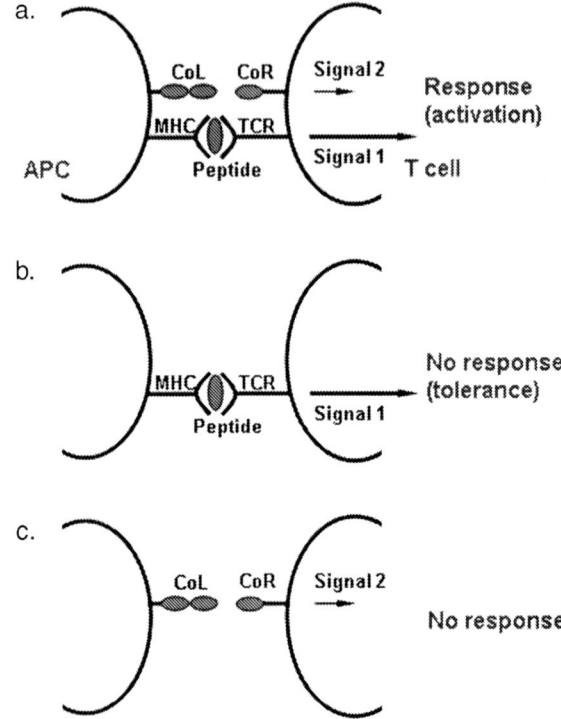

Fig. 1 Two-signal model of T-cell response. **a** For T-cell activation, ligation of T-cell receptor (*TCR*) with MHC–peptide complex provides the primary signal, whereas engagement of co-signaling receptors (*CoR*) and ligands (*CoL*) serves as the second signal. **b** Without TCR–MHC interaction, T-cell response will not be triggered. **c** Without co-signaling, no T-cell response or T-cell anergy/tolerance induction will take place

interaction, a T-cell response will not be triggered. The absence or imbalance of the second signal could result in T-cell anergy/tolerance or over-activation (Chen, 2004).

The co-signaling molecules belong mainly to two families of proteins: the immunoglobulin (Ig) superfamily (including B7-CD28 family) and the tumor necrosis factor (TNF) receptor/ligand superfamily (Fig. 2). The basic structure of CD28-like receptors contains a single Ig variable (IgV) region-like motif as extracellular domain and immunoreceptor tyrosine-based activation motif (ITAM), immunoreceptor tyrosine-based inhibition motif (ITIM) and/or immunoreceptor tyrosine-based switch motif (ITSM) in the intracellular region, which are responsible for signaling transduction. The current members of receptors include CD28, cytotoxic T-lymphocyte antigen-4 (CTLA-4), inducible costimulator (ICOS), program death 1 (PD-1) and B and T-lymphocyte attenuator (BTLA) (Watanabe et al., 2003). Their known ligands all belong to the B7 ligand family with the exception of BTLA, which interacts with a TNFR member, herpes virus entry mediator (HVEM). B7 ligand family has a structural feature of two Ig-like extracellular domains, one IgV and one Ig constant (Ig C) region-like domains (Fig. 1). Crystal structure studies indicate that B7 ligand and receptor interact through their IgV domains (Schwartz et al., 2001; Stamper et al., 2001; Zhang et al., 2004). The majority of so-called ligands could also receive signals to deliver functions and, in a strict sense, should be called "counter-receptors".

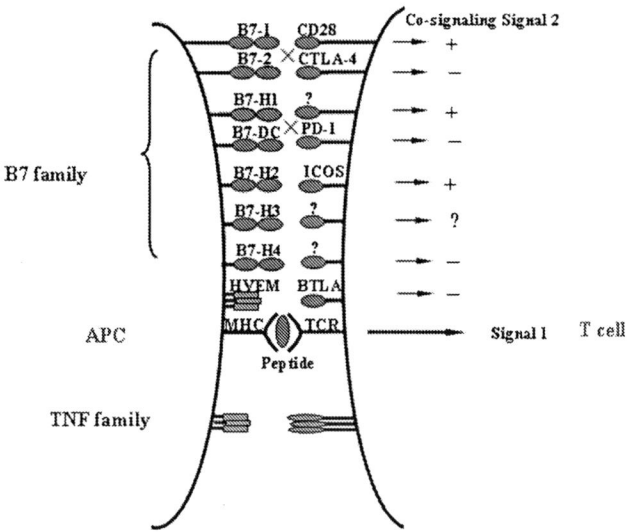

Fig. 2 B7/CD28 family of costimulatory ligands/receptors. Current B7 ligand family includes B7-1 (CD80), B7-2 (CD86), B7-H1 (PD-L1, CD274), B7-H2 (ICOSL, CD275), B7-H3 (CD276) and B7-H4 (B7S1). CD28 receptor family member includes CD28, CTLA-4 (CD152), ICOS (CD278), PD-1 (CD279) and BTLA

The classical B7 family co-signaling pathways B7-CD28/CTLA-4 were discovered in the 1990s (Linsley et al., 1990, 1991). CD28, expressed on naïve T cells, plays an essential role in T-cell priming. It delivers a primary costimulatory signal to T cells by ligation of B7-1 (CD80) or B7-2 (CD86) expressed on antigen presentation cells (APCs) (Lenschow et al., 1996). In contrast, engagement of CTLA-4 on activated T cells by B7-1 and B7-2 will attenuate T-cell response (Chambers et al., 2001; Cross et al., 1995; Krummel and Allison, 1995). In recent years, a series of B7 homologs with similarities in both sequence and extracellular domain structures have been discovered. These molecules have much broader distribution than the classic B7 counter-receptors and they also broadly contribute to the modulation of immune responses (Table 1).

Table 1 Protein sequence identities (%) among human B7 family members

Human	B7-1	B7-2	B7-H1	B7-DC	B7-H2	B7-H3	B7-H4
B7-1	100	25	22	23	23	25	22
B7-2		100	22	20	22	23	23
B7-H1			100	39	21	29	23
B7-DC				100	20	24	25
B7-H2					100	29	21
B7-H3						100	27
B7-H4							100

Extracellular domains of human B7 ligand family members were compared by ClustalW program

3 B7-H1-PD-1 Pathway

B7 homolog 1 (B7-H1), the third member of B7 family in addition to B7-1 and B7-2, was identified by a homology-search method in 1999 (Dong et al., 1999). Both mouse and human B7-H1 messenger RNAs have a broad tissue distribution (Dong et al., 1999). In contrast, constitutive cell surface expression of protein is limited to antigen presentation cells (APC), including dendritic cells, macrophages in lymphoid tissue and macrophage-like cells. Expression of B7-H1, however, could be induced on virtually all stromal cells, in addition to hematopoietic cells, including epithelial and endothelial cells by various pro-inflammatory cytokines, such as interferons (Mazanet and Hughes, 2002; Wiendl et al., 2003).

B7-DC, another B7 family member discovered in 2001 (Latchman et al., 2001; Tseng et al., 2001), shares the highest sequence similarity with B7-H1. Nevertheless, unlike the universally inducible pattern of B7-H1, surface expression of B7-DC is restricted to dendritic cells and activated macrophages.

PD-1, a B7 receptor family member expressed on activated T and B cells (Agata et al., 1996), is identified as a receptor for both B7-H1 and B7-DC (Freeman et al., 2000; Tseng et al., 2001). Cytoplasmic domain of PD-1 contains two signaling transduction motifs, an ITIM and an ITSM (Ishida et al., 1992). The ITSM domain has been implicated to be responsible for the downstream signaling events critical to suppress T- and B-cell responses (Okazaki et al., 2001). Cumulative evidence indicates that PD-1 is important for the maintenance of peripheral tolerance. PD-1-deficient mice bred onto various backgrounds show severe autoimmune diseases. In the BALB/c background, PD-1-deficient mice had dilated cardiomyopathy with a high titer of autoantibody against heart-specific protein (Nishimura et al., 2001). In C57BL/6 background, PD-1-deficient mice spontaneously developed lupus-like arthritis and glomerulonephritis (Nishimura et al., 1999). When backcrossed onto the NOD background, PD-1 deficiency increases the penetrance of spontaneous diabetes (Wang et al., 2005). These data strongly support a negative regulatory function of PD-1 in T and B cells.

The different spatial expression patterns of B7-DC and B7-H1 suggest B7-H1 is a major negative regulatory ligand for PD-1 in peripheral tissues whereas B7-DC is an important ligand in lymphoid organs (Chen, 2004). This hypothesis is supported by several animal models. Blockade of B7-H1, but not B7-DC, by systemic administration of blocking antibody, accelerated experimental autoimmune encephalitis (EAE) (Salama et al., 2003), autoimmune diabetes (Ansari et al., 2003), autoimmune hepatitis (Dong et al., 2004) and graft-versus-host disease (GVHD) (Blazar et al., 2003).

4 Upregulation of B7-H1 in Cancer and Chronic Viral Infections

The central function of B7-H1 to maintain peripheral tolerance is exploited in cancer (Dong et al., 2002). Immunohistochemistry studies demonstrated that B7-H1 was expressed on many types of freshly isolated human cancer tissues of kidney, breast, stomach, colon, lung, ovary, bladder, liver, cervix, esophagus, glioma and melanoma. Clinical studies on renal cell carcinoma (RCC), esophageal, gastric, ovarian and breast cancers have established the correlation of increased B7-H1 expression by tumor tissue with poor prognosis (Dong et al., 2002; Iwai et al., 2002). Long-term (over 10 years) follow-up of RCC (Thompson et al., 2006) and ovarian cancer patients (Hamanishi et al., 2007) also provided a clear correlation of B7-H1 expression with poor patient survival.

Meanwhile, in human HIV (Day et al., 2006; Trautmann et al., 2006), HCV (Urbani et al., 2006) and HBV (Boni et al., 2007) patients, upregulation of PD-1 on viral-specific T cell and systematic upregulation of B7-H1 (Barber et al.,2006; Chen et al., 2007) are associated with T-cell dysfunction and disease progression. In all cases, blockade of B7-H1-PD-1 pathway improved viral-specific T-cell function in vitro, indicating a potentially valuable therapeutic manipulation to sustain immunity to viral infections.

5 Mechanisms Underlying B7-H1-Mediated Suppression

Employing cell culture systems which are reconstituted with B7-H1-expressing cell lines, B7-H1 and PD-1 blocking antibodies and tumor-specific T cells, experimental data unequivocally support the idea that tumor-associated B7-H1 negatively affects T-cell functions in multiple aspects, including inhibition of T-cell proliferation, T-cell production of IL-2 and IFNγ, promotion of IL-10 production and suppression of cytolytic activity of tumor-specific CD8 T cell (CTL) (Dong et al., 2002). In several mouse tumor models in vivo, tumor-associated B7-H1 induces CTL apoptosis or confers resistance to lysis by CTL. In addition, B7-H1+ tumors are shown to be resistant to T-cell adoptive-transfer immunotherapy or therapeutic antibody anti-CD137. Furthermore, B7-H1 was shown to be necessary for the maintenance of T-cell exhaustion in a chronic infection mouse model of lymphocytic choriomeningitis virus (LCMV). Recent in vivo anergy/tolerance models also reveal a critical role of B7-H1-PD-1 pathway in initiation and maintenance of T-cell anergy (Goldberg et al., 2007; Keir et al., 2006; Martin-Orozco et al., 2006; Tsushima et al., 2007). Each of these mechanisms will be discussed in detail (summarized in Fig. 3).

5.1 T-Cell Deletion/Apoptosis

In an early study, cells from a human melanoma cell line 624mel, which express gp100 tumor antigen, were cultured together with a human CD8+ cytolytic T-cell (CTL) clone M15, which is specific for an epitope of gp100. As a result, B7-H1-transfected 624mel increased the apoptosis of M15 in a 5-day culture in vitro. Inclusion of B7-H1 blocking antibody or PD-1 fusion protein significantly inhibited apoptosis (Dong et al., 2002). In line with this in vitro observation, an adoptive transfer experiment using the mouse P815 tumor model further demonstrates that tumor-associated B7-H1 promotes the deletion of activated tumor-specific T cell in vivo. In those studies, the 2C TCR transgenic T cell recognizes the p2Ca peptide

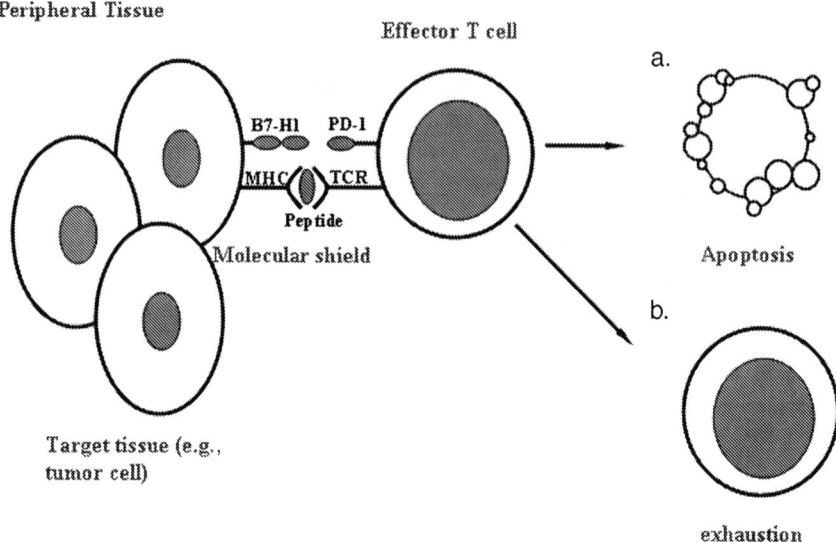
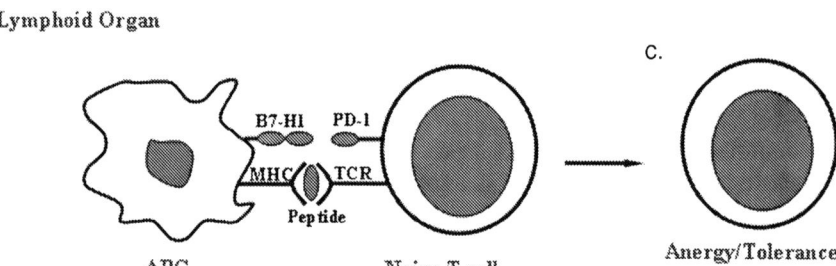

Fig. 3 B7-H1-mediated inhibitory mechanisms. According to the fate and functional outcome of T cell after B7-H1 engagement, we classify B7-H1-mediated suppression mechanisms into three main categories: deletion/apoptosis, resistance to CTL lysis and anergy. B7-H1 induces T-cell deletion/apoptosis, provides resistance to lysis and induces exhaustion in the peripheral, targeting effector T cell (**a, b**), whereas antigen presenting cell associated B7-H1 mediates anergy/tolerance formation in the lymphoid organs (**c**)

presented by P815 and lyses the tumor. However, when activated 2C T cells were adoptively transferred into mice bearing B7-H1-transfected P815 tumor, 2C T cells were quickly deleted. About 35% of 2C cells underwent apoptosis within 8 h post-transfer; in contrast, only 10% of 2C were apoptotic in mice bearing mock-transfected P815 tumor (Dong et al., 2002). Examination of tumor cell number in the peritoneal cavity 24 h after 2C transfer revealed an increase in B7-H1+ P815, but not mock P815. This experiment supports a role for B7-H1 in the induction of apoptosis of effector T cells. Injection of B7-H1 blocking antibody could inhibit the growth of B7-H1-positive tumor in the peritoneal cavity (Dong et al., 2002), further supporting this conclusion.

In addition to tumor models, B7-H1-induced T-cell apoptosis is also observed in a mouse corneal allograft system. B7-H1 is constitutively expressed in the corneal of eye, an immune-privileged site. In the B7-H1-positive corneal allograft, PD-1-positive infiltrating T cell underwent quick apoptosis once it came in contact with the allograft. Blockade of B7-H1 led to allograft rejection due to tissue destruction by infiltrating T cells (Hori et al., 2006). Similarly, B7-H1+ liver stellate cells could also mediate a B7-H1-dependent apoptosis of T cells (Dong et al., 2004). Further evidence supporting the role of B7-H1 in T-cell apoptosis was obtained through studies of B7-H1-deficient mice, in which a significant decrease of CD8+ T-cell apoptosis was observed in the liver (Dong et al., 2004).

5.2 Resistance to CTL Lysis: The Molecular Shield Hypothesis

In addition to the induction of apoptosis upon contacting T cells, B7-H1 on tumor cells is also found to confer resistance to lysis by CTL. In a mouse P815 tumor model, transfer of a tumor-specific TCR transgenic CTL clone P1A eradicates established wild-type P815 tumors. However, B7-H1-transfected P815 tumors are much more resistant to CTL lysis, which is not associated with T-cell apoptosis. Interestingly, when mock-transfected and B7-H1-transfected P815 were mixed in equal number and co-cultured with P1A CTL, 90% of mock-transfected P815 tumor cells were killed; in contrast, only 35% of B7-H1+ P815 were lysed. After engaged by B7-H1-positive P815, no obvious increase of apoptosis of P1A CTL was observed. Similarly, we also found that B7-H1+ P815 cells were also more resistant to 2C transgenic T-cell lysis in vitro (data not shown). Taken together, these findings provide a new interpretation, in addition to apoptosis, for our previous observation that B7-H1+ P815 tumor is resistant to 2C T cell-transfer immunotherapy (Dong et al., 2002). Upon incubation with B7-H1+ tumor cells, CTL remained fully functional, capable of recognizing and destroying B7-H1-negative tumor. Inclusion of B7-H1 or PD-1 blocking antibodies completely eliminated the resistance against CTL killing and led to regression of established B7-H1-positive tumors (Hirano et al., 2005; Iwai et al., 2002). This observation indicates that B7-H1 on tumor cells could confer resistance to lysis by fully activated T cells. While this mechanism clearly operates in cell culture systems, it remains to be addressed whether this immune evasion mechanism also operates in animal cancer models and in cancer patients. We refer to this phenomenon as a molecular shield and believe it may be critical as an immune escaper mechanism in cancer patients.

5.3 Generation and Maintenance of T-Cell Anergy and Exhaustion

T-cell anergy/tolerance and exhaustion are two similar but distinct phenomena. While both depict a hypoactive T-cell response to antigen-specific challenge, T-cell exhaustion is more specifically defined in chronic infection settings with prolonged exposure to antigen stimulation. As a result, a reduction in both number and functionality of antigen-specific CTL was observed. In a chronic-infection mouse model of LCMV, one highly invasive lab-derived LCMV clone will cause prolonged infection with detectable viremia in blood and multiple peripheral organs. PD-1 was dramatically upregulated on CTL in response to the viral infection, and its expression was maintained during chronic infection, while B7-H1 is persistently expressed on spleen cells. These infected mice have not only a significant decrease in the number of antigen-specific memory CD8 T-cell population, but also a functional impairment of their remaining memory T cells. Interestingly, blockade of B7-H1/PD-1 pathway by monoclonal antibodies against B7-H1 or PD-1 increased proliferation of several virus-specific CTL clones, restored CTL function and reduced viral burden (Barber et al.,2006). More importantly, persistent upregulation of PD-1 is also observed on HIV, HCV and HBV viral-specific CTLs in chronically infected patients and correlates with impairment of CTL function and disease progression (Day et al., 2006; Trautmann et al., 2006; Urbani et al., 2006). B7-H1 expression is also elevated on hepatocytes of HCV-infected patients and on circulating myeloid dendritic cells in HBV-infected patients (Chen et al., 2007). B7-H1/PD-1 blockade could revive exhausted CTL, restore HIV-, HCV- and HBV-specific CTL proliferation and cytokine production in vitro, which could potentially translate into better viral control in vivo. However, since the B7-H1-PD-1 pathway is crucial to maintain peripheral tolerance as shown in PD-1-deficient mice, blockade of B7-H1 might lead to autoimmune diseases. Indeed, when B7-H1-deficient mice were infected with the chronic LCMV clone, instead of clearing the virus, they succumbed to LCMV infection during acute infection phase due to severe immunopathology, thus indicating B7-H1 blockade in early infection may cause autoimmunity and even autoimmune diseases. As a result, with any future clinical application using B7-H1 blockade, the timing and frequency of antibody treatment must be carefully evaluated.

Other human viral pathogen, such as rhinovirus (Kirchberger et al., 2005), and bacterial pathogen, such as *Helicobacter pylori* (Das et al., 2006), also target B7-H1-PD-1 pathway to inhibit host immune response to achieve persistent infection. Both pathogens upregulate B7-H1 expression on antigen presenting cells (APC) which negatively affects APC-T-cell interaction.

The B7-H1/PD-1 pathway has also been recently shown to determine the initiation and maintenance of T anergy (Goldberg et al., 2007; Tsushima et al., 2007). In a peptide-induced anergy model, naïve CD8+ OT-1 TCR transgenic T cells, which recognize an $H-2K^b$-restricted epitope of chicken ovalbumin (OVA), are first exposed to excessive amount of soluble OVA peptide. OT-1 cells undergo a rapid expansion in the first week, followed by massive apoptosis of activated OT-1 T cell. The remaining survived cells become anergic, which no longer respond to OVA antigen restimulation (Barber et al.,2006). B7-H1 and PD-1 are quickly upregulated in the first 48 h in the lymphoid organs during the anergy induction phase, then downregulated to basal levels within a week. Because activated OT-1 T cells could only be found 72 h after antigen administration, these data support the idea that the B7-H1/PD-1 pathway determines anergy induction in lymphoid organs prior to T-cell exit into peripheral organs. In this peptide-induced anergy model, ablation or blockade of B7-H1-PD-1 pathway not only prevented anergy induction in the priming phase, but also reversed established anergy at a re-challenge step (Tsushima et al., 2007), suggesting that B7-H1 also plays a role in the maintenance of T-cell anergy.

In a different self-tolerance model, when influenza hemagglutinin (HA)-specific CD8+ T cells were adoptively transferred into HA transgenic mice, they become PD-1 positive and functionally tolerized at the priming phase. B7-H1/PD-1 blockade broke self-antigen-mediated tolerance induction, which led to the formation of functional CTL (Goldberg et al., 2007). Similarly, when naïve OT-1 cells were transferred into RIP-OVA transgenic mice expressing OVA antigen in the pancreatic islet cells, B7-H1/PD-1 blockade induced islet cell destruction and autoimmune diabetes (Martin-Orozco et al., 2006). In spontaneous autoimmune diabetes in nonobese diabetic (NOD) mice, blocking endogenous B7-H1/PD-1 pathway also accelerated disease onset and progression (Ansari et al., 2003). Furthermore, in a diabetes remission model by tolerance induction through FcR-nonbinding anti-CD3 and insulin-pulsed APCs treatment, B7-H1/PD-1 was found to be essential for tolerance induction and maintenance (Fife et al., 2006). Endogenous B7-H1 was also shown to be critical to maintain feto-maternal tolerance with its expression reported in placenta. B7-H1 antibody treatment increased the abortion rate in an abortion-prone allogeneic mating model (Guleria et al., 2005).

6 B7-H1 at the Crossroad of Peripheral Tolerance, Autoimmunity and Infection

It has become clear that the B7-H1/PD-1 pathway suppresses diverse T-cell immune responses through multiple mechanisms and at different tissue locations. The effect of B7-H1 in T-cell apoptosis and as a molecular shield operates mainly in the effector phase and in peripheral organs. Alternatively, antigen presenting cell-associated B7-H1-mediated tolerance induction could happen in the lymphoid organs, in addition to peripheral tissues, and provides protection for peripheral tissue from autoimmunity.

Various pathogens (virus or bacteria), parasites and tumor cells clearly exploit this suppressive pathway to evade host immune attack and to achieve persistent infection or growth. Modulation of the B7-H1 pathway could break tolerance, revive exhausted T cells and protect effector T cells from deletion, thus providing a promising therapeutic target against tumors and chronic viral infection. However, clinical manipulation of B7-H1 pathway must achieve proper balance between breaking pathogen-induced tolerance and minimizing pathogen-induced immunopathology.

Acknowledgments We thank Jennifer Osborne for editing the manuscript. This work is partially supported by the National Institutes of Health grants CA098731, CA106861 and CA113341.

References

Agata, Y., Kawasaki, A., Nishimura, H., Ishida, Y., Tsubata, T., Yagita, H., and Honjo, T. (1996). Expression of the PD-1 antigen on the surface of stimulated mouse T and B lymphocytes. *Int Immunol* 8:765–772.

Ansari, M. J., Salama, A. D., Chitnis, T., Smith, R. N., Yagita, H., Akiba, H., Yamazaki, T., Azuma, M., Iwai, H., Khoury, S. J., et al. (2003). The programmed death-1 (PD-1) pathway regulates autoimmune diabetes in nonobese diabetic (NOD) mice. *J Exp Med* 198:63–69.

Barber, D. L., Wherry, E. J., Masopust, D., Zhu, B., Allison, J. P., Sharpe, A. H., Freeman, G. J., and Ahmed, R. (2006). Restoring function in exhausted CD8 T cells during chronic viral infection. *Nature* 439:682–687.

Blazar, B. R., Carreno, B. M., Panoskaltsis-Mortari, A., Carter, L., Iwai, Y., Yagita, H., Nishimura, H., and Taylor, P. A. (2003). Blockade of programmed death-1 engagement accelerates graft-versus-host disease lethality by an IFN-gamma-dependent mechanism. *J Immunol* 171: 1272–1277.

Boni, C., Fisicaro, P., Valdatta, C., Amadei, B., Di Vincenzo, P., Giuberti, T., Laccabue, D., Zerbini, A., Cavalli, A., Missale, G., et al. (2007). Characterization of Hbv-specific T cell dysfunction in chronic Hbv infection. *J Virol* 81:4215–4225.

Chambers, C. A., Kuhns, M. S., Egen, J. G., and Allison, J. P. (2001). CTLA-4-mediated inhibition in regulation of T cell responses: mechanisms and manipulation in tumor immunotherapy. *Annu Rev Immunol* 19:565–594.

Chen, L. (2004). Co-inhibitory molecules of the B7-CD28 family in the control of T-cell immunity. *Nat Rev Immunol* 4:336–347.

Chen, L., Zhang, Z., Chen, W., Li, Y., Shi, M., Zhang, J., Wang, S., and Wang, F. S. (2007). B7-h1 up-regulation on myeloid dendritic cells significantly suppresses T cell immune function in patients with chronic hepatitis B. *J Immunol* 178:6634–6641.

Cross, A. H., Girard, T. J., Giacoletto, K. S., Evans, R. J., Keeling, R. M., Lin, R. F., Trotter, J. L., and Karr, R. W. (1995). Long-term inhibition of murine experimental autoimmune encephalomyelitis using CTLA-4-Fc supports a key role for CD28 costimulation. *J Clin Invest* 95:2783–2789.

Das, S., Suarez, G., Beswick, E. J., Sierra, J. C., Graham, D. Y., and Reyes, V. E. (2006). Expression of B7-H1 on gastric epithelial cells: its potential role in regulating T cells during *Helicobacter pylori* infection. *J Immunol* 176:3000–3009.

Day, C. L., Kaufmann, D. E., Kiepiela, P., Brown, J. A., Moodley, E. S., Reddy, S., Mackey, E. W., Miller, J. D., Leslie, A. J., DePierres, C., et al. (2006). PD-1 expression on HIV-specific T cells is associated with T-cell exhaustion and disease progression. *Nature* 443:350–354.

Dong, H., Strome, S. E., Salomao, D. R., Tamura, H., Hirano, F., Flies, D. B., Roche, P. C., Lu, J., Zhu, G., Tamada, K., et al. (2002). Tumor-associated B7-H1 promotes T-cell apoptosis: a potential mechanism of immune evasion. *Nat Med* 8:793–800.

Dong, H., Zhu, G., Tamada, K., and Chen, L. (1999). B7-H1, a third member of the B7 family, co-stimulates T-cell proliferation and interleukin-10 secretion. *Nat Med* 5:1365–1369.

Dong, H., Zhu, G., Tamada, K., Flies, D. B., van Deursen, J. M., and Chen, L. (2004). B7-H1 determines accumulation and deletion of intrahepatic CD8(+) T lymphocytes. *Immunity* 20:327–336.

Fife, B. T., Guleria, I., Gubbels Bupp, M., Eagar, T. N., Tang, Q., Bour-Jordan, H., Yagita, H., Azuma, M., Sayegh, M. H., and Bluestone, J. A. (2006). Insulin-induced remission in new-onset NOD mice is maintained by the PD-1-PD-L1 pathway. *J Exp Med* 203:2737–2747.

Freeman, G. J., Long, A. J., Iwai, Y., Bourque, K., Chernova, T., Nishimura, H., Fitz, L. J., Malenkovich, N., Okazaki, T., Byrne, M. C., et al. (2000). Engagement of the PD-1 immunoinhibitory receptor by a novel B7 family member leads to negative regulation of lymphocyte activation. *J Exp Med* 192:1027–1034.

Goldberg, M. V., Maris, C. H., Hipkiss, E. L., Flies, A. S., Zhen, L., Tuder, R. M., Grosso, J. F., Harris, T. J., Getnet, D., Whartenby, K. A., et al. (2007). Role of PD-1 and its ligand, B7-H1, in early fate decisions of CD8 T cells. *Blood* 110.106–192.

Guleria, I., Khosroshahi, A., Ansari, M. J., Habicht, A., Azuma, M., Yagita, H., Noelle, R. J., Coyle, A., Mellor, A. L., Khoury, S. J., and Sayegh, M. H. (2005). A critical role for the programmed death ligand 1 in fetomaternal tolerance. *J Exp Med* 202:231–237.

Hamanishi, J., Mandai, M., Iwasaki, M., Okazaki, T., Tanaka, Y., Yamaguchi, K., Higuchi, T., Yagi, H., Takakura, K., Minato, N., et al. (2007). Programmed cell death 1 ligand 1 and tumor-infiltrating CD8+ T lymphocytes are prognostic factors of human ovarian cancer. *Proc Natl Acad Sci USA* 104:3360–3365.

Hirano, F., Kaneko, K., Tamura, H., Dong, H., Wang, S., Ichikawa, M., Rietz, C., Flies, D. B., Lau, J. S., Zhu, G., et al. (2005). Blockade of B7-H1 and PD-1 by monoclonal antibodies potentiates cancer therapeutic immunity. *Cancer Res* 65:1089–1096.

Hori, J., Wang, M., Miyashita, M., Tanemoto, K., Takahashi, H., Takemori, T., Okumura, K., Yagita, H., and Azuma, M. (2006). B7-H1-induced apoptosis as a mechanism of immune privilege of corneal allografts. *J Immunol* 177:5928–5935.

Ishida, Y., Agata, Y., Shibahara, K., and Honjo, T. (1992). Induced expression of PD-1, a novel member of the immunoglobulin gene superfamily, upon programmed cell death. *EMBO J* 11:3887–3895.

Iwai, Y., Ishida, M., Tanaka, Y., Okazaki, T., Honjo, T., and Minato, N. (2002). Involvement of PD-L1 on tumor cells in the escape from host immune system and tumor immunotherapy by PD-L1 blockade. *Proc Natl Acad Sci USA* 99:12293–12297.

Keir, M. E., Liang, S. C., Guleria, I., Latchman, Y. E., Qipo, A., Albacker, L. A., Koulmanda, M., Freeman, G. J., Sayegh, M. H., and Sharpe, A. H. (2006). Tissue expression of PD-L1 mediates peripheral T cell tolerance. *J Exp Med* 203:883–895.

Kirchberger, S., Majdic, O., Steinberger, P., Bluml, S., Pfistershammer, K., Zlabinger, G., Deszcz, L., Kuechler, E., Knapp, W., and Stockl, J. (2005). Human rhinoviruses inhibit the accessory function of dendritic cells by inducing sialoadhesin and B7-H1 expression. *J Immunol* 175:1145–1152.

Krambeck, A. E., Thompson, R. H., Dong, H., Lohse, C. M., Park, E. S., Kuntz, S. M., Leibovich, B. C., Blute, M. L., Cheville, J. C., and Kwon, E. D. (2006). B7-H4 expression in renal cell carcinoma and tumor vasculature: associations with cancer progression and survival. *Proc Natl Acad Sci USA* 103:10391–10396.

Krummel, M. F., and Allison, J. P. (1995). CD28 and CTLA-4 have opposing effects on the response of T cells to stimulation. *J Exp Med* 182:459–465.

Latchman, Y., Wood, C. R., Chernova, T., Chaudhary, D., Borde, M., Chernova, I., Iwai, Y., Long, A. J., Brown, J. A., Nunes, R., et al. (2001). PD-L2 is a second ligand for PD-1 and inhibits T cell activation. *Nat Immunol* 2:261–268.

Lenschow, D. J., Walunas, T. L., and Bluestone, J. A. (1996). CD28/B7 system of T cell costimulation. *Annu Rev Immunol* 14:233–258.

Linsley, P. S., Brady, W., Urnes, M., Grosmaire, L. S., Damle, N. K., and Ledbetter, J. A. (1991). CTLA-4 is a second receptor for the B cell activation antigen B7. *J Exp Med* 174:561–569.

Linsley, P. S., Clark, E. A., and Ledbetter, J. A. (1990). T-cell antigen CD28 mediates adhesion with B cells by interacting with activation antigen B7/BB-1. *Proc Natl Acad Sci USA* 87:5031–5035.

Martin-Orozco, N., Wang, Y. H., Yagita, H., and Dong, C. (2006). Cutting edge: programmed death (PD) ligand-1/PD-1 interaction is required for CD8+ T cell tolerance to tissue antigens. *J Immunol* 177:8291–8295.

Mazanet, M. M., and Hughes, C. C. (2002). B7-H1 is expressed by human endothelial cells and suppresses T cell cytokine synthesis. *J Immunol* 169:3581–3588.

Mueller, D. L., Jenkins, M. K., and Schwartz, R. H. (1989). Clonal expansion versus functional clonal inactivation: a costimulatory signalling pathway determines the outcome of T cell antigen receptor occupancy. *Annu Rev Immunol* 7:445–480.

Nishimura, H., Nose, M., Hiai, H., Minato, N., and Honjo, T. (1999). Development of lupus-like autoimmune diseases by disruption of the PD-1 gene encoding an ITIM motif-carrying immunoreceptor. *Immunity* 11:141–151.

Nishimura, H., Okazaki, T., Tanaka, Y., Nakatani, K., Hara, M., Matsumori, A., Sasayama, S., Mizoguchi, A., Hiai, H., Minato, N., and Honjo, T. (2001). Autoimmune dilated cardiomyopathy in PD-1 receptor-deficient mice. *Science* 291:319–322.

Okazaki, T., Maeda, A., Nishimura, H., Kurosaki, T., and Honjo, T. (2001). PD-1 immunoreceptor inhibits B cell receptor-mediated signaling by recruiting src homology 2-domain-containing tyrosine phosphatase 2 to phosphotyrosine. *Proc Natl Acad Sci USA* 98:13866–13871.

Salama, A. D., Chitnis, T., Imitola, J., Ansari, M. J., Akiba, H., Tushima, F., Azuma, M., Yagita, H., Sayegh, M. H., and Khoury, S. J. (2003). Critical role of the programmed death-1 (PD-1) pathway in regulation of experimental autoimmune encephalomyelitis. *J Exp Med* 198:71–78.

Schwartz, J. C., Zhang, X., Fedorov, A. A., Nathenson, S. G., and Almo, S. C. (2001). Structural basis for co-stimulation by the human CTLA-4/B7-2 complex. *Nature* 410:604–608.

Stamper, C. C., Zhang, Y., Tobin, J. F., Erbe, D. V., Ikemizu, S., Davis, S. J., Stahl, M. L., Seehra, J., Somers, W. S., and Mosyak, L. (2001). Crystal structure of the B7-1/CTLA-4 complex that inhibits human immune responses. *Nature* 410:608–611.

Thompson, R. H., Kuntz, S. M., Leibovich, B. C., Dong, H., Lohse, C. M., Webster, W. S., Sengupta, S., Frank, I., Parker, A. S., Zincke, H., et al. (2006). Tumor B7-H1 is associated with poor prognosis in renal cell carcinoma patients with long-term follow-up. *Cancer Res* 66: 3381–3385.

Trautmann, L., Janbazian, L., Chomont, N., Said, E. A., Gimmig, S., Bessette, B., Boulassel, M. R., Delwart, E., Sepulveda, H., Balderas, R. S., et al. (2006). Upregulation of PD-1 expression on HIV-specific CD8+ T cells leads to reversible immune dysfunction. *Nat Med* 12:1198–1202.

Tseng, S. Y., Otsuji, M., Gorski, K., Huang, X., Slansky, J. E., Pai, S. I., Shalabi, A., Shin, T., Pardoll, D. M., and Tsuchiya, H. (2001). B7-DC, a new dendritic cell molecule with potent costimulatory properties for T cells. *J Exp Med* 193:839–846.

Tsushima, F., Yao, S., Shin, T., Flies, A., Flies, S., Xu, H., Tamada, K., Pardoll, D. M., and Chen, L. (2007). Interaction between B7-H1 and PD-1 determines initiation and reversal of T-cell anergy. *Blood* 110:180–185.

Urbani, S., Amadei, B., Tola, D., Massari, M., Schivazappa, S., Missale, G., and Ferrari, C. (2006). PD-1 expression in acute hepatitis C virus (HCV) infection is associated with HCV-specific CD8 exhaustion. *J Virol* 80:11398–11403.

Wang, J., Yoshida, T., Nakaki, F., Hiai, H., Okazaki, T., and Honjo, T. (2005). Establishment of NOD-Pdcd1−/− mice as an efficient animal model of type I diabetes. *Proc Natl Acad Sci USA* 102:11823–11828.

Watanabe, N., Gavrieli, M., Sedy, J. R., Yang, J., Fallarino, F., Loftin, S. K., Hurchla, M. A., Zimmerman, N., Sim, J., Zang, X., et al. (2003). BTLA is a lymphocyte inhibitory receptor with similarities to CTLA-4 and PD-1. *Nat Immunol* 4:670–679.

Wiendl, H., Mitsdoerffer, M., Schneider, D., Chen, L., Lochmuller, H., Melms, A., and Weller, M. (2003). Human muscle cells express a B7-related molecule, B7-H1, with strong negative immune regulatory potential: a novel mechanism of counterbalancing the immune attack in idiopathic inflammatory myopathies. *FASEB J* 17:1892–1894.

Zhang, X., Schwartz, J. C., Guo, X., Bhatia, S., Cao, E., Lorenz, M., Cammer, M., Chen, L., Zhang, Z. Y., Edidin, M. A., et al. (2004). Structural and functional analysis of the costimulatory receptor programmed death-1. *Immunity* 20:337–347.

Regulatory T Cells in Cancer

Silvia Piconese and Mario P. Colombo

1 Introduction

It is now widely accepted that tumors actively devise subversion over a variety of immune players (Zitvogel et al., 2006). Counterbalancing such immune escape has become a major goal of immunotherapy today. The "self" nature of several tumor-associated antigens explains their weak immunogenicity and the need to overcome self-tolerance to properly activate a response against them. Immunosuppression accompanying tumor progression renders such response very unlikely. Regulatory T cells (Treg) act on maintaining tolerance and exerting immunosuppression, depending on their relative number in the $CD4^+$ T-cells pool.

The unique feature of regulatory T lymphocytes is represented by the ability to actively suppress immune responses. The characterization of minor features resulted in the classification of Treg into two main subsets: the "naturally arising" Treg that develop in the thymus due to high-affinity TCR triggering and suppress bystander T-cell proliferation by a still unknown mechanism requiring cell-to-cell interaction and the "adaptive" Treg, which develop peripherally following antigenic stimulation in the presence of IL-10 (Tr1 subset) or TGF-β (Th3 subset). Such classification is susceptible to adjustments, since regulatory lymphocytes phenotypically indistinguishable from natural Treg can develop in the periphery following low-dose antigen and TGF-β administration (Kretschmer et al., 2005; Wing, 2006). Therefore, thymic development, suboptimal antigenic stimulation and peripheral conversion all contribute to create the total pool of Treg, whose maintenance involves MHC II molecules (Gavin et al., 2002), the cytokine IL-2 (Malek and Bayer, 2004) and costimulatory molecules such as CD28 (Tang et al., 2003) and CD40 (Guiducci et al., 2005).

In spite of their heterogeneous origin, Treg subtypes share distinctive markers. Sakaguchi's group (Sakaguchi et al., 1995) first associated the regulatory phenotype to CD25, the α-chain of the high-affinity receptor for interleukin 2 (IL-2). Treg

---

S. Piconese
Immunotherapy and Gene Therapy Unit, Department of Experimental Oncology, Fondazione IRCCS Istituto Nazionale dei Tumori, Milan, Italy

depletion by means of CD25 targeting was significantly used to demonstrate their role as a "common basis" among the several aspects of immune regulation (Shimizu et al., 1999). Due to the intrinsic limit of this marker, associated to both activated effectors and regulatory T cells, other molecules have been proposed to identify Treg, such as the glucocorticoid-induced tumor necrosis factor receptor (GITR), OX40, the cytotoxic T-lymphocyte antigen 4 (CTLA-4), whose expression, however, only partially overlaps with the regulatory phenotype. The actual breakthrough in Treg characterization was the discovery that the forkhead box transcription factor Foxp3 was the master gene of Treg lineage (Fontenot et al., 2003; Hori et al., 2003). Very recent molecular studies have provided the renewed interpretation of Foxp3 as a mediator that "amplifies and fixes pre-established molecular features of Treg cells" (Gavin et al., 2007). Still, Foxp3 can be transiently expressed also by activated T cells in humans (Wang et al., 2007) and the very last identified unique signature distinguishing Treg is the downregulation of cyclic nucleotide phosphodiesterase 3B (Gavin et al., 2007).

Although the molecular signature of Treg has been extensively dissected, the fine mechanisms by which they exert suppression are not fully resolved yet. The original finding that cell-to-cell contact is required for in vitro suppression (Takahashi et al., 1998; Thornton and Shevach, 1998) has been challenged by several observations in

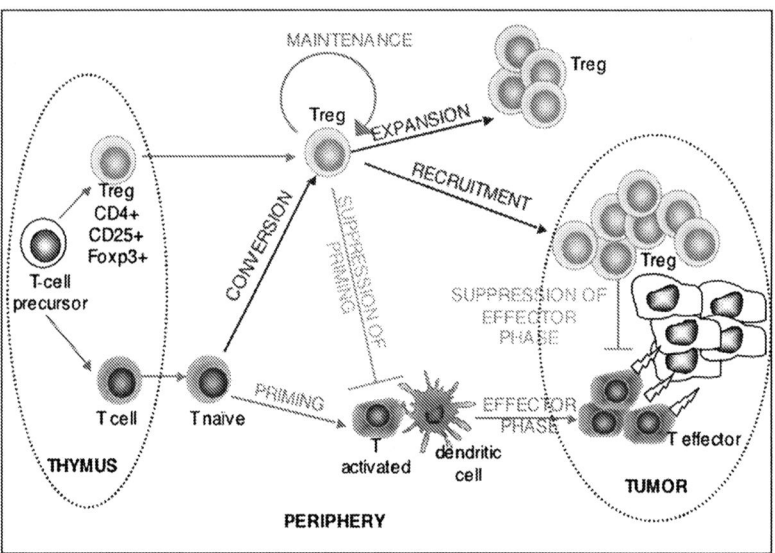

Fig. 1 Tumor fosters Treg toward immunosuppression. Growing tumor breaks Treg homeostasis by actively increasing Treg number in lymphoid organs and within the tumor mass. The total pool of Treg in tumor bearers includes not only thymus-derived Treg but also newly derived Treg originated from peripheral conversion of naïve T cells. Tumor-associated Treg can undergo expansion following tumor-induced signals and can be actively recruited at the tumor site by chemokine gradient. Treg can suppress both the priming and the effector phase of the anti-tumor immune response. Tumor cells directly and indirectly promote conversion, expansion and recruitment of Treg

complex systems, in which both cellular interactions and soluble factors have been demonstrated to participate in overall Treg-mediated suppression. The identification of possible targets of Treg suppression has been likewise submitted to continuous update, since it has been shown that not only T cells, but also B lymphocytes, NK cells and dendritic cells are susceptible to Treg inhibition. Overall, emerging evidences place Treg central to the immunoregulatory network (Fig. 1).

2 Treg and Tumors: of Mice and Men

Immunosuppression physiologically occurs in immune-privileged organs like testis, placenta and eye. Furthermore, some pathological conditions like tumorigenesis can use mechanisms providing immune privilege to escape from the host immune response. Treg are involved in both physiological and pathological suppression of immune reactivity.

Moreover, though natural Treg display a TCR repertoire mainly directed toward self-antigens (Hsieh et al., 2004), they suppress not only auto-reactive but also allogeneic immune responses; therefore, they might counterbalance anti-tumor immunity regardless of the targeted antigens, either tumor-associated self-antigens or tumor-specific non-self-molecules.

Besides exerting suppressive effects in the tumor microenvironment, regulatory lymphocytes are able to recirculate through lymphoid compartment and peripheral tissues, spreading local tolerance systemically. Indeed, systemic ablation of Treg elicits concomitant immunity at sites distant from primary tumor (Turk et al., 2004). Moreover, anti-tumor Treg-targeted therapies have frequently been associated to multiorgan autoimmune manifestations (Phan et al.,2003), as the effects of Treg inhibition spread to non-tumor tissues undermining systemic tolerance.

Treg biology should be taken into account when designing immunotherapeutic approaches. Indeed, the administration of IL-2 in cancer patients resulted in Treg rather than T-helper expansion, since this cytokine is critically involved in Treg proliferation, maintenance and suppressive function. In addition, some experimental vaccination protocols, which were believed to be immunostimulatory, elicited Treg amplification thus attenuating their potential therapeutic efficacy (Zhou et al., 2006).

In recent years, the role of Treg in the intricate network characterizing tumor-associated immune tolerance has been extensively investigated. Despite an incomplete understanding of their biology, the involvement of Treg in tumor immune escape remains undoubted.

2.1 Treg in Murine Tumors

Berendt and North (1980) first described T lymphocytes with suppressive activity in mice bearing an immunogenic fibrosarcoma, which prevented adoptively transferred effector cells from inducing tumor regression. The almost 20 years of silence that followed this pioneer observation were mainly due to the absence of a marker that specifically identified such suppressor T cells. Subsequently, a $CD4^+$ T-cell population similar to North's suppressor T cells was found involved in autoimmune manifestations. These cells were characterized by the expression of the surface marker CD25 and shown to be responsible for the control of autoimmunity. Indeed, removal of CD25-positive T cells led to autoimmune disease in several organs and reconstitution of the eliminated population reverted the pathological status (Powrie and Mason, 1990; Sakaguchi et al., 1985; Sugihara et al., 1988). This population was found to be anergic and suppressive in vitro and proposed to be responsible for autoimmunity induced by thymectomy 3 days postnatal, in accordance to the key role of thymus in central tolerance (Itoh et al., 1999).

The knowledge acquired in the field of autoimmunity was soon translated into that of tumor immunology by Sakaguchi and collaborators in 1999 (Shimizu et al., 1999). Administration of anti-CD25-depleting antibody (PC61 clone) prior to injection of a leukemia cell line induced CTL- and NK-mediated tumor rejection. Similar results were also obtained in other laboratories toward hematological and solid tumor models (Golgher et al., 2002; Jones et al., 2002; Onizuka et al., 1999). Concurrent depletion of effector T cells, activated by concomitant immunity or vaccination, represents the main limit of using the same treatment in a therapeutic rather than preventive setting (Onizuka et al., 1999). Other limitations of the depletion approach will be extensively discussed in Sect. 6.1.

Sakaguchi proposed the term of regulatory T cells as "a common basis between tumor immunity and autoimmunity" (Shimizu et al., 1999). Anti-tumor and anti-self responses are strictly linked. Indeed, mice rejecting the melanoma cell line B16F10 because of Treg depletion also mounted efficient immunity toward the self melanocyte differentiation antigen tyrosinase (Jones et al., 2002). The immune response elicited by Treg depletion was directed against tumor antigens shared among different histotypes, since mice rejecting the colon carcinoma cell line CT26 were protected against tumors of different origin (Golgher et al., 2002). This observation stressed the possibility that Treg removal might allow auto-reactive T cells to target self-antigens expressed by tumor cells, being potentially dangerous for normal tissues.

Synergistic effects of CTLA-4 blockade and Treg depletion in inducing anti-tumor and anti-self responses following melanoma vaccination indicate that CTLA-4 is not the suppressive element on Treg (Sutmuller et al., 2001). The combined treatments were likely acting on different levels of autoimmunity control. The dominant Treg-mediated tolerance was broken by CD25 depletion, while blocking the inhibitory receptor CTLA-4 reversed the recessive tolerance of autoreactive lymphocytes. The protection from melanoma growth was strictly accompanied by depigmentation caused by immune attack of self-antigens expressed by melanocytes. However, the sole Treg depletion is sufficient to induce vitiligo in the course of melanoma immunotherapy. In a setting in which lymphocytes, transgenic for the TCR to the melanoma antigen gp100, were adoptively transferred in B16-bearing mice, naturally occurring Treg were suppressing both T-helper and antigen-specific CD8+ cells. Their removal was also associated with tumor rejection and vitiligo (Antony et al., 2005).

Depletion of CD4+ lymphocytes, either alone or in combination with GM-CSF, leads to complete rejection of B16 melanoma cell line (Turk et al., 2004). Here for the first time an alternative approach was proposed: to target Treg by functionally inhibiting them rather than depleting them. Indeed, Sakaguchi had previously demonstrated that the stimulation of GITR on Treg reversed their suppressive function (Shimizu et al., 2002). When applied to cancer immunotherapy, this approach promoted concomitant immunity. GITR triggering consistently improves vaccine-induced immunity and reverts tolerance to tumor antigens without producing overt autoimmune side effects when administered within the tumor (Cohen et al., 2006; Ko et al., 2005; Ramirez-Montagut, 2006). In all of these models, however, GITR stimulation appeared mainly to boost effector T cells rather than inhibit Treg.

Since Treg heavily infiltrate tumors of different origin, targeting Treg locally rather than systemically could avoid generalized autoimmune manifestations. This issue has been investigated by Yu et al. (2005) who demonstrated that Treg ablation within the tumor mass leads to the complete rejection of advanced highly immunogenic lesions. This result was achieved by administration of anti-CD4 that, instead of anti-CD25-depleting antibody, allowed eliminating CD4+ Treg while sparing CD8+CD25+-activated effector cells. This model highlighted that Treg are capable of actively hindering the endogenous immune response that might be generated against tumors.

In transgenic mice carrying the Her-2/neu under the MMTV promoter, the oncogene is a self-antigen. Although progressing tumor induces Treg expansion (Ambrosino, 2006 and our unpublished observation), natural Treg seem to have no role during earlier immunosurveillance (Chiodoni et al., 2006).

2.2 Treg in Human Cancer

In humans, it was observed that T lymphocytes infiltrating non-small cell lung cancer and late-stage ovarian cancer expressed CD25 and produced the inhibitory cytokine TGF-β, thus suggesting that Treg were actively hindering anti-tumor immunity (Woo et al., 2001). It was later demonstrated that in patients with pancreatic, breast, hepatocellular and gastric carcinoma, Treg were expanded not only in the tumor microenvironment but also in the draining lymph nodes, the ascites and the peripheral blood, associated with the malignancy progression (Liyanage et al., 2002; Ormandy et al., 2005; Sasada et al., 2003). Treg accumulation in human ovarian epithelial cancer has been demonstrated to be linked to active recruitment of CCR4-expressing Treg, from periphery to tumor microenvironment, by the chemokine CCL-22 produced by tumor cells and by tumor-infiltrating macrophages (Curiel et al., 2004). For this tumor, a positive correlation between Treg increase and a poor prognosis has been demonstrated.

The role proposed for natural Treg in human solid tumors is quite different from the indications arising from hematological malignancies. Indeed, the reported Treg accumulation in myeloma and lymphoma patients has been controversially associated with the clinical outcome of the disease, since both anti-tumor effector lymphocytes and tumor cells themselves may be subjected to Treg suppression. Treg are recruited within human Hodgkin and non-Hodgkin lymphomas by the chemokine CCL-22, where they functionally suppress tumor-infiltrating T cells (Ishida et al., 2006; Yang et al., 2006). Unexpectedly, such Treg accumulation results in improved overall survival of follicular lymphoma patients (Carreras et al., 2006), suggesting that Treg can inhibit proliferation of transformed cells.

The adverse effects of Treg have influenced the design of immunotherapy protocols in clinical settings. Attention was given to the potential concurrent Treg expansion during IL-2 administration (Antony and Restifo, 2005) or following dendritic cells-based vaccination (Banerjee et al., 2006). Treg neutralization has been tested to break tolerance to tumor antigens. For instance, the CD25-depleting agent Denileukin diftitox, a fusion protein composed by interleukin 2 conjugated to the diphtheria toxin, showed some efficacy in ovary, breast, lung and renal carcinoma treatments (Barnett et al., 2005; Dannull et al., 2005), but was totally ineffective against melanoma (Attia et al., 2005). On the other hand, treatment of metastatic ovarian carcinoma and melanoma with anti-CTLA-4 antibody, while showing some efficacy in association with vaccination, led to severe autoimmune reactions (Hodi et al., 2003; Phan et al., 2003). Clearly, additional studies are needed to improve Treg neutralization in clinical settings.

3 Targets of Treg Suppression in Anti-tumor Immunity

Treg have been shown to suppress immune cell types belonging to the innate and the adaptive response and involved in both priming and effector phase of anti-tumor immunity.

3.1 Treg Suppression of Anti-tumor Innate Response and Priming

The continuous exposure to microbial and food-derived antigens contributes to the establishment of inflammatory conditions that, in the gut, might lead to carcinogenesis. Erdman et al. (2003) have shown that adoptive transfer of Treg specifically inhibits both the early phase of inflammation and the subsequent dysplasia occurring in RAG knock-out mice upon microbe infection. Treg-mediated suppression was IL-10-dependent as no protection was obtained by transferring IL-10-deficient Treg. In other systems, Treg might help tumor progression through the inhibition of NK and dendritic cells.

Natural killer (NK) cells have a primary role in the innate response and in immunosurveillance. Treg can suppress NK cells both indirectly, inhibiting interleukin-2 production by T lymphocytes, and directly, via TGF-β (Ghiringhelli et al., 2005a). Activated Treg express TGF-β bound to the latency-associated protein (LAP) on the cell surface, while resting NK cells express the TGF-β receptor. Blockade of TGF-β signal reverses Treg suppression over NK function and rescues anti-tumor natural cytotoxicity.

The tumor microenvironment provides a suppressive milieu whose effects spread distally to the bone marrow, inducing the expansion of myeloid-derived suppressor cells (MDSC) (Gabrilovich et al., 2007; Melani et al., 2003). It is presumable that, in the tumor microenvironment, Treg contribute to macrophage skewing toward a tolerogenic phenotype by producing suppressive cytokines and, in the periphery, activate suppressive metabolic pathways in MDSC.

At the interface between innate and adaptive immunity, dendritic cells have a crucial role in determining the fate of anti-tumor response and their suppression is common to several escape mechanisms. Indeed, tumor-infiltrating dendritic cells are usually blocked in a tolerogenic state that is difficult to reverse either in vitro or in vivo. In several conditions, they display low levels of costimulatory molecules, produce low amounts of immunostimulatory cytokines and poorly activate T-cells proliferation in mixed leukocyte reactions (Serra et al., 2003). Dendritic cells infiltrating the colon carcinoma tumor CT26 produce IL-10 and tumor-associated T cells display a suppressor phenotype. Such immunosuppressive microenvironment inhibits the priming toward an unrelated antigen administered concurrently to tumor injection. Treg depletion rescues normal cytokine production and improves tumor outcome (Jarnicki et al., 2006). In this setting, bidirectional interactions between monocytes and Treg might occur: tolerogenic dendritic cells could induce T-cell suppression, while Treg could suppress APC maturation and function, establishing a vicious circle of immunosuppression (Fig. 2).

The direct Treg-suppressive effect on dendritic cells activation has been clearly demonstrated by in vitro studies. Naïve Treg specifically modulate cytokine production by bone marrow-derived dendritic cells, reducing the pro-inflammatory IL-6 and augmenting the suppressive IL-10 upon maturating stimuli. Not only natural Treg, but also Treg derived from tumor-bearing animals display suppressive properties over activation of bone marrow-derived dendritic cells (Larmonier et al., 2007). An elegant demonstration of Treg involvement in tumor-induced dendritic cell suppression has been provided by Dercamp et al. (2005). They showed that in vivo reversal of dendritic cells paralysis by means of CpG administration is greater when coupled to Treg depletion. They further demonstrated that Treg suppression is partly due to IL-10 production; however, Treg seem not to be the only source of IL-10, since their depletion and IL-10 neutralization show synergistic therapeutic effects. The resulting rescue of dendritic cells functions leads to efficient priming of tumor-specific $CD8^+$ T lymphocytes. Considering that STAT3 targeting restores dendritic cells functions in tumor-bearing host, IL-10 is likely to be involved in dendritic cells tolerization, via STAT3 activation (Kortylewski et al., 2005).

Enzymatic alterations, such as IDO activation, have also been associated to the tolerogenic phenotype of antigen-presenting cells. Treg may take part in this process. Fallarino et al. (2003) have demonstrated that Treg activate IDO expression

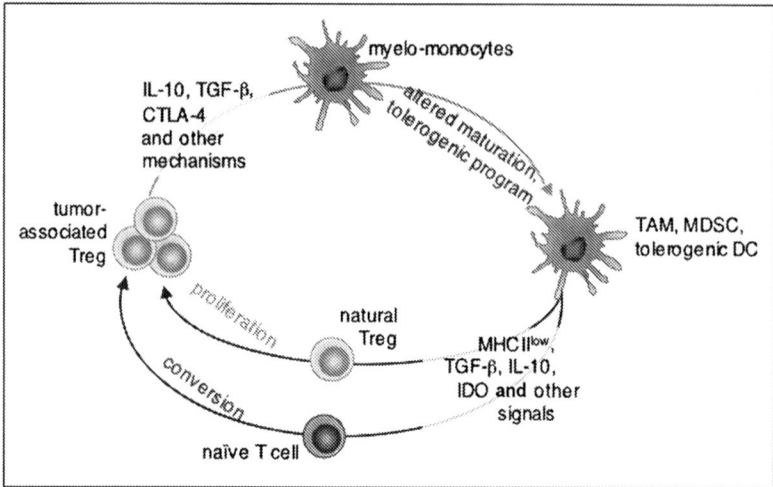

Fig. 2 Treg endorse a vicious circle of immunosuppression. Treg maintain their own activity by bidirectional interaction with the tumor-associated myeloid populations. Treg activate several immunosuppressive pathways in myelo-monocytes, such as IDO activation by CTLA-4/B7 signaling. Moreover, Treg-derived immunosuppressive cytokines induce the differentiation of tolerogenic dendritic cells and suppressive tumor-associated macrophages (TAM) and activates tolerogenic enzymatic pathways in myeloid-derived suppressor cells (MDSC). Conversely, these myeloid populations, characterized by sub-immunogenic presentation, production of immunosuppressive cytokines and activation of tolerogenic pathways such as IDO, can promote both conversion of T cells into Treg and proliferation of already-differentiated Treg. A self-enhancing cycle of suppression establishes increasing tolerance along tumor progression

in dendritic cells in vitro. Indeed CTLA-4, constitutively expressed by Treg and overexpressed upon TCR activation, can activate the B7 signal on dendritic cells, which induces expression of IDO leading to tryptophan depletion and accumulation of pro-apoptotic mediators.

In light of what has been discussed above, the classical model of Treg-mediated direct suppression of lymphocyte proliferation requires partial re-interpretation. Indeed, Treg could inhibit T-cell proliferation by suppressing cognate APC interactions with effector cells. This model was confirmed by in vitro and in vivo imaging analyses of Treg function. By adding differentially labeled Treg and responder T cells to the classical in vitro suppression assay, Tang and Krummel (2006) have shown that possible. The emerging hypothesis is that, while T-helper cells license APC to induce priming, Treg may counteract the cognate activation. Furthermore, in two different models of experimental autoimmune diseases, in vivo imaging showed Treg interfering with APC–effector complexes, leading to the formation of unstable and short interactions at the immunological synapse (Tadokoro et al., 2006; Tang, 2006). A similar hypothesis has been proposed after showing that upon co-culture, human Treg cells are able to induce the expression of the inhibitory receptors ILT-3 and ILT-4 by dendritic cells (Vlad et al., 2005).

3.2 Treg Suppression of Anti-tumor Adaptive Immunity

Although antigen-presenting cells may be involved in Treg-mediated suppression, in vitro, Treg are able to suppress the proliferation of responder T cells in the absence of other cells (Shevach, 2002). Moreover, the regulation of adaptive responses by Treg can occur not only in secondary lymphoid organs, but also in peripheral tissues, where already-primed effectors reside. Indeed, the correct localization of Treg at inflammatory sites is crucial for their in vivo activity (Siegmund et al., 2005).

$CD8^+$ CTL are considered the main cellular subset responsible for tumor elimination in both humans and mice. Transfer of antigen-specific CTL induces tumor rejection, while the cotransfer of $CD4^+$ Treg abolishes such effects (Chen et al., 2005). In this experimental setting, Treg mainly inhibited the cytotoxicity of $CD8^+$ T lymphocytes rather than their proliferation, survival or cytokines secretion. TGF-β signaling seems critical for Treg immunoregulation, since Treg fail to suppress $CD8^+$ cells genetically engineered to lose responsiveness to this cytokine. TGF-β is an important regulator of CTL functions, altering the transcriptional program of $CD8^+$ lymphocytes toward a defective production of cytotoxic granules (Thomas and Massague, 2005). Similar data have been obtained from a human cytotoxic cell line, which kills autologous tumor cells in vitro unless Treg are present. Cytotoxicity is retained if TGF-β signaling is blocked during the co-culture (Somasundaram et al., 2002). This finding encourages the development of Treg-targeted therapies, since Treg neutralization may fully rescue CTL function and lead to effective tumor eradication. By directly visualizing CTL functions in draining lymph nodes of tumor-bearing mice, Mempel et al. (2006) have described the kinetics of Treg-mediated suppression. Even in the presence of Treg, CTL stored normal amounts of lytic mediators, but they reduced the time of contact with target cells, and the release of granules do not persist long enough to kill the target.

Efficient immune response requires activation of $CD4^+$ T cells. Adoptive immunotherapy combining tumor-specific $CD4^+$ and $CD8^+$ lymphocytes is more efficient than the sole CTL transfer (Antony, 2005). $CD4^+$ T lymphocytes were the first target of Treg suppression identified in vitro (Thornton and Shevach, 1998) and are also consistently inhibited in vivo. Casares et al. (2003) have shown in a murine tumor model that Treg depletion restores IFN-γ production by $CD4^+$ T cells and results in both CD8-dependent and CD8-independent protection from tumor growth. The reactivity of $CD4^+$ T lymphocytes in the presence of tumor is detectable only after proper immunostimulation. Otherwise, these lymphocytes are kept in a strict state of anergy by tumor-derived immunosuppressive signals and fail to produce cytokines upon ex vivo restimulation (Cuenca et al., 2003). A consistent portion of such unresponsive $CD4^+$ lymphocytes is phenotypically and functionally regulatory cells and includes not only naturally arising Treg but also tumor-induced adaptive Treg derived from conversion of non-Treg precursors (Valzasina et al., 2006). Conversion of potential T-helper lymphocytes into Treg is an important issue of T-cell suppression in tumor-bearing hosts. Treg themselves may contribute to the conversion of non-Treg cells by producing suppressive mediators locally. Indeed, it has been demonstrated in vitro that Treg induce a suppressive phenotype in $CD4^+$

target cells (Qiao et al., 2007). Conversion mechanisms will be further discussed in Sect. 4.2.

The complex machinery of anti-tumor response includes the humoral response in addition to cellular immunity. The rejection of Her-2-expressing tumor is mainly mediated by an antibody response against the extracellular portion of the receptor (Reilly et al., 2001). This implies that both B cells and Th2 T cells are involved in anti-tumor immunity. Treg can affect the functions of both lymphocyte subsets. Indeed, Treg can directly kill B lymphocytes (Zhao et al., 2006) and indirectly affect Th2 differentiation, thus hampering the induction of humoral response (Wing and Sakaguchi, 2006).

Treg are likely involved in several steps of anti-tumor response. As recently described by Gallimore and colleagues (Simon et al., 2007), they affect both innate and adaptive responses. On one hand, Treg transfer inhibits the NK-dependent rejection of tumor in the absence of T and B lymphocytes. On the other hand, their depletion, by means of PC61 antibody, synergizes with anti-tumor vaccination by unveiling CD4 and antibody responses.

From this picture, some complexity arises that cannot be easily resolved. Besides the cellular targets of Treg suppression, the phase of immune response in which Treg exert inhibition is still controversial. The observation that some human and murine tumors specifically recruit Treg from the draining lymph nodes implies that Treg are needed for immunosuppression at the site where effector phase takes place. Moreover, depletion of intratumor $CD4^+$ T cells results in efficient eradication only at late stages of murine tumor growth, suggesting that Treg suppress mainly after the priming phase (Yu et al., 2005). Conversely, other clinical and pre-clinical studies show that Treg extensively accumulate also at the draining lymph node where effector T cells encounter antigen-presenting cells and are activated. The actual cognate interaction between Treg and T cells has been described in tumor-draining lymph nodes by intravital imaging studies (Mempel et al., 2006). Moreover, the draining lymph nodes of murine fibrosarcoma contain both Treg and effector cells; suppression occurs by means of partial expression of CD86, which acts as a negative regulator of T cells by CTLA-4 interaction (Hiura et al., 2005). In conclusion, the plasticity of Treg-suppressive function likely licenses them to affect a variety of not mutually exclusive stages of the immune response.

4 Mechanisms of Treg Expansion in Tumor Bearers

One of the most intriguing, yet unsolved, questions is the mechanism leading to Treg accumulation in tumor bearers. The main mechanisms proposed to contribute to Treg expansion are recruitment, proliferation and conversion (reviewed by Zou, 2006). It is conceivable that these events are not mutually exclusive, rather they may concurrently contribute to expand Treg pools in tumor settings (Munn and Mellor, 2006) (Fig. 1).

The specific recruitment of Treg at the tumor beds has been clearly described by Curiel et al. (2004) in human ovarian cancer. Both tumor cells and tumor-associated macrophages produce the chemokine CCL22 that attracts Treg through the CCR4 receptor leading to the preferential accumulation, at the tumor site, of suppressive rather than effector T cells. Blockade of this signal efficiently hinders Treg recruitment. In these ovarian carcinomas, the increase of Treg at the tumor site correlates with decreased Treg number in the draining lymph nodes, thus suggesting that recruitment may be mostly responsible for the observed Treg enrichment. In contrast, several different tumor models support Treg expansion in both tumor and draining lymph nodes, thus suggesting that mechanisms other than migration, such as proliferation and conversion, contribute to Treg accumulation.

4.1 Role of Proliferation in Tumor-Associated Treg Enrichment

Robert North described suppressor T cells as cycling cells susceptible to antimitotic treatment (North and Awwad, 1990). A recent evidence for Treg proliferation in tumor-bearing animals has been provided by Ghiringhelli et al. (2005b), who showed that Treg derived from tumors and lymph nodes actively incorporate the intercalating agent 5-bromo-2'-deoxyuridine (BrdU) that, while marking proliferation, does not provide information on the number of cell divisions. In this setting, Treg proliferation was fostered by immature myeloid dendritic cells (IMDC) that accumulate in lymphoid organs in response to unknown tumor-produced factors. Such IMDC promote the proliferation of Treg ex vivo, and their injection in vivo induces BrdU incorporation by otherwise resting Treg. The molecular signal driving IMDC-mediated Treg proliferation was TGF-β, since Treg from mice transgenic for the dominant negative form of the TGF-β receptor II did not proliferate in the same setting. This finding was somehow unexpected, since the main cytokine responsible for Treg expansion was thought to be IL-2. However, neither dendritic cells nor T cells are expected to produce IL-2 in the immunosuppressive tumor microenvironment. IMDC express MHC II and costimulatory molecules at low level, which might be sufficient to induce Treg proliferation but insufficient to prime an effector response. This observation is in line with a model in which steady-state dendritic cells allow self-tolerance by sustaining Treg homeostasis (Lutz and Schuler, 2002), while it differs from the notion of mature dendritic cells and IL-2 requirements for Treg proliferation (Yamazaki et al., 2003).

Proliferation of tumor-associated Treg also occurs in the context of tumor antigen-specific T-cell response (Zhou et al., 2006). Clonotypic anti-HA CD4$^+$ T cells transferred into A20HA-bearing mice show a heterogeneous response in terms of CFSE dilution profile. A fraction of them shows impaired proliferation, while another displays high proliferation upon primary in vivo stimulation. Following in vitro antigen restimulation, the former restores proliferation whereas the latter becomes hyporesponsive and suppressive, indicating that donor lymphocytes recognizing the antigen at first encounter, in vivo, consist of Treg. Therefore, T cells effectively activated in a tumor setting are at most Treg. Such Treg are characterized by GITR, CD25 and Foxp3 expression, and once primed they are able to suppress in the absence of persisting antigen. Tumor-associated Treg suppress in vitro via cell-to-cell contact rather than cytokines production, a feature attributed to naturally arising Treg and not to antigen-induced Treg. An open question is whether the expanded Treg derive from the portion of transferred clonotypic Treg among the CD4 population. Even when highly purified $CD4^+$ $CD25^-$ T cells are injected into tumor-bearing mice, a fraction of Treg can develop, suggesting that conversion mechanism of naïve lymphocytes into Treg can actually occur in tumor-bearing mice, thus contributing to overall Treg amplification. These data outline a complex picture in which both natural Treg and induced Treg can undergo proliferation suggesting that Treg proliferation and conversion are not mutually exclusive.

It is arguable that pre-existing natural Treg may induce newly derived Treg from naïve precursors; such a mechanism would imply that natural Treg may "spread" their suppressive potential to other de novo induced Treg, a concept called "infectious tolerance". This issue has been clearly analyzed recently (Zhou and Levitsky, 2007). When antigen-specific, differentially labeled, Treg and T naïve were transferred into tumor-bearing mice, the former were preferentially expanded, without affecting the rate of conversion of the latter. This means that, in this setting, conversion does not depend on natural Treg, but on other players, most likely antigen-presenting cells or tumor-produced factors.

These results may explain the findings reported by others (Bui et al., 2006): when differentially marked Treg and T naïve are transferred into tumor-bearing hosts, the majority of tumor-infiltrating Treg derive from expansion of natural Treg rather than conversion of naïve T cells. Indeed, natural Treg promptly expand in tumor-bearing mice overcoming in number the converted population. However, in spite of the absolute minority of de novo induced Treg, the conversion rate in tumor bearers is significantly higher than in physiological conditions, showing that tumor can actually force the conversion process independently from pre-existing natural Treg.

4.2 Tumor Promotes Conversion of Naïve Precursors into Newly Derived Treg

The first direct evidence of polyclonal T-cells conversion in tumor-bearing mice came from our group (Valzasina et al., 2006). We have shown that tumor-induced Treg accumulation occurs even in thymectomized mice, which lack natural Treg. In this context, the only source of newly formed Treg is the compartment of naïve T cells. We tested this hypothesis by transferring polyclonal naïve $CD4^+CD25^-$ T cells into mice bearing established tumors. We observed that the rate of conversion, measured as CD25 and Foxp3 gaining by transferred cells, is markedly higher in draining lymph nodes and spleens of tumor-bearing compared to tumor-free mice. In our setting, proliferation is not occurring, nor is it necessary to achieve conversion, since T cells converted into Treg even if pre-treated with an anti-mitotic agent.

Conversion mechanisms could explain why the kinetics of effectors versus Treg priming in tumor-draining nodes is characterized by Treg increase and the concomitant effector decrease (Hiura et al., 2005). It is still unclear in that study whether converted Treg derive from uncommitted naïve precursors or have to pass through an intermediate activation phase before acquiring suppressive ability. Further studies are needed to elucidate this intriguing issue.

The signals leading to conversion in the tumor setting are not fully understood yet. Experimental in vitro models have shown that T-cell activation in the presence of TGF-β induces Foxp3 expression and conversion into Treg (Chen et al., 2003). This cytokine could be provided by tumor-associated immune cells, such as IMDC (Ghiringhelli et al., 2005b), and by tumor cells themselves, as demonstrated by Liu et al. (2007). Indeed, in the presence of tumor-conditioned medium, naïve T cells acquire suppressive function ex vivo in a TGF-β-dependent way, and TGF-β neutralization reduces in vivo the proportion of converted Treg. Two groups have recently identified this cytokine as arbiter of T-cells differentiation. Indeed, while TGF-β alone promotes Treg development, the concurrent availability of IL-6 from inflammatory dendritic cells induces the differentiation of IL-17-producing Th17 cells (Bettelli et al., 2006; Veldhoen et al., 2006). Modulation of the balance of these cytokines could be exploited either to switch on responsiveness in cancer immunotherapy or to switch off exacerbated autoimmunity. TNF-α is another inflammatory cytokine whose modulation affects Treg compartment. Indeed, TNF-α blockade results in the differentiation of $CD4^+CD25^-$ T cells into Treg via TGF-β, leading to an unbalanced proportion of effector and Treg (Goldstein et al., 2007; Nadkarni et al., 2007). This finding, which arose from the study of autoimmune diseases, could be applied to tumor immunotherapy, especially in cases where TNF-α is targeted at the tumor to minimize toxic effect and to favor the generation of a pro-inflammatory microenvironment, unfavorable to Treg induction.

Cytokines are thought not to be the only player in conversion, since the presence of functional antigen-presenting cells is required for in vitro and likely in vivo conversion, determining the fate of T-cell activation. An elegant study from von Boehmer's group (Kretschmer et al., 2005) has provided a clue to understanding the possible mechanisms responsible for conversion. They showed that conversion of antigen-specific T cells occurs in sub-immunogenic antigen-presenting conditions and TGF-β availability; conversely, proliferation takes place in a fully immunogenic context. Strikingly, T cells that cannot respond to TGF-β proliferate more and convert less, while T cells that cannot produce IL-2 proliferate less and convert more. This result highlights an inverse correlation between the two processes and suggests that anergy induction and conversion are strictly connected toward induction of tolerance.

The described experimental conditions of suboptimal antigen availability closely resemble the tumor setting, where antigen-presenting cells are blocked in a tolerogenic state and suppressive cytokines are secreted into the tumor microenvironment. Such tolerogenic milieu could be responsible for Treg induction. Recent findings suggest an important role for specific enzymatic activities in generating a suppressive metabolic context. IDO-expressing dendritic cells are a relevant example. Depleting tryptophan from the milieu and secreting toxic catabolites, IDO impairs $CD8^+$ cell functions and concurrently converts $CD4^+$-naïve T cells into Treg

(Fallarino et al., 2006). This means that conversion can be triggered by the same stimuli that lead to intrinsic cell unresponsiveness and that recessive and dominant forms of tolerance can be co-induced in suppressive environments. Nonetheless, tumor cells directly produce IDO and induce both in vitro and in vivo conversion of naïve precursors into Treg. This is the case of human acute myeloid leukemia (AML), about half of which produce IDO and lead to elevated levels of Treg in peripheral blood. AML cells, co-cultured in vitro with $CD4^+CD25^-$, induce conversion of the latter into Treg via IDO. The same results were obtained with A20 mouse lymphoma cells. Indeed, administration of an IDO inhibitor significantly decreases, in vivo, the conversion rate of transferred $CD4^+CD25^-$ T cells into Treg in A20-bearing mice, while the same treatment has no impact on the conversion in mice bearing IDO-negative tumors that likely exploit other suppressive pathways to achieve conversion (Curti et al., 2007). IDO is not the only suppressive enzyme that is related to Treg induction. For instance, several murine and human tumors express the enzyme cyclooxygenase (COX)-2, whose main product, prostaglandin E2, exerts suppressive functions in the tumor microenvironment, including induction of Treg (Sharma et al., 2005).

In addition to dendritic and tumor cells, other tumor-associated cell populations can be involved in Treg conversion. Huang et al. (2006) demonstrated that a subset of MDSC arising in tumor-bearing mice, following IFN-γ stimulation, release TGF-β and IL-10, thus skewing toward induction of Treg. It is conceivable that MDSC and Treg, two suppressive populations belonging to the myeloid and lymphoid lineage, respectively, are functionally linked in generating an overall state of immunosuppression in tumor bearers. Also, some macrophages can actively promote Treg generation. Indeed, in experimental models of immune privilege and oral tolerance, the lack of F4/80 has been associated with impaired generation of CD8 Treg (Lin et al., 2005). A similar scenario could be predicted for tolerance induced by tumor-infiltrating macrophages, which derive from circulating MDSC and locally produce soluble factors that are favorable to tumor growth.

In summary, tolerogenic monocytes such as MDSC, immature dendritic cells and tumor-associated macrophages can promote Treg induction; on the other hand, Treg promote the development of a suppressive microenvironment in which myeloid cells cannot mature properly (see Sect. 3.1). Therefore, Treg and other suppressive cells contribute to create a vicious circle of immune suppression (Fig. 2). Emblematic of this interaction is the very recent finding that human T cells may convert into Treg by "trogocytosis" (the physical exchange of membrane fragments between cells) of fragments derived from antigen-presenting cells, containing the suppressive MHC molecule HLA-G (LeMaoult et al., 2007).

5 Antigen Specificity of Tumor-Associated Treg

Since TCR triggering is required not only for conversion of T cells into professional suppressors, but also for proliferation of already-committed Treg, presumably the factors driving the two processes are the origin of tumor antigens and the repertoire of responding T cells. Indeed, tumor antigens may encompass tumor-specific molecules, over-represented self-antigens and germ cell-associated antigens. Consequently, the host recognition of such antigens may vary depending on their degree of expression as self-molecules.

On the other hand, naïve T cells, activated T cells and natural Treg express diverse TCR repertoires. Indeed, Treg have a TCR repertoire more similar to autoreactive than to naïve T lymphocytes (Hsieh et al., 2004, 2006). This finding confirms that thymic selection is an inexact process that allows auto-reactive cells to escape deletion, but that, nonetheless, counterbalances potential autoimmunity by positively selecting self-reactive suppressor cells. Overall, the final pool of naïve T cells includes not only allo-reactive but also escaped auto-reactive T cells, making possible the recognition of both non-self and self tumor antigens.

This model predicts that tumor-specific antigens, recognized as non-self, could activate allo-reactive naïve T cells, whereas self-antigens, shared by tumor and normal cells, preferentially prime Treg and auto-reactive non-Treg. Therefore, expansion of pre-existing Treg should be driven by self-antigens, while both self- and non-self-restricted antigens could trigger conversion of naïve T cells into Treg.

Several attempts have been made to clarify the TCR identity of tumor-associated Treg. Vaccination against serologically identified non-mutated tumor antigens, which are also expressed by normal cells, leads to exacerbation of tumor growth, since these self-molecules activate Treg to exert suppression over immune response (Nishikawa et al., 2003). The mechanism of Treg accumulation, either from proliferation or conversion, has not been characterized in this model, but it is very likely that pre-existing natural Treg promptly respond to anti-self priming.

T-cell clones established from lymphocytes infiltrating human melanoma display suppressive properties, resembling Treg. Two antigenic specificities of such tumor-infiltrating Treg were identified: LAGE1 and antigen recognized by Treg cells 1 (ARTC1) (Wang et al., 2004, 2005). In spite of their tissue-restricted expression (only tumor and testicular cells express these antigens), they are tolerated by the host immune system, maybe even more than other peripheral antigens. Indeed, testis is an immune-privileged site and Treg specific for testis-restricted antigens can re-create immune privilege at the tumor site (Munn, 2006). Moreover, ARTC1 can be presented in an APC-independent manner by tumor cells themselves, leading to local suppression.

Some human and murine tumors are characterized by specific overexpression of self-molecules. This is the case of the abnormal expression of Her-2 receptor by human and mouse tumors. Transgenic mice expressing the mutated rat Her-2 under MMTV promoter (called BALB/neuT) develop spontaneous mammary carcinogenesis. DNA vaccination against Her-2 elicits potent anti-tumor immune response in non-transgenic mice, while it achieves much weaker response in transgenic littermates that express Her-2 in the thymus. The T-cell activation in tolerant versus non-tolerant hosts is characterized by different, mutually exclusive, TCR repertoire (Rolla et al., 2006). Indeed, thymic expression of Her-2 and negative selection limit the number of clones generating specific Treg.

6 Treg-Targeted Immunotherapeutic Strategies

The failure of many ongoing immunotherapeutic protocols has been ascribed to the activity of suppressive cells such as MDSC and Treg. Overcoming suppression while promoting activation is the must of immunotherapy today. Achieving Treg neutralization is therefore an absolute requirement.

6.1 Neutralizing Treg by Depletion

Neutralization of Treg suppression was first attempted in tumor-bearing mice by means of the anti-CD25 monoclonal antibody (PC61). The administration of this antibody resulted in the production of generalized autoimmune disease similar to that observed in mice thymectomized at day 3 of age, which cannot develop natural Treg (Taguchi and Takahashi, 1996). It was initially observed that PC61 administration resulted in the disappearance of $CD25^+$ cells, as detected by staining with a different clone of anti-CD25 antibody (7D4) (Onizuka et al., 1999). The depletion of $CD25^+$ T cells resulted in enhancement of immune reactivity and impairment of tumor growth in several tumor models, thus definitively confirming the central role of Treg in anti-tumor immunity (reviewed by Zou, 2006).

However, Treg depletion by anti-CD25 antibody has raised several criticisms. First of all, Treg share CD25 expression with activated $CD4^+$ and $CD8^+$ lymphocytes. The most successful outcome has been achieved when the treatment was applied in naïve mice prior to tumor inoculation since, when given to tumor-bearing mice, PC61 neutralizes both Treg and T cells activated by concomitant immunity. To eliminate Treg without affecting the activated $CD8^+$ cells, the intratumor depletion of $CD4^+$ T cells has been tested (Yu et al., 2005). It is noteworthy that this treatment, given at late stages of tumor progression (when the vast majority of $CD4^+$ tumor-infiltrating cells consist of Treg), induces complete eradication of the growing tumor.

An additional problem associated with Treg depletion is the possibility that newly derived Treg can quickly replenish the lymphoid compartment upon depletion, re-establishing suppression. Indeed, peripheral conversion of naïve precursor into Treg, which occurs in physiological conditions, is accelerated in the presence of growing tumors (Valzasina et al., 2006). This implies that Treg neutralization could be better achieved by functional inactivation rather than by physical depletion.

Recently, the activity of the antibody PC61 has been revised. Kohm et al. (2006) have shown that injection of an anti-CD25 antibody (7D4 clone) results in the downregulation of CD25 on the Treg surface, rather than the physical elimination of CD25-expressing cells, without affecting the percentage of $Foxp3^+$ T cells, which marks the true Treg. Loss of CD25 expression by Treg would indeed lead to their functional inactivation, since Treg require IL-2 responsiveness in order to exert their suppressive function. However, this issue remains controversial, because other authors have shown that administration of the PC61 antibody, more frequently used than 7D4, results not only in the reduction of CD25$^+$ T cells but also in significant reduction of Foxp3$^+$ T cells (Nair et al., 2007; Simon et al., 2007).

Therapeutic depletion of Treg by specific targeting of Foxp3-expressing cells is not feasible due to the intracellular localization of this transcription factor. Some recently developed transgenic mouse models provide new experimental tools suitable to dissect biological processes involving Foxp3 in Treg development and function. For instance, DEREG mice, which express the receptor of diphtheria toxin under control of Foxp3 promoter, display severe autoimmune phenotype upon diphtheria toxin administration, due to the selective ablation of Foxp3-expressing cells (Lahl et al., 2007). Moreover, Rudensky's group has recently created a transgenic model of inducible ablation of *loxP*-flanked Foxp3. The administration of the recombinase *cre* does not eliminate Treg but inactivates their suppressive ability and skews their differentiation toward Th1 (Williams and Rudensky, 2007).

Indirect depletion of Foxp3$^+$ Treg in tumor-bearing mice has been achieved by Gilboa's group (Nair et al., 2007). This original approach consists of eliciting cytotoxic T cells directed against Foxp3-expressing cells to eliminate Treg. Mice have been vaccinated with dendritic cells stably transfected with the Foxp3 mRNA. This treatment synergizes with vaccination against the trp-2 antigen in the B16 melanoma model. Interestingly, for yet unknown reasons, it induces Treg depletion selectively at the tumor site while sparing the peripheral Treg, thus favoring anti-tumor immunity without affecting systemic tolerance.

In order to translate Treg depletion into clinic, and in the absence of any approved antibody for clinical studies, attention was given to the immunotoxin Dinileukin diftitox (Ontak). This compound consists of a fusion protein between interleukin 2 and the diphtheria toxin and selectively kills cells expressing the high-affinity receptor for IL-2, among which are Treg. Several clinical studies have investigated the potential efficacy of this drug as adjuvant of vaccination in cancer patients, providing encouraging although not fully satisfactory results. Indeed, Ontak has shown some efficacy in depleting Treg and promoting tumor rejection during the treatment of ovary, breast, lung and renal-cell carcinoma (reviewed by Zou, 2006). Conversely, immunotherapy of melanoma with Ontak has produced controversial results. Indeed, Rosenberg's group has observed neither Treg depletion nor melanoma regression upon Ontak administration (Attia et al., 2005), whereas Mahnke et al. (2007) have recently reported that this drug is able to elicit significant depletion of Treg and increase anti-tumor reactivity, even though it did not induce complete disease remission in metastatic melanoma patients. Although different experimental conditions could explain the observed discrepancies, the molecule could be refined to obtain better efficiency and consistency of results. For instance, the IL-2 present in Ontak may provide a triggering signal to Treg rather than eliminating them. To avoid this effect, preliminary in vitro studies have shown that the fusion protein LMB-2, composed by a bacterial toxin and the Fv portion of anti-CD25 antibody, can efficiently deplete human Treg without providing any triggering (Attia et al., 2006).

Another strategy to deplete Treg in clinical immunotherapy is cyclophosphamide. Robert North first reported that this compound eliminates cycling suppressor cells, thus allowing anti-tumor responses (Berendt and North, 1980). However, cyclophosphamide administration quickly produces a "rebound" effect of peripheral leukocytes that should be avoided to obtain successful therapy (Proietti et al., 1998). Repetitive low-dose administration of this compound is able to selectively decrease the presence of Treg in cancer patients (Ghiringhelli et al., 2007). However, since cyclophosphamide specifically targets proliferating cells, its overall efficacy in inducing tumor regression could be hampered by the continuous generation of Treg by proliferation-independent conversion.

6.2 Treg Functional Inactivation

Because of the above-described drawbacks in Treg depletion strategies, novel approaches should be developed to inhibit the suppressive function of Treg, especially those expanded in tumor bearers. Several attempts have been made in this direction by targeting molecules that are strictly related to the suppressive ability of Treg.

A molecule constitutively expressed by Treg and associated to their function is CTLA-4. The blockade of CTLA-4 does not result in Treg depletion, but in their expansion in lymph nodes (Quezada et al., 2006); nevertheless, in some instances, it inhibits their suppressive function. In the model of inflammatory bowel disease, Treg suppression of colitogenic T cells is inhibited by CTLA-4 blockade. Since such inhibition is lost in the case CTLA-4 knock-out Treg are targeted, the experiment indicates that Treg-associated CTLA-4 was responsible for immunosuppression in the gut (Read et al., 2006). Conversely, when used as adjuvant in cancer immunotherapy, CTLA-4 blockade has been shown to improve the immune response even in mice previously depleted of Treg (Sutmuller et al., 2001). In line with this evidence, Allison's group has recently demonstrated that CTLA-4 blockade, in conjunction with GM-CSF-based vaccination, modifies the intratumor balance between Treg and T effectors, restoring the normal proliferation and function of the latter (Quezada et al., 2006). The most important side effect produced by anti-CTLA-4 treatment is the concurrent development of severe autoimmunity, such as enterocolitis, hepatitis, uveitis, and even hypophysitis (Blansfield et al., 2005). It is thought that toxicity can be partially due to the short interval between treatments that, overlapping the half-life of the antibody, provokes a dose accumulation. By simply prolonging such interval it could be possible to reduce toxicity.

Glucocorticoid-induced tumor necrosis factor receptor (GITR) (Nocentini et al., 1997) is constitutively expressed by murine Treg. Following the observation that receptor triggering reverses Treg suppression (Shimizu et al., 2002), administration of anti-GITR agonist antibody has been evaluated as a possible cancer treatment. Both systemic and intratumor GITR triggering leads to rejection of established tumors without producing overt autoimmunity (Ko et al., 2005). Subsequent studies have demonstrated that GITR immunostimulatory properties are mainly due to effectors stimulation rather than Treg inhibition. Indeed, GITR triggering displays the same efficacy in both Treg-depleted and non-depleted mice (Cohen et al., 2006; Ramirez-Montagut, 2006) and even induces Treg expansion, which might partially hinder the overall efficacy of such treatment (Ramirez-Montagut, 2006).

OX40 is a costimulatory molecule of the tumor necrosis factor receptor family (Mallett et al., 1990) that is expressed on murine Treg constitutively and on effector T cells upon activation. Triggering OX40 on activated effectors leads to strong costimulation that impedes and reverses tolerance (Bansal-Pakala et al., 2001). On the other hand, stimulation of OX40 on Treg completely abolishes their suppressive function (Takeda et al., 2004; Valzasina et al., 2005). Although distinct, the effects of OX40 on the two populations contribute to the same final outcome of enhanced anti-tumor immunity. Administration of agonist anti-OX40 antibody or OX40L-Ig to tumor-bearing mice has shown significant anti-tumor efficacy in several tumor models (Kjaergaard et al., 2000; Weinberg et al., 2000). Moreover, a tumor cell vaccine engineered to express both GM-CSF and OX40L exhibits higher protection and superior therapeutic efficacy than the vaccine engineered with a single agent (Gri et al., 2003). It is conceivable that in these settings, OX40 stimulation fosters anti-tumor immunity by concurrently boosting effector cells and inhibiting Treg.

Manipulation of Toll-like receptors (TLR) in the tumor microenvironment stimulates an anti-tumor immune response. The expression of such receptors is not restricted to dendritic cells but extended to other cells of the immune system, including Treg, and even to epithelial cells. Triggering of TLR on dendritic cells can indirectly affect Treg by inducing IL-6 production, which skews Treg precursors toward Th17 (Veldhoen et al., 2006) and renders effectors unresponsive to Treg suppression (Pasare and Medzhitov, 2003). Similarly, administration of inactivated Sendai virus particles triggers an unknown receptor on dendritic cells, which in response produces IL-6 leading to Treg inhibition and tumor rejection (Kurooka and Kaneda, 2007).

Also Treg express some Toll-like receptors, whose triggering can exert direct effects on their function. Indeed, Peng et al. (2005) have demonstrated that triggering TLR-8 on human Treg strongly inhibits their suppressive properties. The search for small agonistic molecules to be used as Treg blockers represents a challenge in cancer immunotherapy.

7 Conclusions

The role of Treg in tumor immunology has gained more and more interest, as a consequence of the observation that intrinsic tolerance to tumor antigens is not easily broken unless dominant tolerance is inhibited. Being able to inhibit a variety of cell subsets and immune processes, Treg are central in suppressing anti-tumor immunity. Treg increase in number in the vast majority of both murine and human tumors and this generates a vicious circle of immunosuppression. Treg inactivation, in the absence of autoimmune side effects, could be the key for successful immunotherapy.

References

Ambrosino, E., Spadaro, M., Iezzi, M., Curcio, C., Forni, G., Musiani, P., Wei, W. Z., and Cavallo, F. (2006). Immunosurveillance of Erbb2 carcinogenesis in transgenic mice is concealed by a dominant regulatory T-cell self-tolerance. *Cancer Res* 66(15):7734–7740.

Antony, P. A., Piccirillo, C. A., Akpinarli, A., Finkelstein, S. E., Speiss, P. J., Surman, D. R., Palmer, D. C., Chan, C. C., Klebanoff, C. A., Overwijk, W. W., Rosenberg, S. A., and Restifo, N. P. (2005). CD8+ T cell immunity against a tumor/self-antigen is augmented by CD4+ T helper cells and hindered by naturally occurring T regulatory cells. *J Immunol* 174(5): 2591–2601.

Antony, P. A., and Restifo, N. P. (2005). CD4+CD25+ T regulatory cells, immunotherapy of cancer, and interleukin-2. *J Immunother (1997)* 28(2):120–128.

Attia, P., Maker, A. V., Haworth, L. R., Rogers-Freezer, L., and Rosenberg, S. A. (2005). Inability of a fusion protein of IL-2 and diphtheria toxin (Denileukin Diftitox, DAB389IL-2, ONTAK) to eliminate regulatory T lymphocytes in patients with melanoma. *J Immunother (1997)* 28(6):582–592.

Attia, P., Powell, D. J., Jr, Maker, A. V., Kreitman, R. J., Pastan, I., and Rosenberg, S. A. (2006). Selective elimination of human regulatory T lymphocytes in vitro with the recombinant immunotoxin LMB-2. *J Immunother (1997)* 29(2):208–214.

Banerjee, D. K., Dhodapkar, M. V., Matayeva, E., Steinman, R. M., and Dhodapkar, K. M. (2006). Expansion of FOXP3high regulatory T cells by human dendritic cells (DCs) in vitro and after injection of cytokine-matured DCs in myeloma patients. *Blood* 108(8):2655–2661.

Bansal-Pakala, P., Jember, A. G., and Croft, M. (2001). Signaling through OX40 (CD134) breaks peripheral T-cell tolerance. *Nat Med* 7(8):907–912.

Barnett, B., Kryczek, I., Cheng, P., Zou, W., and Curiel, T. J. (2005). Regulatory T cells in ovarian cancer: biology and therapeutic potential. *Am J Reprod Immunol* 54(6):369–377.

Berendt, M. J., and North, R. J. (1980). T-cell-mediated suppression of anti-tumor immunity. An explanation for progressive growth of an immunogenic tumor. *J Exp Med* 151(1):69–80.

Bettelli, E., Carrier, Y., Gao, W., Korn, T., Strom, T. B., Oukka, M., Weiner, H. L., and Kuchroo, V. K. (2006). Reciprocal developmental pathways for the generation of pathogenic effector TH17 and regulatory T cells. *Nature* 441(7090):235–238.

Blansfield, J. A., Beck, K. E., Tran, K., Yang, J. C., Hughes, M. S., Kammula, U. S., Royal, R. E., Topalian, S. L., Haworth, L. R., Levy, C., Rosenberg, S. A., and Sherry, R. M. (2005). Cytotoxic T-lymphocyte-associated antigen-4 blockage can induce autoimmune hypophysitis in patients with metastatic melanoma and renal cancer. *J Immunother (1997)* 28(6): 593–598.

Bui, J. D., Uppaluri, R., Hsieh, C. S., and Schreiber, R. D. (2006). Comparative analysis of regulatory and effector T cells in progressively growing versus rejecting tumors of similar origins. *Cancer Res* 66(14):7301–7309.

Carreras, J., Lopez-Guillermo, A., Fox, B. C., Colomo, L., Martinez, A., Roncador, G., Montserrat, E., Campo, E., and Banham, A. H. (2006). High numbers of tumor-infiltrating FOXP3-positive regulatory T cells are associated with improved overall survival in follicular lymphoma. *Blood* 108(9):2957–2964.

Casares, N., Arribillaga, L., Sarobe, P., Dotor, J., Lopez-Diaz de Cerio, A., Melero, I., Prieto, J., Borras-Cuesta, F., and Lasarte, J. J. (2003). CD4+/CD25+ regulatory cells inhibit activation of tumor-primed CD4+ T cells with IFN-gamma-dependent antiangiogenic activity, as well as long-lasting tumor immunity elicited by peptide vaccination. *J Immunol* 171(11):5931–5939.

Chen, M. L., Pittet, M. J., Gorelik, L., Flavell, R. A., Weissleder, R., von Boehmer, H., and Khazaie, K. (2005). Regulatory T cells suppress tumor-specific CD8 T cell cytotoxicity through TGF-beta signals in vivo. *Proc Natl Acad Sci USA* 102(2):419–424.

Chen, W., Jin, W., Hardegen, N., Lei, K. J., Li, L., Marinos, N., McGrady, G., and Wahl, S. M. (2003). Conversion of peripheral CD4+CD25- naive T cells to CD4+CD25+ regulatory T cells by TGF-beta induction of transcription factor Foxp3. *J Exp Med* 198(12): 1875–1886.

Chiodoni, C., Iezzi, M., Guiducci, C., Sangaletti, S., Alessandrini, I., Ratti, C., Tiboni, F., Musiani, P., Granger, D. N., and Colombo, M. P. (2006). Triggering CD40 on endothelial cells contributes to tumor growth. *J Exp Med* 203(11):2441–2450.

Cohen, A. D., Diab, A., Perales, M. A., Wolchok, J. D., Rizzuto, G., Merghoub, T., Huggins, D., Liu, C., Turk, M. J., Restifo, N. P., Sakaguchi, S., and Houghton, A. N. (2006). Agonist anti-GITR antibody enhances vaccine-induced CD8(+) T-cell responses and tumor immunity. *Cancer Res* 66(9):4904–4912.

Cuenca, A., Cheng, F., Wang, H., Brayer, J., Horna, P., Gu, L., Bien, H., Borrello, I. M., Levitsky, H. I., and Sotomayor, E. M. (2003). Extra-lymphatic solid tumor growth is not immunologically ignored and results in early induction of antigen-specific T-cell anergy: dominant role of cross-tolerance to tumor antigens. *Cancer Res* 63(24):9007–9015.

Curiel, T. J., Coukos, G., Zou, L., Alvarez, X., Cheng, P., Mottram, P., Evdemon-Hogan, M., Conejo-Garcia, J. R., Zhang, L., Burow, M., Zhu, Y., Wei, S., Kryczek, I., Daniel, B., Gordon, A., Myers, L., Lackner, A., Disis, M. L., Knutson, K. L., Chen, L., and Zou, W. (2004). Specific recruitment of regulatory T cells in ovarian carcinoma fosters immune privilege and predicts reduced survival. *Nat Med* 10(9):942–949.

Curti, A., Pandolfi, S., Valzasina, B., Aluigi, M., Isidori, A., Ferri, E., Salvestrini, V., Bonanno, G., Rutella, S., Durelli, I., Horenstein, A. L., Fiore, F., Massaia, M., Colombo, M. P., Baccarani, M., and Lemoli, R. M. (2007). Modulation of tryptophan catabolism by human leukemic cells results in the conversion of CD25− into CD25+ T regulatory cells. *Blood* 109(7): 2871–2877.

Dannull, J., Su, Z., Rizzieri, D., Yang, B. K., Coleman, D., Yancey, D., Zhang, A., Dahm, P., Chao, N., Gilboa, E., and Vieweg, J. (2005). Enhancement of vaccine-mediated antitumor immunity in cancer patients after depletion of regulatory T cells. *J Clin Invest* 115(12): 3623–3633.

Dercamp, C., Chemin, K., Caux, C., Trinchieri, G., and Vicari, A. P. (2005). Distinct and overlapping roles of interleukin-10 and CD25+ regulatory T cells in the inhibition of antitumor CD8 T-cell responses. *Cancer Res* 65(18):8479–8486.

Erdman, S. E., Rao, V. P., Poutahidis, T., Ihrig, M. M., Ge, Z., Feng, Y., Tomczak, M., Rogers, A. B., Horwitz, B. H., and Fox, J. G. (2003). CD4(+)CD25(+) regulatory lymphocytes require interleukin 10 to interrupt colon carcinogenesis in mice. *Cancer Res* 63(18): 6042–6050.

Fallarino, F., Grohmann, U., Hwang, K. W., Orabona, C., Vacca, C., Bianchi, R., Belladonna, M. L., Fioretti, M. C., Alegre, M. L., and Puccetti, P. (2003). Modulation of tryptophan catabolism by regulatory T cells. *Nat Immunol* 4(12):1206–1212.

Fallarino, F., Grohmann, U., You, S., McGrath, B. C., Cavener, D. R., Vacca, C., Orabona, C., Bianchi, R., Belladonna, M. L., Volpi, C., Santamaria, P., Fioretti, M. C., and Puccetti, P. (2006). The combined effects of tryptophan starvation and tryptophan catabolites down-regulate T cell receptor zeta-chain and induce a regulatory phenotype in naive T cells. *J Immunol* 176(11):6752–6761.

Fontenot, J. D., Gavin, M. A., and Rudensky, A. Y. (2003). Foxp3 programs the development and function of CD4+CD25+ regulatory T cells. *Nat Immunol* 4(4):330–336.

Gabrilovich, D. I., Bronte, V., Chen, S. H., Colombo, M. P., Ochoa, A., Ostrand-Rosenberg, S., and Schreiber, H. (2007). The terminology issue for myeloid-derived suppressor cells. *Cancer Res* 67(1):425; author reply 426.

Gavin, M. A., Clarke, S. R., Negrou, E., Gallegos, A., and Rudensky, A. (2002). Homeostasis and anergy of CD4(+)CD25(+) suppressor T cells in vivo. *Nat Immunol* 3(1):33–41.

Gavin, M. A., Rasmussen, J. P., Fontenot, J. D., Vasta, V., Manganiello, V. C., Beavo, J. A., and Rudensky, A. Y. (2007). Foxp3-dependent programme of regulatory T-cell differentiation. *Nature* 445(7129):771–775.

Ghiringhelli, F., Menard, C., Puig, P. E., Ladoire, S., Roux, S., Martin, F., Solary, E., Le Cesne, A., Zitvogel, L., and Chauffert, B. (2007). Metronomic cyclophosphamide regimen selectively depletes CD4+CD25+ regulatory T cells and restores T and NK effector functions in end stage cancer patients. *Cancer Immunol Immunother* 56(5):641–648.

Ghiringhelli, F., Menard, C., Terme, M., Flament, C., Taieb, J., Chaput, N., Puig, P. E., Novault, S., Escudier, B., Vivier, E., Lecesne, A., Robert, C., Blay, J. Y., Bernard, J., Caillat-Zucman, S., Freitas, A., Tursz, T., Wagner-Ballon, O., Capron, C., Vainchencker, W., Martin, F., and Zitvogel, L. (2005a). CD4+CD25+ regulatory T cells inhibit natural killer cell functions in a transforming growth factor-beta-dependent manner. *J Exp Med* 202(8): 1075–1085.

Ghiringhelli, F., Puig, P. E., Roux, S., Parcellier, A., Schmitt, E., Solary, E., Kroemer, G., Martin, F., Chauffert, B., and Zitvogel, L. (2005b). Tumor cells convert immature myeloid dendritic cells into TGF-beta-secreting cells inducing CD4+CD25+ regulatory T cell proliferation. *J Exp Med* 202(7):919–929.

Goldstein, I., Ben-Horin, S., Koltakov, A., Chermoshnuk, H., Polevoy, V., Berkun, Y., Amariglio, N., and Bank, I. (2007). alpha1beta1 Integrin+ and regulatory Foxp3+ T cells constitute two functionally distinct human CD4+ T cell subsets oppositely modulated by TNFalpha blockade. *J Immunol* 178(1):201–210.

Golgher, D., Jones, E., Powrie, F., Elliott, T., and Gallimore, A. (2002). Depletion of CD25+ regulatory cells uncovers immune responses to shared murine tumor rejection antigens. *Eur J Immunol* 32(11):3267–3275.

Gri, G., Gallo, E., Di Carlo, E., Musiani, P., and Colombo, M. P. (2003). OX40 ligand-transduced tumor cell vaccine synergizes with GM-CSF and requires CD40-Apc signaling to boost the host T cell antitumor response. *J Immunol* 170(1):99–106.

Guiducci, C., Valzasina, B., Dislich, H., and Colombo, M. P. (2005). CD40/CD40L interaction regulates CD4+CD25+ T reg homeostasis through dendritic cell-produced IL-2. *Eur J Immunol* 35(2):557–567.

Hiura, T., Kagamu, H., Miura, S., Ishida, A., Tanaka, H., Tanaka, J., Gejyo, F., and Yoshizawa, H. (2005). Both regulatory T cells and antitumor effector T cells are primed in the same draining lymph nodes during tumor progression. *J Immunol* 175(8):5058–5066.

Hodi, F. S., Mihm, M. C., Soiffer, R. J., Haluska, F. G., Butler, M., Seiden, M. V., Davis, T., Henry-Spires, R., MacRae, S., Willman, A., Padera, R., Jaklitsch, M. T., Shankar, S., Chen, T. C., Korman, A., Allison, J. P., and Dranoff, G. (2003). Biologic activity of cytotoxic T lymphocyte-associated antigen 4 antibody blockade in previously vaccinated metastatic melanoma and ovarian carcinoma patients. *Proc Natl Acad Sci USA* 100(8):4712–4717.

Hori, S., Nomura, T., and Sakaguchi, S. (2003). Control of regulatory T cell development by the transcription factor Foxp3. *Science* 299(5609):1057–1061.

Hsieh, C. S., Liang, Y., Tyznik, A. J., Self, S. G., Liggitt, D., and Rudensky, A. Y. (2004). Recognition of the peripheral self by naturally arising CD25+ CD4+ T cell receptors. *Immunity* 21(2):267–277.

Hsieh, C. S., Zheng, Y., Liang, Y., Fontenot, J. D., and Rudensky, A. Y. (2006). An intersection between the self-reactive regulatory and nonregulatory T cell receptor repertoires. *Nat Immunol* 7(4):401–410.

Huang, B., Pan, P. Y., Li, Q., Sato, A. I., Levy, D. E., Bromberg, J., Divino, C. M., and Chen, S. H. (2006). Gr-1+CD115+ immature myeloid suppressor cells mediate the development of tumor-induced T regulatory cells and T-cell anergy in tumor-bearing host. *Cancer Res* 66(2): 1123–1131.

Ishida, T., Ishii, T., Inagaki, A., Yano, H., Komatsu, H., Iida, S., Inagaki, H., and Ueda, R. (2006). Specific recruitment of CC chemokine receptor 4-positive regulatory T cells in Hodgkin lymphoma fosters immune privilege. *Cancer Res* 66(11):5716–5722.

Itoh, M., Takahashi, T., Sakaguchi, N., Kuniyasu, Y., Shimizu, J., Otsuka, F., and Sakaguchi, S. (1999). Thymus and autoimmunity: production of CD25+CD4+ naturally anergic and suppressive T cells as a key function of the thymus in maintaining immunologic self-tolerance. *J Immunol* 162(9):5317–5326.

Jarnicki, A. G., Lysaght, J., Todryk, S., and Mills, K. H. (2006). Suppression of antitumor immunity by IL-10 and TGF-beta-producing T cells infiltrating the growing tumor: influence of tumor environment on the induction of CD4+ and CD8+ regulatory T cells. *J Immunol* 177(2): 896–904.

Jones, E., Dahm-Vicker, M., Simon, A. K., Green, A., Powrie, F., Cerundolo, V., and Gallimore, A. (2002). Depletion of CD25+ regulatory cells results in suppression of melanoma growth and induction of autoreactivity in mice. *Cancer Immun* 2:1.

Kjaergaard, J., Tanaka, J., Kim, J. A., Rothchild, K., Weinberg, A., and Shu, S. (2000). Therapeutic efficacy of OX-40 receptor antibody depends on tumor immunogenicity and anatomic site of tumor growth. *Cancer Res* 60(19):5514–5521.

Ko, K., Yamazaki, S., Nakamura, K., Nishioka, T., Hirota, K., Yamaguchi, T., Shimizu, J., Nomura, T., Chiba, T., and Sakaguchi, S. (2005). Treatment of advanced tumors with agonistic anti-GITR mAb and its effects on tumor-infiltrating Foxp3+CD25+CD4+ regulatory T cells. *J Exp Med* 202(7):885–891.

Kohm, A. P., McMahon, J. S., Podojil, J. R., Begolka, W. S., DeGutes, M., Kasprowicz, D. J., Ziegler, S. F., and Miller, S. D. (2006). Cutting edge: anti-CD25 monoclonal antibody injection results in the functional inactivation, not depletion, of CD4+CD25+ T regulatory cells. *J Immunol* 176(6):3301–3305.

Kortylewski, M., Kujawski, M., Wang, T., Wei, S., Zhang, S., Pilon-Thomas, S., Niu, G., Kay, H., Mule, J., Kerr, W. G., Jove, R., Pardoll, D., and Yu, H. (2005). Inhibiting Stat3 signaling in the hematopoietic system elicits multicomponent antitumor immunity. *Nat Med* 11(12):1314–1321.

Kretschmer, K., Apostolou, I., Hawiger, D., Khazaie, K., Nussenzweig, M. C., and von Boehmer, H. (2005). Inducing and expanding regulatory T cell populations by foreign antigen. *Nat Immunol* 6(12):1219–1227.

Kurooka, M., and Kaneda, Y. (2007). Inactivated Sendai virus particles eradicate tumors by inducing immune responses through blocking regulatory T cells. *Cancer Res* 67(1):227–236.

Lahl, K., Loddenkemper, C., Drouin, C., Freyer, J., Arnason, J., Eberl, G., Hamann, A., Wagner, H., Huehn, J., and Sparwasser, T. (2007). Selective depletion of Foxp3+ regulatory T cells induces a scurfy-like disease. *J Exp Med* 204(1):57–63.

Larmonier, N., Marron, M., Zeng, Y., Cantrell, J., Romanoski, A., Sepassi, M., Thompson, S., Chen, X., Andreansky, S., and Katsanis, E. (2007). Tumor-derived CD4(+)CD25(+) regulatory T cell suppression of dendritic cell function involves TGF-beta and IL-10. *Cancer Immunol Immunother* 56(1):48–59.

LeMaoult, J., Caumartin, J., Daouya, M., Favier, B., Le Rond, S., Gonzalez, A., and Carosella, E. D. (2007). Immune regulation by pretenders: cell-to-cell transfers of HLA-G make effector T cells act as regulatory cells. *Blood* 109(5):2040–2048.

Lin, H. H., Faunce, D. E., Stacey, M., Terajewicz, A., Nakamura, T., Zhang-Hoover, J., Kerley, M., Mucenski, M. L., Gordon, S., and Stein-Streilein, J. (2005). The macrophage F4/80 receptor is required for the induction of antigen-specific efferent regulatory T cells in peripheral tolerance. *J Exp Med* 201(10):1615–1625.

Liu, V. C., Wong, L. Y., Jang, T., Shah, A. H., Park, I., Yang, X., Zhang, Q., Lonning, S., Teicher, B. A., and Lee, C. (2007). Tumor evasion of the immune system by converting CD4+CD25– T cells into CD4+CD25+ T regulatory cells: role of tumor-derived TGF-beta. *J Immunol* 178(5):2883–2892.

Liyanage, U. K., Moore, T. T., Joo, H. G., Tanaka, Y., Herrmann, V., Doherty, G., Drebin, J. A., Strasberg, S. M., Eberlein, T. J., Goedegebuure, P. S., and Linehan, D. C. (2002). Prevalence of regulatory T cells is increased in peripheral blood and tumor microenvironment of patients with pancreas or breast adenocarcinoma. *J Immunol* 169(5):2756–2761.

Lutz, M. B., and Schuler, G. (2002). Immature, semi-mature and fully mature dendritic cells: which signals induce tolerance or immunity? *Trends Immunol* 23(9):445–449.

Mahnke, K., Schonfeld, K., Fondel, S., Ring, S., Karakhanova, S., Wiedemeyer, K., Bedke, T., Johnson, T. S., Storn, V., Schallenberg, S., and Enk, A. H. (2007). Depletion of CD4+CD25+ human regulatory T cells in vivo: kinetics of Treg depletion and alterations in immune functions in vivo and in vitro. *Int J Cancer* 120(12):2723–2733.

Malek, T. R., and Bayer, A. L. (2004). Tolerance, not immunity, crucially depends on IL-2. *Nat Rev Immunol* 4(9):665–674.

Mallett, S., Fossum, S., and Barclay, A. N. (1990). Characterization of the MRC OX40 antigen of activated CD4 positive T lymphocytes—a molecule related to nerve growth factor receptor. *EMBO J* 9(4):1063–1068.

Melani, C., Chiodoni, C., Forni, G., and Colombo, M. P. (2003). Myeloid cell expansion elicited by the progression of spontaneous mammary carcinomas in c-erbB-2 transgenic BALB/c mice suppresses immune reactivity. *Blood* 102(6):2138–2145.

Mempel, T. R., Pittet, M. J., Khazaie, K., Weninger, W., Weissleder, R., von Boehmer, H., and von Andrian, U. H. (2006). Regulatory T cells reversibly suppress cytotoxic T cell function independent of effector differentiation. *Immunity* 25(1):129–141.

Munn, D. H., and Mellor, A. L. (2006). The tumor-draining lymph node as an immune-privileged site. *Immunol Rev* 213:146–158.

Nadkarni, S., Mauri, C., and Ehrenstein, M. R. (2007). Anti-TNF-alpha therapy induces a distinct regulatory T cell population in patients with rheumatoid arthritis via TGF-beta. *J Exp Med* 204(1):33–39.

Nair, S., Boczkowski, D., Fassnacht, M., Pisetsky, D., and Gilboa, E. (2007). Vaccination against the forkhead family transcription factor Foxp3 enhances tumor immunity. *Cancer Res* 67(1):371–380.

Nishikawa, H., Kato, T., Tanida, K., Hiasa, A., Tawara, I., Ikeda, H., Ikarashi, Y., Wakasugi, H., Kronenberg, M., Nakayama, T., Taniguchi, M., Kuribayashi, K., Old, L. J., and Shiku, H. (2003). CD4+ CD25+ T cells responding to serologically defined autoantigens suppress anti-tumor immune responses. *Proc Natl Acad Sci USA* 100(19):10902–10906.

Nocentini, G., Giunchi, L., Ronchetti, S., Krausz, L. T., Bartoli, A., Moraca, R., Migliorati, G., and Riccardi, C. (1997). A new member of the tumor necrosis factor/nerve growth factor receptor family inhibits T cell receptor-induced apoptosis. *Proc Natl Acad Sci USA* 94(12):6216–6221.

North, R. J., and Awwad, M. (1990). Elimination of cycling CD4+ suppressor T cells with an anti-mitotic drug releases non-cycling CD8+ T cells to cause regression of an advanced lymphoma. *Immunology* 71(1):90–95.

Onizuka, S., Tawara, I., Shimizu, J., Sakaguchi, S., Fujita, T., and Nakayama, E. (1999). Tumor rejection by in vivo administration of anti-CD25 (interleukin-2 receptor alpha) monoclonal antibody. *Cancer Res* 59(13):3128–3133.

Ormandy, L. A., Hillemann, T., Wedemeyer, H., Manns, M. P., Greten, T. F., and Korangy, F. (2005). Increased populations of regulatory T cells in peripheral blood of patients with hepatocellular carcinoma. *Cancer Res* 65(6):2457–2464.

Pasare, C., and Medzhitov, R. (2003). Toll pathway-dependent blockade of CD4+CD25+ T cell-mediated suppression by dendritic cells. *Science* 299(5609):1033–1036.

Peng, G., Guo, Z., Kiniwa, Y., Voo, K. S., Peng, W., Fu, T., Wang, D. Y., Li, Y., Wang, H. Y., and Wang, R. F. (2005). Toll-like receptor 8-mediated reversal of CD4+ regulatory T cell function. *Science* 309(5739):1380–1384.

Phan, G. Q., Yang, J. C., Sherry, R. M., Hwu, P., Topalian, S. L., Schwartzentruber, D. J., Restifo, N. P., Haworth, L. R., Seipp, C. A., Freezer, L. J., Morton, K. E., Mavroukakis, S. A., Duray, P. H., Steinberg, S. M., Allison, J. P., Davis, T. A., and Rosenberg, S. A. (2003). Cancer regression and autoimmunity induced by cytotoxic T lymphocyte-associated antigen 4 blockade in patients with metastatic melanoma. *Proc Natl Acad Sci USA* 100(14):8372–8377.

Powrie, F., and Mason, D. (1990). OX-22high CD4+ T cells induce wasting disease with multiple organ pathology: prevention by the OX-22low subset. *J Exp Med* 172(6):1701–1708.

Proietti, E., Greco, G., Garrone, B., Baccarini, S., Mauri, C., Venditti, M., Carlei, D., and Belardelli, F. (1998). Importance of cyclophosphamide-induced bystander effect on T cells for a successful tumor eradication in response to adoptive immunotherapy in mice. *J Clin Invest* 101(2):429–441.

Qiao, M., Thornton, A. M., and Shevach, E. M. (2007). CD4+ CD25+ [corrected] regulatory T cells render naive CD4+ CD25− T cells anergic and suppressive. *Immunology* 120(4):447–455.

Quezada, S. A., Peggs, K. S., Curran, M. A., and Allison, J. P. (2006). CTLA4 blockade and GM-CSF combination immunotherapy alters the intratumor balance of effector and regulatory T cells. *J Clin Invest* 116(7):1935–1945.

Ramirez-Montagut, T., Chow, A., Hirschhorn-Cymerman, D., Terwey, T. H., Kochman, A. A., Lu, S., Miles, R. C., Sakaguchi, S., Houghton, A. N., and van den Brink, M. R. (2006). Glucocorticoid-induced TNF receptor family related gene activation overcomes tolerance/ignorance to melanoma differentiation antigens and enhances antitumor immunity. *J Immunol* 176(11):6434–6442.

Read, S., Greenwald, R., Izcue, A., Robinson, N., Mandelbrot, D., Francisco, L., Sharpe, A. H., and Powrie, F. (2006). Blockade of CTLA-4 on CD4+CD25+ regulatory T cells abrogates their function in vivo. *J Immunol* 177(7):4376–4383.

Reilly, R. T., Machiels, J. P., Emens, L. A., Ercolini, A. M., Okoye, F. I., Lei, R. Y., Weintraub, D., and Jaffee, E. M. (2001). The collaboration of both humoral and cellular HER-2/neu-targeted immune responses is required for the complete eradication of HER-2/neu-expressing tumors. *Cancer Res* 61(3):880–883.

Rolla, S., Nicolo, C., Malinarich, S., Orsini, M., Forni, G., Cavallo, F., and Ria, F. (2006). Distinct and non-overlapping T cell receptor repertoires expanded by DNA vaccination in wild-type and HER-2 transgenic BALB/c mice. *J Immunol* 177(11):7626–7633.

Sakaguchi, S., Fukuma, K., Kuribayashi, K., and Masuda, T. (1985). Organ-specific autoimmune diseases induced in mice by elimination of T cell subset. I. Evidence for the active participation of T cells in natural self-tolerance; deficit of a T cell subset as a possible cause of autoimmune disease. *J Exp Med* 161(1):72–87.

Sakaguchi, S., Sakaguchi, N., Asano, M., Itoh, M., and Toda, M. (1995). Immunologic self-tolerance maintained by activated T cells expressing IL-2 receptor alpha-chains (CD25). Breakdown of a single mechanism of self-tolerance causes various autoimmune diseases. *J Immunol* 155(3):1151–1164.

Sasada, T., Kimura, M., Yoshida, Y., Kanai, M., and Takabayashi, A. (2003). CD4+CD25+ regulatory T cells in patients with gastrointestinal malignancies: possible involvement of regulatory T cells in disease progression. *Cancer* 98(5):1089–1099.

Serra, P., Amrani, A., Yamanouchi, J., Han, B., Thiessen, S., Utsugi, T., Verdaguer, J., and Santamaria, P. (2003). CD40 ligation releases immature dendritic cells from the control of regulatory CD4+CD25+ T cells. *Immunity* 19(6):877–889.

Sharma, S., Yang, S. C., Zhu, L., Reckamp, K., Gardner, B., Baratelli, F., Huang, M., Batra, R. K., and Dubinett, S. M. (2005). Tumor cyclooxygenase-2/prostaglandin E2-dependent promotion of FOXP3 expression and CD4+ CD25+ T regulatory cell activities in lung cancer. *Cancer Res* 65(12):5211–5220.

Shevach, E. M. (2002). CD4+ CD25+ suppressor T cells: more questions than answers. *Nat Rev Immunol* 2(6):389–400.

Shimizu, J., Yamazaki, S., and Sakaguchi, S. (1999). Induction of tumor immunity by removing CD25+CD4+ T cells: a common basis between tumor immunity and autoimmunity. *J Immunol* 163(10):5211–5218.

Shimizu, J., Yamazaki, S., Takahashi, T., Ishida, Y., and Sakaguchi, S. (2002). Stimulation of CD25(+)CD4(+) regulatory T cells through GITR breaks immunological self-tolerance. *Nat Immunol* 3(2):135–142.

Siegmund, K., Feuerer, M., Siewert, C., Ghani, S., Haubold, U., Dankof, A., Krenn, V., Schon, M. P., Scheffold, A., Lowe, J. B., Hamann, A., Syrbe, U., and Huehn, J. (2005). Migration matters: regulatory T-cell compartmentalization determines suppressive activity in vivo. *Blood* 106(9):3097–3104.

Simon, A. K., Jones, E., Richards, H., Wright, K., Betts, G., Godkin, A., Screaton, G., and Gallimore, A. (2007). Regulatory T cells inhibit Fas ligand-induced innate and adaptive tumour immunity. *Eur J Immunol* 37(3):758–767.

Somasundaram, R., Jacob, L., Swoboda, R., Caputo, L., Song, H., Basak, S., Monos, D., Peritt, D., Marincola, F., Cai, D., Birebent, B., Bloome, E., Kim, J., Berencsi, K., Mastrangelo, M., and Herlyn, D. (2002). Inhibition of cytolytic T lymphocyte proliferation by autologous CD4+/CD25+ regulatory T cells in a colorectal carcinoma patient is mediated by transforming growth factor-beta. *Cancer Res* 62(18):5267–5272.

Sugihara, S., Izumi, Y., Yoshioka, T., Yagi, H., Tsujimura, T., Tarutani, O., Kohno, Y., Murakami, S., Hamaoka, T., and Fujiwara, H. (1988). Autoimmune thyroiditis induced in mice depleted of particular T cell subsets. I. Requirement of Lyt-1 dull L3T4 bright normal T cells for the induction of thyroiditis. *J Immunol* 141(1):105–113.

Sutmuller, R. P., van Duivenvoorde, L. M., van Elsas, A., Schumacher, T. N., Wildenberg, M. E., Allison, J. P., Toes, R. E., Offringa, R., and Melief, C. J. (2001). Synergism of cytotoxic T lymphocyte-associated antigen 4 blockade and depletion of CD25(+) regulatory T cells in antitumor therapy reveals alternative pathways for suppression of autoreactive cytotoxic T lymphocyte responses. *J Exp Med* 194(6):823–832.

Tadokoro, C. E., Shakhar, G., Shen, S., Ding, Y., Lino, A. C., Maraver, A., Lafaille, J. J., and Dustin, M. L. (2006). Regulatory T cells inhibit stable contacts between CD4+ T cells and dendritic cells in vivo. *J Exp Med* 203(3):505–511.

Taguchi, O., and Takahashi, T. (1996). Administration of anti-interleukin-2 receptor alpha antibody in vivo induces localized autoimmune disease. *Eur J Immunol* 26(7):1608–1612.

Takahashi, T., Kuniyasu, Y., Toda, M., Sakaguchi, N., Itoh, M., Iwata, M., Shimizu, J., and Sakaguchi, S. (1998). Immunologic self-tolerance maintained by CD25+CD4+ naturally anergic and suppressive T cells: induction of autoimmune disease by breaking their anergic/suppressive state. *Int Immunol* 10(12):1969–1980.

Takeda, I., Ine, S., Killeen, N., Ndhlovu, L. C., Murata, K., Satomi, S., Sugamura, K., and Ishii, N. (2004). Distinct roles for the OX40–OX40 ligand interaction in regulatory and nonregulatory T cells. *J Immunol* 172(6):3580–3589.

Tang, Q., Adams, J. Y., Tooley, A. J., Bi, M., Fife, B. T., Serra, P., Santamaria, P., Locksley, R. M., Krummel, M. F., and Bluestone, J. A. (2006). Visualizing regulatory T cell control of autoimmune responses in nonobese diabetic mice. *Nat Immunol* 7(1):83–92.

Tang, Q., Henriksen, K. J., Boden, E. K., Tooley, A. J., Ye, J., Subudhi, S. K., Zheng, X. X., Strom, T. B., and Bluestone, J. A. (2003). Cutting edge: CD28 controls peripheral homeostasis of CD4+CD25+ regulatory T cells. *J Immunol* 171(7):3348–3352.

Tang, Q., and Krummel, M. F. (2006). Imaging the function of regulatory T cells in vivo. *Curr Opin Immunol* 18(4):496–502.

Thomas, D. A., and Massague, J. (2005). TGF-beta directly targets cytotoxic T cell functions during tumor evasion of immune surveillance. *Cancer Cell* 8(5):369–380.

Thornton, A. M., and Shevach, E. M. (1998). CD4+CD25+ immunoregulatory T cells suppress polyclonal T cell activation in vitro by inhibiting interleukin 2 production. *J Exp Med* 188(2):287–296.

Turk, M. J., Guevara-Patino, J. A., Rizzuto, G. A., Engelhorn, M. E., Sakaguchi, S., and Houghton, A. N. (2004). Concomitant tumor immunity to a poorly immunogenic melanoma is prevented by regulatory T cells. *J Exp Med* 200(6):771–782.

Valzasina, B., Guiducci, C., Dislich, H., Killeen, N., Weinberg, A. D., and Colombo, M. P. (2005). Triggering of OX40 (CD134) on CD4(+)CD25+ T cells blocks their inhibitory activity: a novel regulatory role for OX40 and its comparison with GITR. *Blood* 105(7): 2845–2851.

Valzasina, B., Piconese, S., Guiducci, C., and Colombo, M. P. (2006). Tumor-induced expansion of regulatory T cells by conversion of CD4+CD25− lymphocytes is thymus and proliferation independent. *Cancer Res* 66(8):4488–4495.

Veldhoen, M., Hocking, R. J., Atkins, C. J., Locksley, R. M., and Stockinger, B. (2006). TGFbeta in the context of an inflammatory cytokine milieu supports de novo differentiation of IL-17-producing T cells. *Immunity* 24(2):179–189.

Vlad, G., Cortesini, R., and Suciu-Foca, N. (2005). License to heal: bidirectional interaction of antigen-specific regulatory T cells and tolerogenic APC. *J Immunol* 174(10):5907–5914.

Wang, H. Y., Lee, D. A., Peng, G., Guo, Z., Li, Y., Kiniwa, Y., Shevach, E. M., and Wang, R. F. (2004). Tumor-specific human CD4+ regulatory T cells and their ligands: implications for immunotherapy. *Immunity* 20(1):107–118.

Wang, H. Y., Peng, G., Guo, Z., Shevach, E. M., and Wang, R. F. (2005). Recognition of a new ARTC1 peptide ligand uniquely expressed in tumor cells by antigen-specific CD4+ regulatory T cells. *J Immunol* 174(5):2661–2670.

Wang, J., Ioan-Facsinay, A., van der Voort, E. I., Huizinga, T. W., and Toes, R. E. (2007). Transient expression of FOXP3 in human activated nonregulatory CD4+ T cells. *Eur J Immunol* 37(1):129–138.
Weinberg, A. D., Rivera, M. M., Prell, R., Morris, A., Ramstad, T., Vetto, J. T., Urba, W. J., Alvord, G., Bunce, C., and Shields, J. (2000). Engagement of the OX-40 receptor in vivo enhances antitumor immunity. *J Immunol* 164(4):2160–2169.
Williams, L. M., and Rudensky, A. Y. (2007). Maintenance of the Foxp3-dependent developmental program in mature regulatory T cells requires continued expression of Foxp3. *Nat Immunol* 8(3):277–284.
Wing, K., Fehervari, Z., and Sakaguchi, S. (2006). Emerging possibilities in the development and function of regulatory T cells. *Int Immunol* 18(7):991–1000.
Wing, K., and Sakaguchi, S. (2006). Regulatory T cells as potential immunotherapy in allergy. *Curr Opin Allergy Clin Immunol* 6(6):482–488.
Woo, E. Y., Chu, C. S., Goletz, T. J., Schlienger, K., Yeh, H., Coukos, G., Rubin, S. C., Kaiser, L. R., and June, C. H. (2001). Regulatory CD4(+)CD25(+) T cells in tumors from patients with early-stage non-small cell lung cancer and late-stage ovarian cancer. *Cancer Res* 61(12): 4766–4772.
Yamazaki, S., Iyoda, T., Tarbell, K., Olson, K., Velinzon, K., Inaba, K., and Steinman, R. M. (2003). Direct expansion of functional CD25+ CD4+ regulatory T cells by antigen-processing dendritic cells. *J Exp Med* 198(2):235–247.
Yang, Z. Z., Novak, A. J., Stenson, M. J., Witzig, T. E., and Ansell, S. M. (2006). Intratumoral CD4+CD25+ regulatory T-cell-mediated suppression of infiltrating CD4+ T cells in B-cell non-Hodgkin lymphoma. *Blood* 107(9):3639–3646.
Yu, P., Lee, Y., Liu, W., Krausz, T., Chong, A., Schreiber, H., and Fu, Y. X. (2005). Intratumor depletion of CD4+ cells unmasks tumor immunogenicity leading to the rejection of late-stage tumors. *J Exp Med* 201(5):779–791.
Zhao, D. M., Thornton, A. M., DiPaolo, R. J., and Shevach, E. M. (2006). Activated CD4+CD25+ T cells selectively kill B lymphocytes. *Blood* 107(10):3925–3932.
Zhou, G., Drake, C. G., and Levitsky, H. I. (2006). Amplification of tumor-specific regulatory T cells following therapeutic cancer vaccines. *Blood* 107(2):628–636.
Zhou, G., and Levitsky, H. I. (2007). Natural regulatory T cells and de novo-induced regulatory T cells contribute independently to tumor-specific tolerance. *J Immunol* 178(4):2155–2162.
Zitvogel, L., Tesniere, A., and Kroemer, G. (2006). Cancer despite immunosurveillance: immunoselection and immunosubversion. *Nat Rev Immunol* 6(10):715–727.
Zou, W. (2006). Regulatory T cells, tumour immunity and immunotherapy. *Nat Rev Immunol* 6(4):295–307.

Cancer-Induced Signaling Defects in Antitumor T Cells

Alan B. Frey

1 Introduction

Immune responses to tumor antigens are seen in almost all cancers but, since growth is usually progressive, are clearly ineffective in elimination of the tumor (Parmiani et al., 2003). Understanding the biochemical basis for escape from antitumor immunity guides multiple avenues of research endeavor. Failure to eliminate tumors can be potentially explained by a variety of reasons, any or all of which may permit escape from immune-mediated eradication (Rabinovich et al., 2007). The possibilities can be broadly grouped into two categories: inhibition of the antitumor priming phase or inhibition of the effector phase. Antitumor Ig and T cells are found in patients and in animal models proving that some measurable priming occurs during tumor growth (Frey and Monu, 2006; Radoja et al., 2000). That the magnitude of a given antitumor response is clearly less robust in comparison to that detected after microbial infection (where, for example, up to 10 % of circulating T cells can be reactive with Epstein–Barr virus proteins) has been interpreted to mean that tumors grow faster than the immune response can contend with. This notion is overly simplistic since human tumors grow quite slowly and there is clearly inhibition of both expansion in number and function of antitumor T cells (see below). The notion that priming of antitumor T cells is deficient during tumor growth is supported by findings that the frequency of tumor-specific T cells can be considerably increased in patients receiving experimental immunotherapy. Unfortunately, in spite of greater numbers of antitumor effector T cells in the blood of vaccinated patients (that secrete IFN-γ in response to recognition of cognate antigen in vitro), tumors are often not eliminated (Rosenberg et al., 2004). This observation suggests that experimental vaccination in human cancer is imperfect/inadequate, a reasonable interpretation considering the advanced state of

---

A.B. Frey
Department of Cell Biology-MSB623, New York University School of Medicine, 550 First Avenue, New York, NY 10016, USA.
e-mail: freya01@med.nyu.edu, alanbfrey@yahoo.com

disease which patients must demonstrate in order to qualify for experimental therapy and which motivates the pursuit of better candidate vaccines. Alternatively there may be immunesuppression and/or immune tolerance in the tumor microenvironment, the draining LN or systemically which impedes the function of the antitumor T cells induced by vaccination, also a reasonable interpretation (Frey and Monu, 2006).

Antigen-specific tolerance to tumor antigens can be induced by tumor growth (Cuenca et al., 2003; Lee et al., 1999; Overwijk et al., 2003; Willimsky and Blankenstein, 2005) and, since many tumor antigens have been shown to be self-derived differentiation antigens, has focused much attention on attempts to overcome anergy (Pardoll, 2003). The anergic phenotype of antitumor T cells is complex: adoptive transfer of naive T cells reactive to a model xenogenic tumor antigen could expand in tumor-bearing mice implying priming in vivo (Staveley-O'Carroll et al., 1998). However, the T cells did not respond to subsequent in vivo vaccination, a finding that was interpreted to mean that initial encounter with antigen induces anergy. Induced anergy is not a function of T-cell differentiation status since similar results using memory or effector T cells in adoptive transfer protocols have been reported (Horna et al., 2006). A prominent role for $CD4^+CD25^+$ T_{Reg} in anergy induction following vaccination has been shown (Gattinoni et al., 2006) and abundant T_{Reg} are found in some tumors (Curiel et al., 2004), inspiring efforts that attempt to either eliminate or block T_{Reg} activity concurrent with vaccination (Zou, 2005).

Although prominent, the role of T_{Reg} in causing antitumor T-cell dysfunction in vivo is possibly not exclusive since considerable antigen-specific cells are found not activated in situ (Zhou et al., 2006). This observation supports the notion that dendritic cell function (antigen processing and/or cell activation) is likely to be deficient, a topic both recently reviewed (Bronte et al., 2006; Gabrilovich, 2004) and vigorously pursued. Another observation suggestive that priming is defective in tumor-bearing hosts is that several HLA Class I-restricted tumor antigens have been identified (using antitumor TIL as the basis of tumor cDNA expression systems) that are products of tumor-specific mishandling of genetic information—faulty mRNA transcription, splicing or translation (Coulie et al., 1995; Dolcetti et al., 1999; Guilloux et al., 1996; Ishikawa et al., 2003; Lupetti et al., 1998). Originally proposed by T. Boon in 1989 (Boon and Van Pel, 1989), to considerable skepticism (Lindahl, 1991), these antigens are expressed from alternative reading frames or intronic sequences. Clearly these antigens are not "self" and their cognate TCR are therefore likely to be of relatively high affinity. Thus, those T cells probably are not subject to the regulatory constraints imposed by the clonal selection theory on self-reactive T cells in the periphery whose TCR affinity for cognate antigen is modest. Although not directly assessed, the abundance of this class of antitumor T cell is probably low in situ which implies either that the precursor frequency is low or priming of the cognate T cells is deficient. (Their effector-phase functions are also clearly suppressed since tumors grow, implying a second level of defect, discussed in detail below. However, two reports showed enhanced death of cancer cells proximal to $CD8^+$ T cells in tumors having increased microsatellite instability encouraging the notion that antitumor TIL in situ can in some cases express lytic function (Dolcetti et al., 1999; Ishikawa et al., 2003).)

Evidence for defective T-cell priming in tumor-bearing hosts is abundant and it undoubtedly contributes to modest (or failed) antitumor T-cell immune response (discussed in other chapters); nevertheless, since there is demonstrable antitumor immune responses some priming must occur. Antitumor T cells are found in patient blood, LN and tumor tissue but are defective in response to stimulation through the TCR: cytokine release, lytic function and proliferation are often deficient (Whiteside and Parmiani, 1994); the extent of deficiency is dependent upon tumor type and stage, and the type of assay employed. These defects may be explained by several potential mechanisms that can be generalized as (1) the absence of a positive activation signal, i.e., antigen (or costimulation) or (2) the presence of a negative signal, i.e., enhanced (and dominant) activity of an inhibitory signaling receptor or soluble mediator (e.g., PD-1 or TGFβ-1). This latter notion is reminiscent of the phenotype of T cells in response to agonist plus antagonist peptides, or perhaps only partial agonist peptide (Sloan-Lancaster et al., 1996; Sloan-Lancaster et al., 1996).

There is clearly tumor antigen present in the body, since tumors are usually clinically discernible, thus the absence of antigen is not the basis for the non-activated state of antitumor T cells (although processing and presentation by dendritic cells may be impaired). Also, deficient costimulation (by APC) possibly contributes to the limited magnitude of priming. Therefore, albeit not maximally expanded in number, antitumor T cells in the circulation and tumor tissue have suppressed immune responses likely due to some form of inhibition. This review analyzes the data concerning tumor-induced inhibition of antitumor T-cell function.

2 Systemic Defects in Antitumor T-Cell Function

The observation that patients with late-stage cancers can have defective systemic DTH and proliferative T-cell responses, and diminished (or skewed) cytokine production (Stutman, 1975), prompted biochemical analyses of signaling in peripheral and tumor-infiltrating T cells (reviewed in Whiteside and Parmiani, 1994). The subject has been vigorously pursued in both animal models and patients and there is general acceptance of the notion that tumor antigen-specific proliferative responses and cytokine production from PBL (and in splenocytes in rodent models) are deficient and that the magnitude of defects expands as tumor burden increases (Wick et al., 1997). One of the pioneering publications which strove to provide a biochemical basis for defective immunity in tumor-bearing hosts reported analysis of in vivo and in vitro lytic function of spleen-derived, IL-2 plus anti-CD3ε-activated killer cells from mice bearing tumors (Loeffler et al., 1992). Loeffler and colleagues showed a decrease in cytolysis if total spleen T cells were analyzed from mice bearing tumors grown for more than 21 days, implying a tumor-induced defect in the development of lytic potential. The precise nature of the lytic cells assayed cannot be readily determined since the enriched T cells were cultured in IL-2 for an indeterminate time period in vitro before use, and lytic function was not totally abrogated, only severely reduced in comparison to cells originally obtained from control mice. Curiously, production of several cytokines from enriched T cells activated in vitro was enhanced compared to control cells implying that T cells could be activated in vitro, and stated (but not shown) was that proliferation of $CD4^+$ cells in vitro was normal. Conditioned medium from the tumor cells (MCA38, a colon adenocarcinoma) partially blocked the in vitro development of lytic activity in control killer cells and granzyme B and TNF RNA levels were decreased (although protein levels were not quantified and RNA levels for IL-2, IFN-γ and IL-6 were unchanged). That observation was interpreted to mean that tumor growth could cause enhanced degradation or inhibition of transcription of mRNAs encoding selected effector-phase proteins, thus explaining lytic T-cell dysfunction. This paper also suggested that tumor cells secrete a factor(s) that impacts negatively on the development of lytic function in T cells, although further work on that factor has not been published.

Following closely the initial description of a tumor-elaborated factor, which presumptively inhibited development of T-cell lytic function in tumor-bearing mice, a paper was published in which unmanipulated purified spleen T cells of tumor-bearing mice were analyzed for both calcium flux in vitro and levels of some components of the proximal TCR signaling complex (Mizoguchi et al., 1992). Calcium flux was reduced ($\sim$30%) in T cells from tumor-bearing mice after stimulation with anti-CD3ε Ab implying defective TCR-mediated signaling. (Stimulation of T cells with cognate tumor cells was not performed, a technical necessity later shown to dramatically impact upon signaling in antitumor T cells (Koneru et al., 2005).) Stated (but not shown) was that activation of cells with calcium ionophore resulted in normal levels of calcium flux implying that the signaling block was in the proximal TCR pathway. Assessment of total protein tyrosine phosphorylation by immunoblotting of non-ionic detergent cell extracts from in vitro activated T cells showed a dramatic reduction in tyrosine kinase activity. Immunoblotting further showed absent $p56^{lck}$ and $p59^{fyn}$, the two tyrosine kinases most proximal in the TCR signaling cascade (see below). Finally, in order to characterize the structural composition of the TCR, T cells were surface labeled by lactoperoxidase-catalyzed iodination, solubilized with non-ionic detergent, immuneprecipitated with anti-CD3ε and radiolabeled proteins that associated with CD3ε were analyzed by electrophoresis and blotting. TCRζ, an adapter protein that is the target of $p56^{lck}$ (see below) which when phosphorylated recruits ZAP70 permitting activation of downstream signaling pathways, was determined to be absent from the TCR complex in T cells isolated from tumor-bearing mice. Further, and perhaps most surprising, the FcR γ chain normally found associated with CD16 in non-$CD8^+$ T cells (NKT cells, granulocytes, myeloid cells, $\gamma\delta$T cells and NK cells (Ravetch and Kinet, 1991)) was robustly detected in T cells isolated from tumor-bearing but not control mice. Stated, but not shown, was that other aspects of activation in vitro were normal (proliferation, secretion of cytokines and IL-2R upregulation) implying that the T-cell deficit is primarily limited to cytolytic function, a concept since termed "split anergy" (Mescher et al., 2006), although lytic function of T cells was not assessed. (Expression of

FcR γ chain "in exchange" for TCRζ in $CD8^+$ T cells has not been subsequently reported.)

Several aspects of this work bear reflection. Most importantly, this paper ushered to the forefront an important biochemical perspective on the study of immune dysfunction in cancer: the notion that structural or functional defects in components of the proximal TCR signaling machinery are caused by tumor growth and that these defects underlie the defective functional phenotype of T cells in cancer therein permitting growth of antigenic tumor. However, perhaps because of the enduring effect this work has had on motivating subsequent years of research in multiple laboratories, some of the original findings and interpretations can be critically re-evaluated. Crucial is the contradiction that T cells lacking both $p59^{fyn}$ and $p56^{lck}$ can transmit any activation signal when CD3ε is crosslinked by Ab in vitro: if T cells had normal cytokine secretion, upregulation of IL-2R and proliferation, as stated by Mizoguchi and colleagues, these cells must have functional proximal TCR machinery in order to have transmitted an activation signal. That granted, the apparent absence of both proximal kinases ($p59^{fyn}$ and $p56^{lck}$) and TCRζ, assessed after detergent solubilization and immunoblotting (with or without immuneprecipitation or in later publications by flow cytometry of complex mixtures of cells, e.g., Whiteside, 2004), is likely a false-negative finding. This supposition is supported by the authors' data that calcium flux is intact in T cells from tumor-bearing mice, albeit diminished compared to controls. Collectively considered, one interpretation of those results is that biochemical and functional assay of signaling in T cells *before* permeabilization is correct (diminished calcium flux, sometimes depressed cytokine secretion, proliferation) and that apparent loss of selected signaling components ($p56^{lck}$, TCRζ) reflects artifactual degradation of protein attendant to cell lysis conditions. This notion is supported by subsequent results from multiple laboratories that showed TCRζ is particularly sensitive to proteolysis (Bronte et al., 2005; Franco et al., 1995; Gastman et al., 1999; Levey and Srivastava, 1995; Wang et al., 1995).

This conclusion can also be drawn from a subsequent paper from that lab in which the kinetics of loss of TCRζ and $p56^{lck}$ as a function of time of tumor growth was determined (Correa et al., 1997). In contrast to previous publications where TCRζ was "missing" from peripheral T cells after 21 days of tumor growth, Correa et al. showed TCRζ loss only after ~32 days of growth. However, $p56^{lck}$ was found absent at much earlier times of growth (ca. 18 days) raising several questions: first, what is the basis for the disparate results concerning the kinetics of TCRζ loss? The loss of a component of the TCR complex most proximal in the cascade probably is causal to any additional effects on downstream components, so the basis for the most proximal defect should be the primary focus. The focus on the apparent loss of TCRζ, which is downstream of $p56^{lck}$ in the signaling machinery, was not explained. In any event, levels of cytokine secretion in vitro from "late-stage" tumor-bearing mouse T cells after TCR ligation were found to be either diminished (IL-2, IFN-γ) or enhanced (IL-4, IL-10) relative to control T cells, again arguing that T cells could receive and respond to external signals. Importantly, when T cells were lysed under "harsh conditions" (boiling in SDS) TCRζ levels were approximately equal to control samples ($p56^{lck}$ levels were not assessed under harsh lysis conditions). This finding was interpreted to mean the TCRζ localizes to a subcellular compartment that is resistant to non-ionic detergent solubilization in T cells of late-stage tumor-bearing mice. Immunocytochemistry or immunofluorescence microscopy localization of TCRζ, which would validate that notion, was not performed.

Considered in light of multiple publications that show downregulation of proximal TCR signaling molecules as a normal consequence of antigen recognition (Marth et al., 1987; Olszowy et al., 1995; Valitutti et al., 1997; Veillette et al., 1988), it seems possible that the observed changes in levels of p56^{lck} and TCRζ may reflect normal physiological variation after antigen recognition. Furthermore, the consequences to the host of absent proximal TCR signaling in peripheral T cells are expected to be dramatic in that responses to environmental pathogens would be obviated. That would result in systemic pathology akin to acquired immunodeficiency, a condition not seen in rodent models or patients. In addition, using the same tumor model as used in the original findings (MCA38), among multiple other tumor models, our lab showed that mice bearing comparatively large tumors had both normal immune responses to various soluble and cellular antigens in vivo and normal in vitro responses to stimulation if the responding T cells were sufficiently purified. These data argue that there is no systemic T-cell dysfunction in tumor-bearing mice (Radoja et al., 2000). Finally, there have been reports that TCRζ levels are not significantly reduced in either PBL or TIL in renal cell carcinoma (Cardi et al., 1997) or in TIL (in frozen thin sections) in Hodgkin's disease (Dukers et al., 2000) arguing against this notion. Despite the concerns that these data illustrate in consideration of the biochemical basis of systemic T-cell dysfunction caused by tumor growth, tumor-induced defective antitumor-specific immune response is probably common in cancer and is causally related to tumor escape from immune response (Cochran et al., 2006; Shu et al., 2006).

We feel that close attention devoted to the issue of TCRζ levels in systemic or tumor-infiltrating T cells is warranted because, since the original findings made in rodent models, in an effort to determine the basis of dysfunctional antitumor immune responses in patients, other laboratories have published similar findings in patient PBL (Whiteside, 2004) or TIL (Uzzo et al., 1999b), whose conclusions may similarly justify re-evaluation.

3 Defective Proliferation of Peripheral Blood T Cells in Cancer Patients

The year following the original reports of TCRζ loss in systemic T cells in rodent models, two papers were published which described similar observations in human TIL and PBL (Finke et al., 1993; Nakagomi et al., 1993). Proliferation of T cells in vitro was found to be deficient in TIL and PBL isolated from patients with various types of cancers, in much the same manner as was previously determined for spleen cells in rodent models. Typically mononuclear cells were isolated from blood by density gradient fractionation and stimulated with mitogen or anti-CD3ε. In some cases cytokine secretion was also determined to be decreased in comparison to non-stimulated patient PBL. Where analyzed in historical comparison, defective in vitro responses of TIL are usually more pronounced than for PBL (Alexander et al., 1993; Whiteside and Parmiani, 1994). These sorts of data have been interpreted to mean that systemic T-cell immunity is compromised in cancer patients that may reflect tumor-induced immune suppression. The basis of patient defective proliferative response has been pursued and is considered herein.

To concerns about similar experiments using rodent cells (Radoja et al., 2000), we suggest that interpretation of these data be tempered by several caveats. If the patient PBL preparations are not purified they possibly are contaminated by blood cells that have been shown to inhibit T-cell proliferation or cause apoptosis of activated cells in vitro (Schmielau and Finn, 2001). Additional concerns are that proliferation of patient PBL in vitro in some cases was not calculated as a percentage of $CD3^+$ cells within the population and is also not extinguished, rather; the percentage increase in proliferation of patient PBL is diminished relative to PBL from control non-patients (Whiteside, 2004). Absolute levels of incorporation of radiolabeled thymidine per cell is sometimes not revealed nor is non-patient PBL always used in comparison (Reichert et al., 2002) making assessment of results between different labs and tumor types impossible. In addition, the percentage of individual patients within a cohort with diminished responses is sometimes not provided making the "penetrance" of the phenomenon among the larger patient population difficult to establish.

One early paper showed significant inhibition of renal cell carcinoma TIL in vitro proliferation but, curiously, autologous PBL were not affected (Alexander et al., 1993). Four years later the same lab published that culture supernatant from a renal cell cancer cell line could dramatically inhibit proliferation in vitro of normal T cells (Kolenko et al., 1997). Biochemical analysis showed that while $p56^{lck}$ and TCRζ levels were normal, T-cell levels of IL-2R-associated kinase JAK3 were dramatically reduced. (Over the years additional publications from that lab de-emphasized TCRζ loss from T cells or found the majority of patient PBL to have normal levels (Bukowski et al., 1998), instead focusing on other systemic signaling defects induced by factors released from tumor cells which appear to induce a "pre-apoptotic state" in TIL or PBL, see below. Further reports of a role for JAK3 in systemic T-cell defects have not been published.)

4 Enhanced Sensitivity to AICD in Systemic T Cells in Cancer Patients Due To Tumor Secretion of Gangliosides

Over the years there have been many reports of antiproliferative or proapoptotic activities in supernatants of tumor cell cultures which induce T-cell apoptosis in vitro including, but not limited to: TGFβ, IL-10, PGE and soluble reactive oxygen and nitrogen molecules. In at least one type of cancer, glioma patients are known to be lymphopenic, a condition that could be due to high levels of systemic T-cell apoptosis (Dix et al., 1999). In other cancers, perhaps in early stages of tumor progression, T-cell apoptosis may be initiated and yet not show robust DNA fragmentation, but be nonetheless functionally deficient.

Assessment of the apoptotic status of patient purified PBL T cells determined that a significant percentage of $CD3^+$ PBL T cells are induced to apoptosis after 24 h exposure to anti-Fas or tumor cells (ca. 40%), or PMA/ionomycin (ca. 22%, which increases to 80–100% after 48 h) (Uzzo et al., 1999b). Apoptosis of unstimulated patient T cells was low, as was apoptosis of one control T-cell sample obtained from a non-patient. It was concluded that systemic T cells in renal cell carcinoma patients are in a heightened (or partial) activation state, perhaps due to chronic tumor antigen exposure or to a tumor-elaborated factor(s) (see below). It was hypothesized that if the in vitro phenotype reflected T-cell status in vivo, antigen activation (i.e., strong) would induce AICD in antitumor T cells therein impeding effective antitumor immunity in the patient. Heightened sensitivity to AICD in vitro was subsequently shown to be mediated by impairment of activation of NFκB in patient PBL (Li et al., 1994; Uzzo et al., 1999a). Tumor cell culture supernatant contained an activity that induced AICD in Jurkat cells and control PBL upon stimulation that was used as the basis to purify gangliosides, then used in some studies in purified form to study the biochemical basis of the tumor-elaborated activity.

Elevated expression of gangliosides by tumor cells has been long noted (Ladisch et al., 1984); in fact experimental immunotherapy targeting gangliosides has been attempted for many years for several different tumor types (without compelling success) (Blackhall and Shepherd, 2007). Gangliosides can inhibit various immune cells including: maturation of DC (Wolfl et al., 2002), macrophage function (Bennaceur et al., 2006; Bharti and Singh, 2001, 2003), NK cell activity (Ladisch et al., 1984) and cytokine secretion from T cells (Dumontet et al., 1994) and are thought to be shed in exosomes as precursor molecules reflective of alteration of biosynthetic enzymes in tumor cells (Hakomori, 1996). Exactly how elevated gangliosides impact on NFκB signaling is not clear: inhibition of activation of the inhibitor IκBα was shown in one report (Ling et al., 1998), but subsequent papers from the same lab suggest IκBα is not involved, instead degradation of NFκB is enhanced (Thornton et al., 2004). It is conceivable that both the biochemical mechanism of action of gangliosides and the nature of the effect on target T cells (e.g., inhibition of cytokine secretion or proliferation, or induction of apoptosis) are concentration dependent. In addition, it is possible that responses to gangliosides are influenced by the nature of activation signals received by the T cell after (or during) ganglioside exposure (Biswas et al., 2006). Therefore, perhaps the contradictory mechanistic data can be reconciled by these considerations.

The details of ganglioside action are likely to be important since the component of the T-cell signaling cascade affected by gangliosides could be a potential point of intervention for immunotherapy. In this regard, a ganglioside receptor in T cells (or any immune cell) has not yet been described which would be a potential target for therapy. However, galectin-1 in a human neuroblastoma cell line was shown in 1998 to be a receptor for GM_1 (Kopitz et al., 1998). When expressed in tumor cells, galectins (endogenous lectins, reviewed in Hernandez and Baum, 2002) have been shown to inhibit antitumor T-cell immunity although the mechanism remains unclear. One report showed that exogenous galectin-1 could inhibit full antigen-dependent triggering of T cells (revealed as only partial TCRζ phosphorylation) permitting limited signaling (resulting in expression of CD69, IFN-γ secretion and apoptosis, but not proliferation) (Chung et al., 2000). Exactly how regulation of T-cell signaling by galectins occurs is not clear since proximal signaling appears to be impacted without need for additional cofactors. We introduce this topic only to illustrate by analogy that a cell surface receptor for gangliosides may be expressed on T cells which would help guide our thinking about how exogenous gangliosides (presumably tumor-derived) inhibit T-cell signaling: receptor-dependent or receptor-independent. If a ganglioside receptor is expressed on antitumor T cells, then the absolute serum concentration of gangliosides to cause functional inhibition need not be as high as if there were no receptor in order for the concentration to be increased in T cells.

Gangliosides may not require a receptor in order to impact on signaling in T cells. For example, if high concentrations, existing presumably as micelles in the blood or in protein-containing exosomes (although monomeric serum gangliosides have been suggested to exist, Kong et al., 1998), gain proximity to a hydrophobic environment (i.e., the plasma membrane of a T cell), the ganglioside could simply partition into the lipid bilayer for thermodynamic reasons. A question arises here: once fused with the plasma membrane, how would signaling be affected (apparently being reduced as is the phenotype in systemic T cells in patients)? One possibility, yet untested as far as we are aware, is that since the T-cell plasma membrane lipid composition would now be altered by the incorporated gangliosides, formation of the higher order lipid raft upon T-cell activation could be inhibited. Since the TCR complex assembles and is maintained in the lipid raft during activation, the requirements for induction or maintenance of signaling could be adversely affected. Modification of lipid rafts which impact on signal transduction has been reported: in one report exogenous polyunsaturated fatty acids, known to have suppressive effects on signaling in immune cells, were shown to regulate Jurkat cell signaling (Stulnig et al., 1998), and Magee et al. (2005) recently showed that signaling was greatly influenced simply by changing the temperature of the immune cells, which in turn impacted on raft formation. A related notion was postulated by the Ladisch lab who showed that exogenous gangliosides enhanced EGFR aggregation in fibroblasts and resulted in heightened sensitivity to ligand (Liu et al., 2004), paradoxically the obverse phenotype of systemic T cells in cancer patients.

Before gangliosideologists embark on the study of proximal signaling in T cells, considering how many red herrings have been enthusiastically pursued in the past, perhaps it would be prudent to first ascertain the physiological relevance of the notion of ganglioside-influenced T-cell signaling. In doing so the notion of a causal relationship between tumor production of gangliosides and defective T cells in patients would be supported. An important question to answer is: do T cells in cancer patients have tumor-derived gangliosides incorporated in their plasma membrane? Since the identity of several gangliosides made in abundance in selected cancers has long been identified (Ritter and Livingston, 1991), it should be straightforward to answer this question in systemic T cells of cancer patients. If systemic T cells in patients are susceptible to AICD because of hypersecretion of tumor gangliosides then those molecules should be present in patient cells. Furthermore, the concentration of putative tumor-shed gangliosides should be maximal in TIL or other host-derived non-TIL cells present in tumors (fibroblasts or MDSC), also a testable prediction.

5 T-Cell Signaling Affected by Soluble Reactive Oxygen or Nitrogen Metabolites

Several laboratories have pursued experiments suggesting that soluble mediators of T-cell dysfunction are produced by either the tumor cell or host MDSC that accumulate to high abundance in many tumors. These topics will be covered by appropriate experts in other chapters in this monograph. We will comment on those candidate mechanisms only to suggest that a similar level of concern be applied to data implicating those candidate mediators. Either reactive oxygen or nitrogen metabolites (peroxynitrites, hydrogen peroxide or nitric oxide) have been suggested to be involved in induction of various immune cell defects, for example, lytic dysfunction in T cells (Bronte et al., 2005) or NK cells (Kono et al., 1996). Specific biochemical components of the TCR signaling pathway that are targeted by these reactive molecules have not been identified, with the exception of TCRζ postulated to be selectively degraded after T-cell exposure to H_2O_2 and other soluble factors (Kiessling et al., 1999).

6 Inhibition of TIL Signal Transduction by Tumor

In contrast to the somewhat contradictory data concerning the physiology and function of systemic T cells in tumor-bearing hosts, there is almost universal acceptance of the notion that TIL are severely inhibited in lytic function. This makes conceptual sense since if the tumor, or host response to the tumor, produces immune cell signaling inhibitory factors, then those factors should be in the highest concentration closest to the tumor. Since TCR signaling is required for cytolysis, the presence in tumor tissue of T cells whose antigen-specific functions can be demonstrated in vitro upon purification but whose lytic function in situ is not readily detected (Bronte et al., 2005) implies tumor-induced inhibition of TCR signaling in situ. Freshly isolated $CD8^+$ TIL are, like TIL in situ, lytic-defective and when analyzed in in vitro signaling assays using cognate tumor cells as stimulus have been shown to have a blockade in the most proximal portion of the TCR signaling pathway (Koneru et al., 2005; Radoja et al., 2001). TIL receive initial antigen signals upon contact with cognate tumor cells but are unable to perpetuate the signal downstream past $p56^{lck}$ activation. (A detailed description of the regulation of proximal TCR signaling in T cells is provided in the following section.) Thus, PKC translocation/activation,

ERK activation and calcium flux are blocked in spite of having been triggered by contact with antigen-expressing target cells.

We have addressed one unanswered question concerning TIL lytic dysfunction: are the signaling and lytic defects which characterize TIL induced in *any* memory/effector T cell which enters the tumor environment or does the in vivo induction of lytic defects require antigen-specific interaction between the T cell and the tumor cell? We took advantage of the observation of others that after infection with *Listeria monocytogenes* (genetically modified to express ovalbumin) has been cleared in mice, antigen-specific $CD8^+$ T cells populate the non-lymphoid peripheral tissues (Masopust et al., 2001; Pope et al., 2001). As opposed to TIL which are lytic defective until briefly cultured in vitro, these cells are lytic towards the model antigen expressed by the recombinant *L. monocytogenes* immediately upon isolation (Masopust et al., 2001). (Like antitumor TIL, $CD8^+$ anti-ova T cells are memory/effector cells as determined by several criteria, with the obvious difference being lytic capability.) We reasoned if anti-ova T cells isolated from non-lymphoid tissues of tumor-bearing mice are lytic but if isolated from *tumor tissue* are non-lytic then that would indicate that the tumor microenvironment induces the lytic defect. Furthermore, induction of lytic defects would not require antigen-specific interactions in the tumor.

C57BL/6 mice were infected with a sublethal dose of *L. monocytogenes* engineered to express chicken ovalbumin ("*L. monocytogenes^{ova}*"). When infection cleared 10 days post-infection, mice were injected with MCA38 tumor cells and 20 days later total $CD8^+$ T cells were isolated from tumors. In mice recovering from *L. monocytogenes* infection a reasonable percentage of non-lymphoid $CD8^+$ T cells are anti-ova CTL (Masopust et al., 2001); therefore, we anticipated that TIL would contain *both* anti-MCA38-reactive T cells and anti-ova-reactive T cells. Immediately upon isolation TIL were tested for lytic function using as target cells either EL-4 cells, EL-4 cells pulsed with the synthetic peptide comprising the ovalbumin epitope presented by K^b (SIINFEKL) or cognate MCA38 tumor cells. There was robust SIINFEKL-specific killing of pulsed EL-4 cells but not of either non-pulsed EL-4 targets or cognate tumor cells. We interpret this result to mean that simple presence in the tumor microenvironment does not induce lytic dysfunction; infiltrating T cells must be (somehow) rendered susceptible to the tumor-derived inhibition of signaling. Since anti-MCA38 TIL are inhibited by cognate tumor cells in cytolysis this finding suggests that antigen-specific recognition is needed to induce proximal signaling defects. This observation also argues against the production of a soluble factor by the tumor that inhibits T-cell signaling.

The TIL proximal signaling block was mapped by a combination of biochemical and functional assays to reveal that, upon initial contact with cognate tumor target cells, $p56^{lck}$ was activated but becomes rapidly inactivated and therefore unable to phosphorylate one of its targets, kinase ZAP70. $p56^{lck}$ inactivation, by dephosphorylation of the activation motif containing Y^{394}, was shown to be mediated by Shp-1 (Monu, in press, 2007). Those data show that (1) TCRζ, previously suggested to be absent in TIL—in the same tumor model (Mizoguchi et al., 1992)—is in fact not absent and becomes phosphorylated upon tumor contact, (2) MDSC, which in other models have shown a dramatic phenotype in terms of inhibition of antitumor T-cell functions (Gabrilovich, 2004), do not mediate the lytic defect in TIL (Frey and Monu, 2006) and (3) antigen-specific interaction between the TIL and tumor is required to induce lytic dysfunction.

Since cognate MCA38 tumor cells cause induction of proximal TCR signaling defects in TIL (Frey and Monu, 2006; Koneru et al., 2005; Radoja et al., 2001) but it is not certain that the mechanism for signaling inhibition which we describe will be utilized for all tumor types, in the following sections we present (to the best of our ability) a synopsis of proximal TCR signaling in T cells in the hope that other models of dysfunctional antitumor T-cell function may be understood in this context. It should be apparent that the complexity of regulation of proximal TCR-mediated signaling provides robust opportunities for interference by tumor-induced factors.

7 Signal Transduction in Cytolytic T Cells

Lytic function of effector $CD8^+$ T cells is dependent upon oligomerization of molecules involved in signal transduction and is initiated by interaction of the antigen receptor (TCRαβ) with cognate peptide ligand:MHC Class I proteins (Kagi et al., 1996) (Fig. 1). Proteins of the TCR complex are synthesized individually and have been shown to assemble in the endoplasmic reticulum or Golgi and, upon binding of the TCRζ chain, the complex moves en bloc to the plasma membrane (Alarcon et al., 2003). Other proteins that associate with the TCR complex on

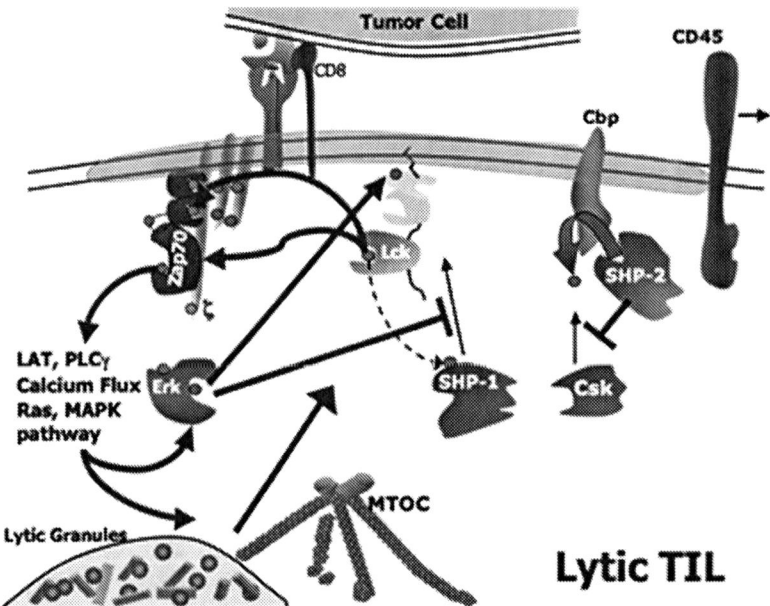

Fig. 1 Schematic of proximal TCR signaling in lytic TIL.

activation are recruited upon TCR binding cognate ligand, such as ZAP70 and LAT (Fig. 1). Non-antigen receptor components of the signaling complex either are located in the cytosolic face of endomembranes (Kabouridis et al., 1997), as soluble proteins in the cytoplasm (Cherukuri et al., 2001; Clements et al., 1999; Rudd, 1999), or are integral plasma membrane proteins (Schraven et al., 1999), but after recruitment to the TCR complex enter into close functional association with one another (Cherukuri et al., 2001; Kane et al., 2000). Antigen recognition induces either a conformational change in, or low-order aggregation of, the TCR (Fernandez-Miguel et al., 1999). The coreceptor CD8, an integral membrane protein heterodimer of α and β chains, is next recruited to the TCR complex wherein it functions to enhance the sensitivity of antigen recognition by stabilizing the low-affinity TCR-MHC/peptide interaction (Zamoyska, 1998). Prior to antigen recognition-induced recruitment of CD8 to the TCR, CD8 is associated with the src-family kinase $p56^{lck}$. Thus, recruitment of $CD8/p56^{lck}$ to the antigen receptor serves several functions: it permits the establishment of a stable higher order oligomeric signaling complex thought to be necessary for accumulation of sufficient information to signify an authentic activation signal, and it causes the association of the kinase most proximal in the signaling cascade with its substrates. Phosphorylation of the TCRζ chain by $p56^{lck}$ upon ligand binding is a very early biochemical consequence of antigen recognition and initiates T-cell "triggering", which results from tyrosine phosphorylation of certain consensus amino acid motifs (ITAMs). After recruitment of $p56^{lck}$, phosphorylation of several associated molecules within the TCR complex rapidly occurs and also the recruitment of additional proteins that are requisite components in perpetuating signal transduction, which are in turn activated by phosphorylation (Samelson, 2002).

In addition to causing recruitment into proximity to its immediate downstream target (TCRζ), antigen recognition causes activation of the kinase function of $p56^{lck}$. In T cells whose TCR is not engaged with cognate antigen (aka, "resting" cells), $p56^{lck}$ exists poised for activation without a requirement for synthesis of any additional factors, a design that permits rapid cell activation (Fig. 2). Activation of $p56^{lck}$ is thought to be by autophosphorylation of a specific tyrosine (residue 394 in mouse, although there may be serine/threonine phosphorylation associated with activation), an event that is prevented by restraint of intermolecular folding, in turn caused by phosphorylation of a tyrosine residue (position 505 in mouse) located near the carboxyl terminus. Regulation of Y^{505} phosphorylation is discussed in detail in the next section and regulation of phosphorylation of the activation motif containing Y^{394} is discussed here.

In T cells at rest, the $p56^{lck}$ inhibitory motif (containing Y^{505}) is phosphorylated (by a negative regulatory kinase Csk), which as mentioned above prevents autophosphorylation. Upon antigen recognition, through its association with CD8, $p56^{lck}$ is brought into proximity with the TCR complex. Coincident with recruitment of $p56^{lck}$, the positive regulatory phosphatase CD45 dephosphorylates $p56^{lck}$ Y^{505} permitting $p56^{lck}$ autophosphorylation. (Activating signals which enable CD45 to act on $p56^{lck}$ Y^{505} are unknown but may be provided *in trans* by counterligands on antigen-bearing target cells.) A second substrate for CD45 is the integral membrane

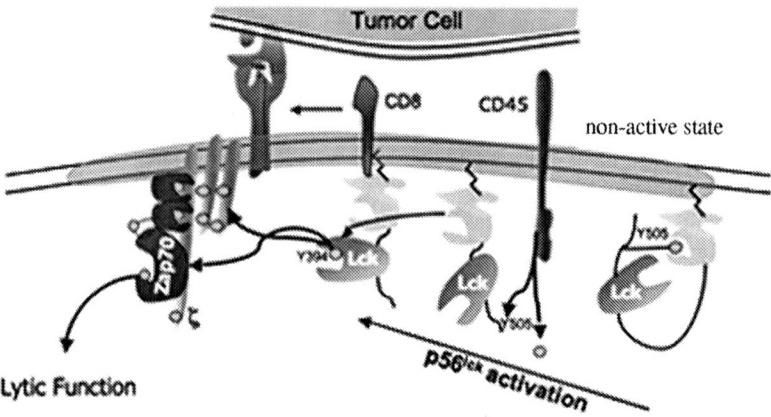

Fig. 2 Schematic of activation of p56^{lck}

adapter protein that recruits Csk to the plasma membrane, Cbp. (Regulation of the p56^{lck} Y^{505} phosphorylation cycle is discussed in detail below.) Thus, at the initiation of T-cell activation, CD45 relieves tonic inhibition of p56^{lck} permitting autophosphorylation coincident with coreceptor-mediated recruitment of p56^{lck} into proximity with its substrate TCRζ and other substrate components of the TCR complex, e.g., CD3. The next biochemical event is recruitment of ZAP70 to phosphorylated TCRζ and CD3ε chains (Samelson, 2002; Weiss and Littman, 1994) whereupon ZAP70 is phosphorylated by p56^{lck}. The nascent lipid raft now contains the TCR complex and a raft-associated adapter protein LAT which is available for phosphorylation by ZAP70. Then additional kinases and adapter proteins (e.g., SLP-76, Gads, Grb2, Vav, Nck, phosphatidylinositol 3-kinase and PLCγ-1) are recruited to phosphorylated ZAP70 in a specific spatio-temporal manner ultimately resulting in calcium flux. TCR-mediated activation of effector-phase functions in CD8$^+$ cells requires calcium flux in that all downstream effector-phase functions are abrogated by events that impede calcium flux.

Following this early phase of signaling (which is quite rapid, ca. within seconds after antigen recognition), subsequent effector functions (such as reorganization of the cytoskeleton, affinity maturation of LFA-1 and subsequent tight adhesion to target cell (Koneru et al., 2006; Morgan et al., 2001), cytokine secretion, mobilization of the MTOC to the immunological synapse (IS) and vectoral discharge of lytic granules) result from coordinated activation of additional downstream signaling pathways (involving other kinases: PI 3-kinase, protein kinase C and MAP kinase (Radoja et al., 2006)), occurring after 1–5 min of contact with antigen-expressing target cells. Thus, the lytic phenotype of an effector CD8$^+$ T cell does not follow a strict linear progression of activation/signal transduction, instead results from a complex interrelated program involving several signaling cascades. Importantly, tonic inhibition of p56^{lck}, the kinase most proximal in the signaling cascade which mediates the initiation of TCR-mediated signaling, is pivotally poised to permit fine control of immune response modulation (Torgersen et al., 2002).

8 Regulation of p56^{lck} Activity by Inhibitory Phosphatases

In addition to control of phosphorylation of p56^{lck} Y^{394} by relief of tonic autoinhibition, phosphorylation of Y^{394} is negatively regulated by the action of inhibitory phosphatases, primarily Shp-1 and PEP (Mustelin et al., 2005). p56^{lck} is a substrate for Shp-1 (Chiang and Sefton, 2001), and other components of the proximal signaling complex are possible substrates (ZAP70, LAT, Vav, Grb2, SLP-76 and PLCγ-1 (Kautz et al., 2001; Plas et al., 1996; Zhang et al., 2003)). Shp-1 is a nonmembrane protein tyrosine phosphatase that contains two amino terminal-located SH2 domains, a single phosphatase domain and a carboxyl terminus containing two sites for tyrosine phosphorylation (Siminovitch and Neel, 1998). The role of Shp-1 in negative regulation of immune cell signaling has been widely studied using the mutant mouse, motheaten (*me/me*). These mice express defective Shp-1 and exhibit a panoply of hematopoietic defects suggesting that Shp-1 plays an essential role in regulating signal transduction in hematopoietic cells (Shultz et al., 1993). Defects include hyperproliferation of macrophages and neutrophils but also abnormal B- and T-cell hyperresponsiveness and development (Hayashi et al., 1988; Lorenz et al., 1996). Shp-1-mediated downregulation of activation of lytic T cells involves both single and double negative and positive feedback circuits which serve to ensure that commitment of a cell to lytic function is restrained unless positive feedback signals are sufficiently robust to overcome the endogenous restraint (Mustelin et al., 2005).

Prior to activation, one of the Shp-1 SH2 domains interacts with and shields the catalytic domain from binding to substrate (Pei et al., 1994). (It is not known whether there is phosphorylation of specific residues associated with maintenance of the sterically inhibited state as there is for p56^{lck}—see below). Upon activation, reflected in phosphorylation at several amino acids, especially Y^{564}, which is accomplished by p56^{lck} (itself a target of Shp-1 therein illustrating a complex regulation program involving multiple negative feedback loops), intermolecular interactions are relaxed and its SH2 domain can now bind to the SH2 domain of binding partners and substrates. Tyrosine phosphorylation of Shp-1 is required for enzymatic activity, but occurs coincident with—or immediately following—localization at the plasma membrane in proximity with its substrates. Shp-1 movement from the cytosol to the membrane is by recruitment to an integral membrane protein, one of a family of "inhibitory signaling receptors" (or adapter proteins) which contain in their cytoplasmic tails a motif for binding Shp-1, the ITIM (D'Ambrosio et al., 1995; Neel, 1997) (Fig. 3).

Activation of Shp-1 enzymatic activity is dramatically enhanced upon binding to inhibitory receptors, ~10–100 fold (Chiang and Sefton, 2001), implying that recruitment to inhibitory receptors may be the major activation mechanism for Shp-1. Many inhibitory signaling receptors are expressed in different hematopoietic cells and have been shown to be able to recruit Shp-1 to the membrane including NK immunoglobulin-like receptors (Vivier and Anfossi, 2004), siglecs (Varki and Angata, 2006) and leukocyte immunoglobulin-like receptors (Ly49 molecules) (Vivier and Daeron, 1997). An example of the dependence of Shp-1 activity upon

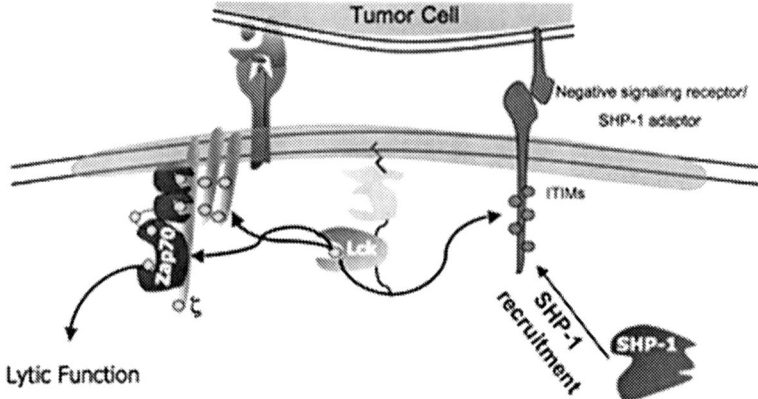

Fig. 3 Recruitment of Shp-1 to the plasma membrane into proximity with p56^{lck} is mediated by activation (tyrosine phosphorylation) of an inhibitory signaling receptor containing ITIM motifs

recruitment to the inhibitory signaling receptor is shown by the phenotype of mice deficient for CD22, an inhibitory signaling receptor involved in regulation of signaling in B cells (O'Keefe et al., 1996). These mice have B-cell hyperproliferation (with attendant increased calcium flux) and decreased Shp-1 recruitment after ligation of the antigen receptor, although Shp-1 can be activated. ITIMs resemble in mechanism of action and function the ITAM motif expressed in adapter proteins in the immunoreceptor signaling pathway (e.g., TCRζ) in that the ITIM becomes tyrosine phosphorylated upon antigen recognition by the immune cell, thought by the kinase that is ultimately targeted for inhibition (Fig. 4). In addition, the extracellular portion of the inhibitory receptor interacts with a counterligand expressed on

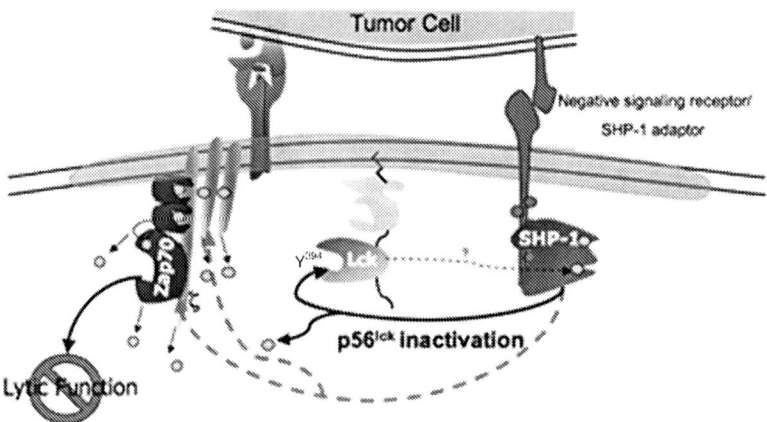

Fig. 4 ITIM motifs are phosphorylated by tyrosine kinases requiring interaction of the inhibitory signaling receptor with its cognate ligand

target cells. The molecular function of this interaction is not known but may serve to stabilize the inhibitory receptor in proximity with the antigen receptor as well as to influence the conformation of the cytoplasmic domain permitting phosphorylation of the ITIM.

Control of Shp-1 activity is complex having multiple different but interacting positive and negative feedback regulatory circuits. Perhaps regulation of Shp-1 can be categorized into two general programs: direct and indirect regulation, each of which has two (approximate) kinetic classes of potential mechanisms, rapid and late-acting. For example, phosphatases can be directly regulated by reactive oxygen species (H_2O_2) which has been shown to be produced in neutrophils, and potentially other hematopoietic cells (Brumell et al., 1996; Meng et al., 2002). Hydrogen peroxide rapidly oxidizes an active site cysteine residue resulting in inhibition of activity (and thus the inability to inactivate target kinases). (In this regard, as discussed in an earlier section, the H_2O_2 reported to be copiously produced by MDSC that accumulate in tumor and peripheral lymph tissues (Kono et al., 1996) possibly is in fact not expressed in situ since proximal signaling in antitumor T cells (especially TIL) is defective. If Shp-1 were rendered inactive by elevated local H_2O_2 levels, enhanced, not suppressed, kinase activity would be expected.) Another mechanism for Shp-1 inactivation involves displacement from proximity to candidate targets. This involves multiple additional protein:protein interactions and is likely slower to instigate since it requires detachment from its binding partner (the inhibitory signaling receptor) that in turn requires either an active dephosphorylation of the receptor ITIM or a diminution of a kinase activity required to maintain ITIM phosphorylation. Alternatively, interaction of the inhibitory signaling receptor with its counterligand may be disrupted which may permit either the inhibitory signaling receptor to move out of proximity to the Shp-1 target kinase or to cause a change in conformation of the domain containing the ITIM therein blocking its phosphorylation.

Indirect regulation would include mechanisms that do not directly affect Shp-1 activity or proximity to substrate. For example, modification of target $p56^{lck}$ to obviate accessibility to Shp-1 could prevent kinase inactivation in spite of activation and recruitment of Shp-1. Although such a function for known $p56^{lck}$-binding proteins (e.g., LIME (Brdickova et al., 2003; Hur et al., 2003), CAML (Tran et al., 2005)) has not been described, shielding from Shp-1 access by a non-$p56^{lck}$ protein is conceivable. However, an induced conformational change in a Shp-1 binding domain (SH2) in $p56^{lck}$ has been reported which results from phosphorylation of $p56^{lck}$ Ser^{59} by ERK (Stefanova et al., 2003). Phosphorylation of $p56^{lck}$ on serine/threonine residues has been long known and associated with enhanced kinase activity (Joung et al., 1995; Watts et al., 1993). Phosphorylation on $p56^{lck}$ Ser^{59} is now hypothesized to reflect robust TCR-mediated signaling with attendant ERK activity. According to this notion, when the quality or duration of antigen stimulation is insufficient or diminished, ERK activity (or its localization proximal to $p56^{lck}$) decreases and thus cannot phosphorylate $p56^{lck}$ Ser^{59} permitting a conformation that is accessible to Shp-1 (Stefanova et al., 2003). Another example of indirect regulation of Shp-1 activity is that the inhibitory receptor itself can be a substrate for Shp-1 (Blasioli et al., 1999) which, when the ITIM is dephosphorylated, no longer binds Shp-1. Perhaps this occurs after the preferred substrate ($p56^{lck}$) has been sufficiently dephosphorylated and would be therefore operative at later times following inhibition of signaling.

9 cAMP-Dependent Modulation of Proximal TCR Signal Transduction by Control of Csk Activity

As mentioned above, control of signaling in T cells can be regulated by the activity of various proteins which function at several different levels in the TCR signaling cascade: positive regulatory kinases ($p56^{lck}$, ZAP70, ERK, phospholipase C, protein kinase C), adaptor molecules (Csk binding protein, Cbp, and linker for activation of T cells, LAT), positive regulatory phosphatases (CD45 and Shp-2), negative regulatory kinases (Csk and PKA), negative regulatory phosphatases (Shp-1 and PEP) and other proteins which function to regulate the activity of signaling components. This latter category includes enzymes involved in posttranslational modification of kinases and phosphatases, e.g., acylation (Mor and Philips, 2006; Resh, 1994), and/or proteins responsible for the subcellular localization of kinases and phosphatases, e.g., Lad (Choi et al., 1999), LIME (Brdickova et al.,2003; Hur et al., 2003), Cbp (Brdicka et al., 2000) and β-arrestin (Tedoldi et al., 2006), and proteins involved in downregulation of signaling enzymes, e.g., ubiquitin ligases (Hawash et al., 2002; Liu et al., 2005).

The most proximal kinase responsible for control of T-cell activation is $p56^{lck}$ whose major structural features were described above. Two characteristics of $p56^{lck}$ suggest it is an ideal candidate for negative regulation by extrinsic factors: it is positioned at the apex of the TCR cascade, thus if inhibited will effectively disable all T-cell function, and its inherent complex regulation—which is designed to permit rapid change in kinase activity—provides multiple potential targets for negative regulation. $p56^{lck}$ activity is influenced by phosphorylation of Tyr^{394} (in the murine enzyme, contained within the activation motif) whose regulation by inhibitory phosphatases was discussed above. In addition to the phosphorylation status of the activation motif, $p56^{lck}$ function is also regulated by phosphorylation of the inhibitory motif containing Tyr^{505}. Regulation of the $p56^{lck}$ inhibitory motif is complex, being affected by the activity of several enzymes, and is directly mediated by the opposing activities of the inhibitory kinase Csk and the positive regulatory phosphatase CD45. $p56^{lck}$ Tyr^{505} is a major target for phosphorylation by Csk, which when phosphorylated causes $p56^{lck}$ to fold such that autocatalytic phosphorylation of Tyr^{394} (and enzyme activation) is prevented (Nada et al., 1991). Therefore, since TIL are blocked in proximal signal transduction (Frey and Monu, 2006; Koneru et al., 2005; Radoja et al., 2001), tumor-induced enhanced Csk activity may be a candidate mechanism for inhibition of antitumor T-cell activation and therein lytic function. As such, analysis of the activation status of $p56^{lck}$ in TIL (to determine if the target of Csk—$p56^{lck}$ Tyr^{505}—is phosphorylated) will implicate Csk in the defective lytic phenotype of antitumor T cells. To date, however, only one publication has examined the phosphorylation status of Csk in TIL (Koneru et al., 2005). Thus, understanding the mechanism by which Csk activity is regulated may reveal if and how the tumor microenvironment might regulate p56^{lck} activity and therein effector T-cell function (Fig. 5).

Activation of primary T cells by *only* TCR crosslinking ("signal 1") results in recruitment of G-proteins into lipid rafts which precedes elevation of cAMP levels (Oh and Schnitzer, 2001). Although cAMP has multiple effects on cell metabolism, it has been long known to inhibit T-cell activation (Kammer, 1988), most likely due to activation of PKA (Tasken and Aandahl, 2004) that in turn activates Csk (by phosphorylation of Ser^{364} (Vang et al., 2001)) which then dampens p56^{lck} activity (Bergman et al., 1992). (In addition, PKA can also directly phosphorylate p56^{lck} at Ser^{42} which may affect the binding specificity of the adjacent SH2 domain (Winkler et al., 1993). Also, PKA activation-dependent negative regulation of signaling in NK cells and B cells has also been reported suggesting a common mechanism of inhibition of src-family kinase function (Levy et al., 1996; Torgersen et al., 1997).) Coincident with, or closely linked to, activation by PKA, Csk is recruited from the cytoplasm to the lipid raft by phosphorylation of the raft-associated Csk binding protein (Cbp) on Tyr^{314}. Phosphorylation of Cbp is regulated by a kinase, probably p56^{lck} (Brdicka et al., 2000), the target of Csk-mediated inactivation of proximal TCR signaling. Since Csk is localized to the membrane in resting T cells, reflecting Cbp phosphorylation (Davidson et al., 2003), in order for the T cell to productively

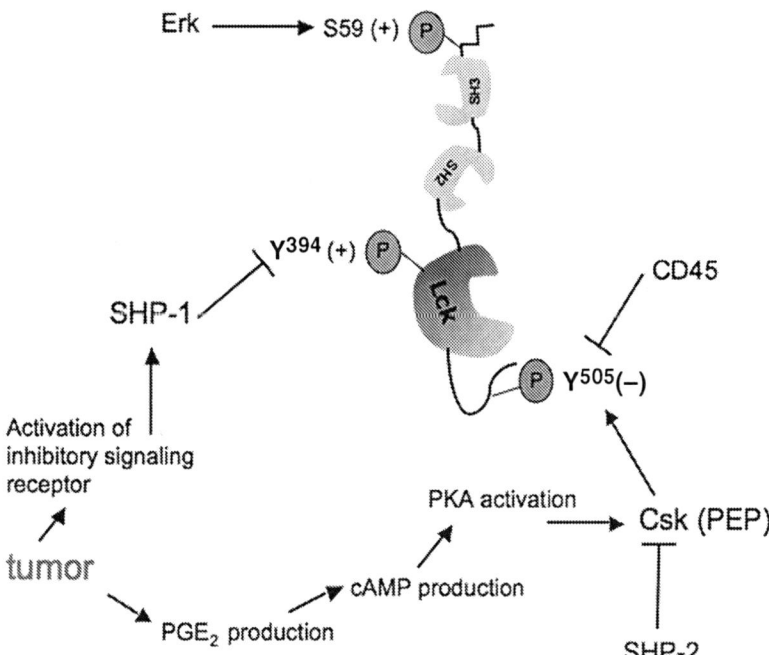

Fig. 5 Schematic of tonic balance of p56^{lck} activation

signal upon antigen recognition, Csk must be either inactivated or displaced from proximity to its substrate (p56^{lck}). This latter mechanism is known to occur upon activation of primary cells and is achieved by dephosphorylation of Cbp by CD45, an event that is one of the very earliest biochemical manifestations of T-cell triggering (Torgersen et al., 2001). CD45 is also able to directly dephosphorylate p56^{lck} Tyr^{505} (Birkeland et al., 1989; Ostergaard et al., 1989), therein facilitating immediate activation of p56^{lck} function as well as restricting Csk-mediated inactivation.

Conversely, activation of T cells by crosslinking *both* the TCR and the costimulatory receptor CD28 results in a reduction of cAMP levels and concomitant diminished activation of PKA, therein avoiding inactivation of p56^{lck} since Csk is not robustly activated (or is sequestered away from its substrate, see below) (Abrahamsen, 2004). Inhibition of cAMP is mediated by phosphodiesterases (PDE), which in T cells is largely accounted for by the PDE type 4 family ("PDE4", containing four genes and multiple isoforms) (Oh and Schnitzer, 2001). Upon TCR and C28 co-activation PDE4 is recruited to lipid rafts simultaneously with G-proteins, implying that the localization of PDE4 in proximity with its substrate (cAMP) under conditions of coordinate TCR and CD28 stimulation permits obviation of cAMP-mediated inhibition of proximal signaling. Under suboptimal conditions of TCR triggering (e.g., without T-cell costimulation) PDE4 is hypothesized to fail to localize to the lipid raft and is therefore unable to inactivate cAMP (Tasken and Stokka, 2006). In addition, in the case of TIL activation in situ, the activity of PDE4 may be overwhelmed by vigorous production of cAMP which may occur upon Ag stimulation in the presence of significant levels of PGE as may potentially accumulate in the microenvironment. Elevated PGE can result from the activity of cyclooxygenase 2 made either in certain types of tumors or in host inflammatory cells recruited to the tumor site (Riedl et al., 2004; Rodriguez et al., 2005) and has been shown to cause elevation of cAMP in lymphocytes (Goodwin et al., 1981). Thus, PKA will be activated which in turn activates Csk-mediated inactivation of p56^{lck}.

As discussed above, in addition to positive regulation of T-cell signaling by activation of a kinase enzymatic activity (often via phosphorylation, e.g., p56^{lck} Tyr^{394}, ZAP70 Tyr^{493}, PLCγ-1 Tyr^{783}), T-cell signaling can also be negatively regulated by phosphorylation events, e.g., phosphatase Shp-1 or kinases Csk and PKA. Similarly, both positive and negative regulation of signaling can be influenced by the subcellular location or compartmentalization, of regulatory enzymes, usually mediated by phosphorylation-dependent alteration of the binding affinity of lipid raft-associated adaptor proteins. As mentioned above, activation of the PKA–Csk negative regulatory path is dependent upon recruitment of Csk to the TCR signaling complex (which is localized in lipid rafts after the TCR is triggered) by phosphorylation of Cbp; in turn PKA is recruited to the raft by phosphorylation of its adapter AKAP (Michel and Scott, 2002). Thus, control of this arm of the inhibitory cascade is mediated by a combination of cAMP-induced activation of PKA activity and unknown events that result in phosphorylation of the adapter responsible for PKA recruitment into proximity to its substrate (AKAP).

Counterbalancing p56^{lck}-dependent recruitment of Csk to Cbp is the activation of two positive regulatory phosphatases which also target Cbp: CD45 and Shp-2.

Dephosphorylation of Cbp by CD45 or Shp-2 results in positive regulation of TCR signaling since Csk cannot colocalize with its substrate. As mentioned above, CD45 dephosphorylation of p56^{lck} and Cbp are probably the earliest biochemical events in TCR signaling, a function that is required for release from tonic inhibition of p56^{lck} activation. After initial activation of T cells, coincident with formation of the higher order antigen receptor signaling complex in the lipid raft (ca. <30 s post-triggering), CD45 is physically excluded from the nascent raft which accumulates at the surface of the T cell at the point of contact with the APC. Thus, very rapidly after T-cell activation CD45 is unable to interact with its substrates (Cbp and p56^{lck}) and no longer participates in regulation of signaling. At intermediate-to-later stages of activation, p56^{lck} can phosphorylate Cbp (both being localized in the now large-sized lipid raft), thus initiating the cycle of Csk-mediated downregulation of proximal signaling by facilitating recruitment of Csk into proximity with its target. Opposing Cbp phosphorylation at this stage of T-cell activation is the activity of Shp-2 that can dephosphorylate Cbp therein interrupting Csk-mediated inhibition of sustained p56^{lck} activation (Zhang et al., 2004). Factors that regulate activation and recruitment of Shp-2 are not well understood but possibly reflect downstream kinases (ERK?) whose activation in turn is dependent upon sustained proximal signaling. In other models, a cytoplasmic scaffolding adaptor protein, Gab1, has been implicated in Shp-2 recruitment to the membrane (Itoh et al., 2000; Sachs et al., 2000; Takahashi-Tezuka et al., 1998), although in T cells this subject is relatively unexplored. The regulatory pathway involving Shp-2 illustrates the complex nature of regulation of T-cell activation involving both positive and negative regulatory feedback mechanisms.

In addition, as mentioned previously, cAMP levels are increased in T cells upon activation (Ledbetter et al., 1986) which if unopposed leads to PKA activation and inhibition of TCR signaling (Abrahamsen et al., 2004). Elevated cAMP, and thus PKA activity, may be part of the normal homeostatic mechanism to dampen signaling and therefore restrict T-cell activation, especially in the absence of costimulation (see below). A counter-regulatory mechanism exists which functions to limit cAMP-induced PKA-mediated inactivation of p56^{lck} and is envisioned to be operative under conditions of robust T-cell activation, i.e., TCR plus CD28 costimulation (Abrahamsen, 2004). As mentioned above, lipid raft-associated PDE4 activity is enhanced upon concomitant ligation of the TCR and CD28. PDE4 is recruited to the lipid raft in a complex with a cytosolic adaptor protein, β-arrestin, using an unknown mechanism for raft association (Perry et al., 2002). (β-arrestin has several interacting partners and is best characterized in regulation of G-protein receptor signaling (Perry and Lefkowitz, 2002).) It is reasonable to presume that signals which influence the binding affinity of the β-arrestin/PDE4 complex to the membrane (potentially phosphorylation events generated from TCR and CD28 signaling) regulate recruitment of PDE4 to the TCR signaling complex and therein blockade of cAMP-mediated dysregulation of TCR signaling. Regulation of β-arrestin binding of PDE4 and subsequent association with lipid rafts is at present unknown but may be speculated to be mediated by distal TCR-mediated signaling (Tasken and Stokka, 2006). Thus there are two levels of control of PKA activity: its recruitment (to AKAP) and its activation (by cAMP in turn counterbalanced by PDE4, itself controlled by recruitment to the signaling complex).

What factors determine the kinetics of activation of the Csk/PKA-mediated inhibitory regulatory pathway relative to productive TCR signaling is probably dictated by the strength (number of ligands per T cell as well as the binding affinity for a given TCR) and quality (being either agonist, partial agonist or antagonist) of antigen-mediated TCR triggering. In this regard, Tasken and colleagues have hypothesized that signals generated by TCR ligation appear to result in a constitutively active "default pathway" of increased cAMP levels leading to inhibition of signaling (Abrahamsen, 2004). Only when TCR ligation is accompanied by CD28 costimulation are cAMP levels reduced therein relieving PKA/Csk activity and permitting sustained T-cell activation. This notion is conceptually appealing since it provides a mechanism that inherently restricts the potentially excessive activity of $CD8^+$ effector CTL since most target cells (especially epithelia-derived tumor cells) do not express CD28 ligands. According to this line of thinking, upon contact with target cells lacking CD28 costimulatory ligands, CTL receive Signal 1 and are activated to kill but the amount of time a given CTL can degranulate is limited since $p56^{lck}$ activity would be curtailed soon after activation. In addition, activation of the Csk/PKA inhibitory pathway involves additional signaling pathways such as PGE_2 which influence the activity of PKA after signaling through non-antigen receptors (Tasken and Stokka, 2006). Ultimately the combination of positive signals (derived from TCR plus costimulation) and negative signals (cAMP) influences the functional status of T cells.

10 Summary

The host immune system is one of the most important elements for protection from tumor development and control of tumor growth. Macfarlane Burnet postulated that a mechanism of immunological character is an evolutionary necessity for protection from neoplastic disease, and the importance of immunosurveillance of cancer has been thoroughly proven. However, an intact immune system often fails to eliminate antigenic tumors and it has been extensively shown that antigen-specific $CD8^+$ TIL are present in most human cancers but are typically non-lytic and unable to mediate tumor eradication (Radoja et al., 2000; Whiteside, 1998).

We feel insufficient research has been focused on the lytic dysfunction of TIL. Apart from research performed by the Whiteside, Finke, Ochoa labs and others demonstrating that some TIL show abnormalities in terms of expression of TCRζ or other TCR-associated signaling proteins (Whiteside, 1999), much of tumor immunology research has focused on enhancement of priming of the immune response. Consideration of the observation that human TIL are antigen-specific, but non-lytic, together with our description of defective lytic function of murine TIL (Koneru et al., 2005; Radoja et al., 2001) and lack of systemic suppression of the immune system in tumor-bearing mice (Radoja et al., 2000), supports the notion that tumor-induced inhibition of TIL lytic function is a common characteristic that may contribute to tumor growth in the presence of antitumor immune response.

The emphasis of our lab to uncover the biochemical basis for TIL lytic dysfunction was predicated on the considerations that because lytic function is dependent upon TCR-mediated signaling and TIL are unable to exocytose lytic granules, residence of antitumor T cells in the tumor microenvironment may induce defective signal transduction. Supporting this hypothesis, our recent work has demonstrated that when conjugated with cognate tumor cells in vitro, signal transduction in non-lytic TIL is blocked, such that proximal tyrosine kinases are not activated, and purified TIL are unable to flux calcium (Koneru et al., 2005). This conclusion is supported by the observation that ZAP70 is only modestly activated and p56^{lck} appears to be inactivated, deficiencies that undoubtedly underlie lytic dysfunction (Fig. 6). In sum, the phenotype of non-lytic TIL appears to result from tumor-induced, Shp-1-mediated, rapid down-modulation of proximal TCR-mediated signaling, which prevents effector-phase function in situ. The factors that cause enhancement of Shp-1 activity in TIL are unknown at present.

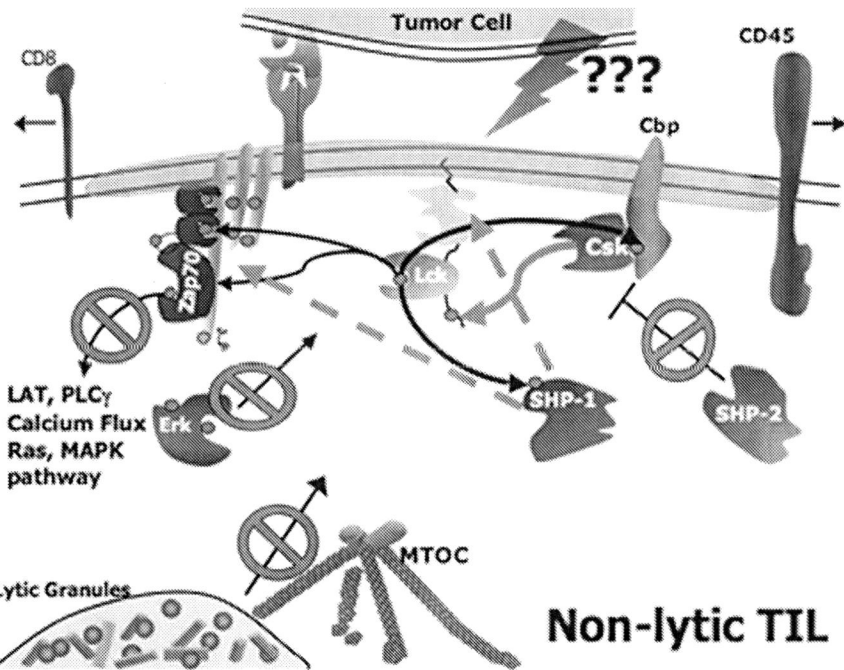

Fig. 6 Schematic of proximal TCR signaling in non-lytic TIL emphasizing inhibition of p56^{lck} activity by Shp-1

Abbreviations

AICD	Activation-induced cell death
AKAP	A-kinase-anchoring protein
APC	Antigen presenting cell
CAML	Calcium modulating cyclophilin ligand
cAMP	Cyclic adenosine monophosphate
Cbp	Csk binding protein
Csk	Carboxy terminal Src kinase
CTL	Cytotoxic T lymphocyte
DTH	Delayed-type hypersensitivity
ERK	Extracellular signal-regulated kinase
Ig	Immunoglobulin
IS	Immunological synapse
ITAM	Immunoreceptor tyrosine-based activation motif
ITIM	Immunoreceptor tyrosine-based inhibition motif
LAT	Linker for activation of T cells
LFA-1	Leukocyte function-associated antigen-1 (CD11a/CD18)
LIME	LCK-interacting molecule
LN	Lymph node(s)
MAPK	mitogen-activated protein kinase
MDSC	Myeloid-derived suppressor cells
MTOC	Microtubule organizing center
PAG	Protein associated with glycosphingolipid-enriched microdomains
PDE	Phosphodiesterase
PEP	Proline, glutamic acid, serine, threonine domain-enriched tyrosine phosphatase
PGE	Prostaglandin E_2
PKA	Protein kinase A
PLCγ-1	Phospholipase C gamma-1
SH2	Src homology 2 domain
Shp-1	SH2-containing tyrosine phosphatase-1
Shp-2	SH2-containing tyrosine phosphatase-2
Slp-76	SH2-domain-containing leukocyte protein of 76 kD
TCR	T-cell receptor
TGFβ-1	Transforming growth factor beta-1
TIL	Tumor-infiltrating lymphocytes
T_{reg}	Regulatory T cells
TUNEL	Terminal deoxynucleotidyl transferase biotin-dUTP nick end labeling
ZAP70	Zeta-chain-associated protein kinase of 70 kD

Acknowledgments Work in the author's laboratory is currently supported by NIH grant CA108573. I am indebted to my students for their dedication and inquisitiveness: I hope that their training proved as pleasurable for them as it was for me.

References

Abrahamsen, H., Baillie, G., Ngai, J., et al. (2004). TCR- and CD28-mediated recruitment of phosphodiesterase 4 to lipid rafts potentiates TCR signaling. *J Immunol* 173:4847–4858.

Alarcon, B., Gil, D., Delgado, P., and Schamel, W. W. (2003). Initiation of TCR signaling: regulation within CD3 dimers. *Immunol Rev* 191:38–46.

Alexander, J. P., Kudoh, S., Melsop, K. A., et al. (1993). T-cells infiltrating renal cell carcinoma display a poor proliferative response even though they can produce interleukin 2 and express interleukin 2 receptors. *Cancer Res* 53:1380–1387.

Bennaceur, K., Popa, I., Portoukalian, J., Berthier-Vergnes, O., and Peguet-Navarro, J. (2006). Melanoma-derived gangliosides impair migratory and antigen-presenting function of human epidermal Langerhans cells and induce their apoptosis. *Int Immunol* 18: 879–886.

Bergman, M., Mustelin, T., Oetken, C., et al. (1992). The human p50csk tyrosine kinase phosphorylates p56lck at Tyr-505 and down regulates its catalytic activity. *EMBO J* 11: 2919–2924.

Bharti, A. C., and Singh, S. M. (2001). Gangliosides derived from a T cell lymphoma inhibit bone marrow cell proliferation and differentiation. *Int Immunopharmacol* 1:155–165.

Bharti, A. C., and Singh, S. M. (2003). Inhibition of macrophage nitric oxide production by gangliosides derived from a spontaneous T cell lymphoma: the involved mechanisms. *Nitric Oxide* 8:75–82.

Birkeland, M. L., Johnson, P., Trowbridge, I. S., and Pure, E. (1989). Changes in CD45 isoform expression accompany antigen-induced murine T-cell activation. *Proc Natl Acad Sci USA* 86:6734–6738.

Biswas, K., Richmond, A., Rayman, P., et al. (2006). GM2 expression in renal cell carcinoma: potential role in tumor-induced T-cell dysfunction. *Cancer Res* 66:6816–6825.

Blackhall, F. H., and Shepherd, F. A. (2007). Small cell lung cancer and targeted therapies. *Curr Opin Oncol* 19:103–108.

Blasioli, J., Paust, S., and Thomas, M. L. (1999). Definition of the sites of interaction between the protein tyrosine phosphatase SHP-1 and CD22. *J Biol Chem* 274:2303–2307.

Boon, T., and Van Pel, A. (1989). T cell-recognized antigenic peptides derived from the cellular genome are not protein degradation products but can be generated directly by transcription and translation of short subgenic regions. A hypothesis. *Immunogenetics* 29:75–79.

Brdicka, T., Pavlistova, D., Leo, A., et al. (2000). Phosphoprotein associated with glycosphingolipid-enriched microdomains (PAG), a novel ubiquitously expressed transmembrane adaptor protein, binds the protein tyrosine kinase csk and is involved in regulation of T cell activation. *J Exp Med* 191:1591–1604.

Brdickova, N., Brdicka, T., Angelisova, P., et al. (2003). LIME: a new membrane Raft-associated adaptor protein involved in CD4 and CD8 coreceptor signaling. *J Exp Med* 198: 1453–1462.

Bronte, V., Cingarlini, S., Marigo, I., et al. (2006). Leukocyte infiltration in cancer creates an unfavorable environment for antitumor immune responses: a novel target for therapeutic intervention. *Immunol Invest* 35:327–357.

Bronte, V., Kasic, T., Gri, G., et al. (2005). Boosting antitumor responses of T lymphocytes infiltrating human prostate cancers. *J Exp Med* 201:1257–1268.

Brumell, J. H., Burkhardt, A. L., Bolen, J. B., and Grinstein, S. (1996). Endogenous reactive oxygen intermediates activate tyrosine kinases in human neutrophils. *J Biol Chem* 271: 1455–1461.

Bukowski, R. M., Rayman, P., Uzzo, R., et al. (1998). Signal transduction abnormalities in T lymphocytes from patients with advanced renal carcinoma: clinical relevance and effects of cytokine therapy. *Clin Cancer Res* 4:2337–2347.

Cardi, G., Heaney, J. A., Schned, A. R., Phillips, D. M., Branda, M. T., and Ernstoff, M. S. (1997). T-cell receptor zeta-chain expression on tumor-infiltrating lymphocytes from renal cell carcinoma. *Cancer Res* 57:3517–3519.

Cherukuri, A., Dykstra, M., and Pierce, S. K. (2001). Floating the raft hypothesis: lipid rafts play a role in immune cell activation. *Immunity* 14:657–660.

Chiang, G. G., and Sefton, B. M. (2001). Specific dephosphorylation of the Lck tyrosine protein kinase at Tyr-394 by the SHP-1 protein-tyrosine phosphatase. *J Biol Chem* 276: 23173–23178.

Choi, Y. B., Kim, C. K., and Yun, Y. (1999). Lad, an adapter protein interacting with the SH2 domain of p56lck, is required for T cell activation. *J Immunol* 163:5242–5249.

Chung, C. D., Patel, V. P., Moran, M., Lewis, L. A., and Miceli, M. C. (2000). Galectin-1 induces partial TCR zeta-chain phosphorylation and antagonizes processive TCR signal transduction. *J Immunol* 165:3722–3729.

Clements, J. L., Boerth, N. J., Lee, J. R., and Koretzky, G. A. (1999). Integration of T cell receptor-dependent signaling pathways by adapter proteins. *Annu Rev Immunol* 17:89–108.

Cochran, A. J., Huang, R. R., Lee, J., Itakura, E., Leong, S. P., and Essner, R. (2006). Tumour-induced immune modulation of sentinel lymph nodes. *Nat Rev Immunol* 6:659–670.

Correa, M. R., Ochoa, A. C., Ghosh, P., Mizoguchi, H., Harvey, L., and Longo, D. L. (1997). Sequential development of structural and functional alterations in T cells from tumor-bearing mice. *J Immunol* 158:5292–5296.

Coulie, P. G., Lehmann, F., Lethe, B., et al. (1995). A mutated intron sequence codes for an antigenic peptide recognized by cytolytic T lymphocytes on a human melanoma. *Proc Natl Acad Sci USA* 92:7976–7980.

Cuenca, A., Cheng, F., Wang, H., et al. (2003). Extra-lymphatic solid tumor growth is not immunologically ignored and results in early induction of antigen-specific T-cell anergy: dominant role of cross-tolerance to tumor antigens. *Cancer Res* 63:9007–9015.

Curiel, T. J., Coukos, G., Zou, L., et al. (2004). Specific recruitment of regulatory T cells in ovarian carcinoma fosters immune privilege and predicts reduced survival. *Nat Med* 10:942–949.

D'Ambrosio, D., Hippen, K. L., Minskoff, S. A., et al. (1995). Recruitment and activation of PTP1C in negative regulation of antigen receptor signaling by Fc gamma RIIB1. *Science* 268:293–297.

Davidson, D., Bakinowski, M., Thomas, M. L., Horejsi, V., and Veillette, A. (2003). Phosphorylation-dependent regulation of T-cell activation by PAG/Cbp, a lipid raft-associated transmembrane adaptor. *Mol Cell Biol* 23:2017–2028.

Dix, A. R., Brooks, W. H., Roszman, T. L., and Morford, L. A. (1999). Immune defects observed in patients with primary malignant brain tumors. *J Neuroimmunol* 100:216–232.

Dolcetti, R., Viel, A., Doglioni, C., et al. (1999). High prevalence of activated intraepithelial cytotoxic T lymphocytes and increased neoplastic cell apoptosis in colorectal carcinomas with microsatellite instability. *Am J Pathol* 154:1805–1813.

Dukers, D. F., Oudejans, J. J., Jaspars, E. H., et al. (2000). All infiltrating T-lymphocytes in Hodgkin's disease express immunohistochemically detectable T-cell receptor zeta-chains in situ. *Histopathology* 36:544–550.

Dumontet, C., Rebbaa, A., Bienvenu, J., and Portoukalian, J. (1994). Inhibition of immune cell proliferation and cytokine production by lipoprotein-bound gangliosides. *Cancer Immunol Immunother* 38:311–316.

Fernandez-Miguel, G., Alarcon, B., Iglesias, A., et al. (1999). Multivalent structure of an alpha-betaT cell receptor. *Proc Natl Acad Sci USA* 96:1547–1552.

Finke, J. H., Zea, A. H., Stanley, J., et al. (1993). Loss of T-cell receptor zeta chain and p56lck in T-cells infiltrating human renal cell carcinoma. *Cancer Res* 53:5613–5616.

Franco, J. L., Ghosh, P., Wiltrout, R. H., et al. (1995). Partial degradation of T-cell signal transduction molecules by contaminating granulocytes during protein extraction of splenic T cells from tumor-bearing mice. *Cancer Res* 55:3840–3846.

Frey, A. B., and Monu, N. (2006). Effector-phase tolerance: another mechanism of how cancer escapes antitumor immune response. *J Leukoc Biol* 79:652–662.

Gabrilovich, D. (2004). Mechanisms and functional significance of tumour-induced dendritic-cell defects. *Nat Rev Immunol* 4:941–952.

Gastman, B. R., Johnson, D. E., Whiteside, T. L., and Rabinowich, H. (1999). Caspase-mediated degradation of T-cell receptor zeta-chain. *Cancer Res* 59:1422–1427.

Gattinoni, L., Powell, D. J., Jr, Rosenberg, S. A., and Restifo, N. P. (2006). Adoptive immunotherapy for cancer: building on success. *Nat Rev Immunol* 6:383–393.

Goodwin, J. S., Bromberg, S., and Messner, R. P. (1981). Studies on the cyclic AMP response to prostaglandin in human lymphocytes. *Cell Immunol* 60:298–307.

Guilloux, Y., Lucas, S., Brichard, V. G., et al. (1996). A peptide recognized by human cytolytic T lymphocytes on HLA-A2 melanomas is encoded by an intron sequence of the N-acetylglucosaminyltransferase V gene. *J Exp Med* 183:1173–1183.

Hakomori, S. (1996). Tumor malignancy defined by aberrant glycosylation and sphingo(glyco)lipid metabolism. *Cancer Res* 56:5309–5318.

Hawash, I. Y., Kesavan, K. P., Magee, A. I., Geahlen, R. L., and Harrison, M. L. (2002). The Lck SH3 domain negatively regulates localization to lipid rafts through an interaction with c-Cbl. *J Biol Chem* 277:5683–5691.

Hayashi, S., Witte, P. L., Shultz, L. D., and Kincade, P. W. (1988). Lymphohemopoiesis in culture is prevented by interaction with adherent bone marrow cells from mutant viable motheaten mice. *J Immunol* 140:2139–2147.

Hernandez, J. D., and Baum, L. G. (2002). Ah, sweet mystery of death! Galectins and control of cell fate. *Glycobiology* 12:127R–136R.

Horna, P., Cuenca, A., Cheng, F., et al. (2006). In vivo disruption of tolerogenic cross-presentation mechanisms uncovers an effective T-cell activation by B-cell lymphomas leading to antitumor immunity. *Blood* 107:2871–2878.

Hur, E. M., Son, M., Lee, O. H., et al. (2003). LIME, a novel transmembrane adaptor protein, associates with p56lck and mediates T cell activation. *J Exp Med* 198:1463–1473.

Ishikawa, T., Fujita, T., Suzuki, Y., et al. (2003). Tumor-specific immunological recognition of frameshift-mutated peptides in colon cancer with microsatellite instability. *Cancer Res* 63:5564–5572.

Itoh, M., Yoshida, Y., Nishida, K., Narimatsu, M., Hibi, M., and Hirano, T. (2000). Role of Gab1 in heart, placenta, and skin development and growth factor- and cytokine-induced extracellular signal-regulated kinase mitogen-activated protein kinase activation. *Mol Cell Biol* 20:3695–3704.

Joung, I., Kim, T., Stolz, L. A., et al. (1995). Modification of Ser59 in the unique N-terminal region of tyrosine kinase p56lck regulates specificity of its Src homology 2 domain. *Proc Natl Acad Sci USA* 92:5778–5782.

Kabouridis, P. S., Magee, A. I., and Ley, S. C. (1997). S-acylation of LCK protein tyrosine kinase is essential for its signalling function in T lymphocytes. *EMBO J* 16:4983–4998.

Kagi, D., Ledermann, B., Burki, K., Zinkernagel, R. M., and Hengartner, H. (1996). Molecular mechanisms of lymphocyte-mediated cytotoxicity and their role in immunological protection and pathogenesis in vivo. *Annu Rev Immunol* 14:207–232.

Kammer, G. M. (1988). The adenylate cyclase-cAMP-protein kinase A pathway and regulation of the immune response. *Immunol Today* 9:222–229.

Kane, L. P., Lin, J., and Weiss, A. (2000). Signal transduction by the TCR for antigen. *Curr Opin Immunol* 12:242–249.

Kautz, B., Kakar, R., David, E., and Eklund, E. A. (2001). SHP1 protein-tyrosine phosphatase inhibits gp91PHOX and p67PHOX expression by inhibiting interaction of PU.1, IRF1, interferon consensus sequence-binding protein, and CREB-binding protein with homologous Cis elements in the CYBB and NCF2 genes. *J Biol Chem* 276:37868–37878.

Kiessling, R., Wasserman, K., Horiguchi, S., et al. (1999). Tumor-induced immune dysfunction. *Cancer Immunol Immunother* 48:353–362.

Kolenko, V., Wang, Q., Riedy, M. C., et al. (1997). Tumor-induced suppression of T lymphocyte proliferation coincides with inhibition of Jak3 expression and IL-2 receptor signaling: role of soluble products from human renal cell carcinomas. *J Immunol* 159:3057–3067.

Koneru, M., Monu, N., Schaer, D., Barletta, J., and Frey, A. B. (2006). Defective adhesion in tumor infiltrating CD8+ T cells. *J Immunol* 176:6103–6111.

Koneru, M., Schaer, D., Monu, N., Ayala, A., and Frey, A. B. (2005). Defective proximal TCR signaling inhibits CD8+ tumor-infiltrating lymphocyte lytic function. *J Immunol* 174:1830–1840.

Kong, Y., Li, R., and Ladisch, S. (1998). Natural forms of shed tumor gangliosides. *Biochim Biophys Acta* 1394:43–56.

Kono, K., Salazar-Onfray, F., Petersson, M., et al. (1996). Hydrogen peroxide secreted by tumor-derived macrophages down-modulates signal-transducing zeta molecules and inhibits tumor-specific T cell-and natural killer cell-mediated cytotoxicity. *Eur J Immunol* 26: 1308–1313.

Kopitz, J., von Reitzenstein, C., Burchert, M., Cantz, M., and Gabius, H. J. (1998). Galectin-1 is a major receptor for ganglioside GM1, a product of the growth-controlling activity of a cell surface ganglioside sialidase, on human neuroblastoma cells in culture. *J Biol Chem* 273: 11205–11211.

Ladisch, S., Ulsh, L., Gillard, B., and Wong, C. (1984). Modulation of the immune response by gangliosides. Inhibition of adherent monocyte accessory function in vitro. *J Clin Invest* 74:2074–2081.

Ledbetter, J. A., Parsons, M., Martin, P. J., Hansen, J. A., Rabinovitch, P. S., and June, C. H. (1986). Antibody binding to CD5 (Tp67) and Tp44 T cell surface molecules: effects on cyclic nucleotides, cytoplasmic free calcium, and cAMP-mediated suppression. *J Immunol* 137: 3299–3305.

Lee, P. P., Yee, C., Savage, P. A., et al. (1999). Characterization of circulating T cells specific for tumor-associated antigens in melanoma patients. *Nat Med* 5:677–685.

Levey, D. L., and Srivastava, P. K. (1995). T cells from late tumor-bearing mice express normal levels of p56lck, p59fyn, ZAP-70, and CD3 zeta despite suppressed cytolytic activity. *J Exp Med* 182:1029–1036.

Levy, F. O., Rasmussen, A. M., Tasken, K., et al. (1996). Cyclic AMP-dependent protein kinase (cAK) in human B cells: co-localization of type I cAK (RI alpha 2 C2) with the antigen receptor during anti-immunoglobulin-induced B cell activation. *Eur J Immunol* 26:1290–1296.

Li, X., Liu, J., Park, J. K., et al. (1994). T cells from renal cell carcinoma patients exhibit an abnormal pattern of kappa B-specific DNA-binding activity: a preliminary report. *Cancer Res* 54:5424–5429.

Lindahl, K. F. (1991). Do we need a pepton hypothesis? *Immunogenetics* 34:1–4.

Ling, W., Rayman, P., Uzzo, R., et al. (1998). Impaired activation of NFkappaB in T cells from a subset of renal cell carcinoma patients is mediated by inhibition of phosphorylation and degradation of the inhibitor, IkappaBalpha. *Blood* 92:1334–1341.

Liu, Y., Li, R., and Ladisch, S. (2004). Exogenous ganglioside GD1a enhances epidermal growth factor receptor binding and dimerization. *J Biol Chem* 279:36481–36489.

Liu, Y. C., Penninger, J., and Karin, M. (2005). Immunity by ubiquitylation: a reversible process of modification. *Nat Rev Immunol* 5:941–952.

Loeffler, C. M., Smyth, M. J., Longo, D. L., et al. (1992). Immunoregulation in cancer-bearing hosts. Down-regulation of gene expression and cytotoxic function in CD8+ T cells. *J Immunol* 149:949–956.

Lorenz, U., Ravichandran, K. S., Burakoff, S. J., and Neel, B. G. (1996). Lack of SHPTP1 results in src-family kinase hyperactivation and thymocyte hyperresponsiveness. *Proc Natl Acad Sci USA* 93:9624–9629.

Lupetti, R., Pisarra, P., Verrecchia, A., et al. (1998). Translation of a retained intron in tyrosinase-related protein (TRP) 2†mRNA generates a new cytotoxic T lymphocyte (CTL)-defined and shared human melanoma antigen not expressed in normal cells of the melanocytic lineage. *J Exp Med* 188:1005–1016.

Magee, A. I., Adler, J., and Parmryd, I. (2005). Cold-induced coalescence of T-cell plasma membrane microdomains activates signalling pathways. *J Cell Sci* 118:3141–3151.

Marth, J. D., Lewis, D. B., Wilson, C. B., Gearn, M. E., Krebs, E. G., and Perlmutter, R. M. (1987). Regulation of pp56lck during T-cell activation: functional implications for the src-like protein tyrosine kinases. *EMBO J* 6:2727–2734.

Masopust, D., Vezys, V., Marzo, A. L., and Lefrancois, L. (2001). Preferential localization of effector memory cells in nonlymphoid tissue. *Science* 291:2413–2417.

Meng, T. C., Fukada, T., and Tonks, N. K. (2002). Reversible oxidation and inactivation of protein tyrosine phosphatases in vivo. *Mol Cell* 9:387–399.

Mescher, M. F., Curtsinger, J. M., Agarwal, P., et al. (2006). Signals required for programming effector and memory development by CD8+ T cells. *Immunol Rev* 211:81–92.
Michel, J. J., and Scott, J. D. (2002). AKAP mediated signal transduction. *Annu Rev Pharmacol Toxicol* 42:235–257.
Mizoguchi, H., O'Shea, J. J., Longo, D. L., Loeffler, C. M., McVicar, D. W., and Ochoa, A. C. (1992). Alterations in signal transduction molecules in T lymphocytes from tumor-bearing mice. *Science* 258:1795–1798.
Mor, A., and Philips, M. R. (2006). Compartmentalized Ras/MAPK signaling. *Annu Rev Immunol* 24:771–800.
Morgan, M. M., Labno, C. M., Van Seventer, G. A., Denny, M. F., Straus, D. B., and Burkhardt, J. K. (2001). Superantigen-induced T cell:B cell conjugation is mediated by LFA-1 and requires signaling through Lck, but not ZAP-70. *J Immunol* 167:5708–5718.
Mustelin, T., Vang, T., and Bottini, N. (2005). Protein tyrosine phosphatases and the immune response. *Nat Rev Immunol* 5:43–57.
Nada, S., Okada, M., MacAuley, A., Cooper, J. A., and Nakagawa, H. (1991). Cloning of a complementary DNA for a protein-tyrosine kinase that specifically phosphorylates a negative regulatory site of p60c-src. *Nature* 351:69–72.
Nakagomi, H., Petersson, M., Magnusson, I., et al. (1993). Decreased expression of the signal-transducing zeta chains in tumor-infiltrating T-cells and NK cells of patients with colorectal carcinoma. *Cancer Res* 53:5610–5612.
Neel, B. G. (1997). Role of phosphatases in lymphocyte activation. *Curr Opin Immunol* 9:405–420.
Oh, P., and Schnitzer, J. E. (2001). Segregation of heterotrimeric G proteins in cell surface microdomains. G(q) binds caveolin to concentrate in caveolae, whereas G(i) and G(s) target lipid rafts by default. *Mol Biol Cell* 12:685–698.
O'Keefe, T. L., Williams, G. T., Davies, S. L., and Neuberger, M. S. (1996). Hyperresponsive B cells in CD22-deficient mice. *Science* 274:798–801.
Olszowy, M. W., Leuchtmann, P. L., Veillette, A., and Shaw, A. S. (1995). Comparison of p56lck and p59fyn protein expression in thymocyte subsets, peripheral T cells, NK cells, and lymphoid cell lines. *J Immunol* 155:4236–4240.
Ostergaard, H. L., Shackelford, D. A., Hurley, T. R., et al. (1989). Expression of CD45 alters phosphorylation of the lck-encoded tyrosine protein kinase in murine lymphoma T-cell lines. *Proc Natl Acad Sci USA* 86:8959–8963.
Overwijk, W. W., Theoret, M. R., Finkelstein, S. E., et al. (2003). Tumor regression and autoimmunity after reversal of a functionally tolerant state of self-reactive CD8+ T cells. *J Exp Med* 198:569–580.
Pardoll, D. (2003). Does the immune system see tumors as foreign or self? *Annu Rev Immunol* 21:807–839.
Parmiani, G., Pilla, L., Castelli, C., and Rivoltini, L. (2003). Vaccination of patients with solid tumours. *Ann Oncol* 14:817–824.
Pei, D., Lorenz, U., Klingmuller, U., Neel, B. G., and Walsh, C. T. (1994). Intramolecular regulation of protein tyrosine phosphatase SH-PTP1: a new function for Src homology 2 domains. *Biochemistry* 33:15483–15493.
Perry, S. J., Baillie, G. S., Kohout, T. A., et al. (2002). Targeting of cyclic AMP degradation to beta 2-adrenergic receptors by beta-arrestins. *Science* 298:834–836.
Perry, S. J., and Lefkowitz, R. J. (2002). Arresting developments in heptahelical receptor signaling and regulation. *Trends Cell Biol* 12:130–138.
Plas, D. R., Johnson, R., Pingel, J. T., et al. (1996). Direct regulation of ZAP-70 by SHP-1 in T cell antigen receptor signaling. *Science* 272:1173–1176.
Pope, C., Kim, S. K., Marzo, A., et al. (2001). Organ-specific regulation of the CD8 T cell response to *Listeria monocytogenes* infection. *J Immunol* 166:3402–3409.
Rabinovich, G., Gabrilovich, D., and Sotomayer, E. (2007). Immunosuppressive strategies that are mediated by tumor cells. *Annu Rev Immunol* 25:267–296.
Radoja, S., Frey, A. B., and Vukmanovic, S. (2006). T-cell receptor signaling events triggering granule exocytosis. *Crit Rev Immunol* 26:265–290.

Radoja, S., Rao, T. D., Hillman, D., and Frey, A. B. (2000). Mice bearing late-stage tumors have normal functional systemic T cell responses in vitro and in vivo. *J Immunol* 164: 2619–2628.

Radoja, S., Saio, M., Schaer, D., Koneru, M., Vukmanovic, S., and Frey, A. B. (2001). CD8(+) tumor-infiltrating T cells are deficient in perforin-mediated cytolytic activity due to defective microtubule-organizing center mobilization and lytic granule exocytosis. *J Immunol* 167: 5042–5051.

Ravetch, J. V., and Kinet, J. P. (1991). Fc receptors. *Annu Rev Immunol* 9:457–492.

Reichert, T. E., Strauss, L., Wagner, E. M., Gooding, W., and Whiteside, T. L. (2002). Signaling abnormalities, apoptosis, and reduced proliferation of circulating and tumor-infiltrating lymphocytes in patients with oral carcinoma. *Clin Cancer Res* 8:3137–3145.

Resh, M. D. (1994). Myristylation and palmitylation of Src family members: the fats of the matter. *Cell* 76:411–413.

Riedl, K., Krysan, K., Pold, M., et al. (2004). Multifaceted roles of cyclooxygenase-2 in lung cancer. *Drug Resist Updat* 7:169–184.

Ritter, G., and Livingston, P. O. (1991). Ganglioside antigens expressed by human cancer cells. *Semin Cancer Biol* 2:401–409.

Rodriguez, P. C., Hernandez, C. P., Quiceno, D., et al. (2005). Arginase I in myeloid suppressor cells is induced by COX-2 in lung carcinoma. *J Exp Med* 202:931–939.

Rosenberg, S. A., Yang, J. C., and Restifo, N. P. (2004). Cancer immunotherapy: moving beyond current vaccines. *Nat Med* 10:909–915.

Rudd, C. E. (1999). Adaptors and molecular scaffolds in immune cell signaling. *Cell* 96:5–8.

Sachs, M., Brohmann, H., Zechner, D., et al. (2000). Essential role of Gab1 for signaling by the c-Met receptor in vivo. *J Cell Biol* 150:1375–1384.

Samelson, L. E. (2002). Signal transduction mediated by the T cell antigen receptor: the role of adapter proteins. *Annu Rev Immunol* 20:371–394.

Schmielau, J., and Finn, O. J. (2001). Activated granulocytes and granulocyte-derived hydrogen peroxide are the underlying mechanism of suppression of t-cell function in advanced cancer patients. *Cancer Res* 61:4756–4760.

Schraven, B., Marie-Cardine, A., Hubener, C., Bruyns, E., and Ding, I. (1999). Integration of receptor-mediated signals in T cells by transmembrane adaptor proteins. *Immunol Today* 20:431–434.

Shu, S., Cochran, A. J., Huang, R. R., Morton, D. L., and Maecker, H. T. (2006). Immune responses in the draining lymph nodes against cancer: implications for immunotherapy. *Cancer Metastasis Rev* 25:233–242.

Shultz, L. D., Schweitzer, P. A., Rajan, T. V., et al. (1993). Mutations at the murine motheaten locus are within the hematopoietic cell protein-tyrosine phosphatase (Hcph) gene. *Cell* 73: 1445–1454.

Siminovitch, K. A., and Neel, B. G. (1998). Regulation of B cell signal transduction by SH2-containing protein-tyrosine phosphatases. *Semin Immunol* 10:329–347.

Sloan-Lancaster, J., and Allen, P. M. (1996). Altered peptide ligand-induced partial T cell activation: molecular mechanisms and role in T cell biology. *Annu Rev Immunol* 14:1–27.

Sloan-Lancaster, J., Steinberg, T. H., and Allen, P. M. (1996). Selective activation of the calcium signaling pathway by altered peptide ligands. *J Exp Med* 184:1525–1530.

Staveley-O'Carroll, K., Sotomayor, E., Montgomery, J., et al. (1998). Induction of antigen-specific T cell anergy: an early event in the course of tumor progression. *Proc Natl Acad Sci USA* 95:1178–1183.

Stefanova, I., Hemmer, B., Vergelli, M., Martin, R., Biddison, W. E., and Germain, R. N. (2003). TCR ligand discrimination is enforced by competing ERK positive and SHP-1 negative feedback pathways. *Nat Immunol* 4:248–254.

Stulnig, T. M., Berger, M., Sigmund, T., Raederstorff, D., Stockinger, H., and Waldhausl, W. (1998). Polyunsaturated fatty acids inhibit T cell signal transduction by modification of detergent-insoluble membrane domains. *J Cell Biol* 143:637–644.

Stutman, O. (1975). Immunodepression and malignancy. *Adv Cancer Res* 22:261–422.
Takahashi-Tezuka, M., Yoshida, Y., Fukada, T., et al. (1998). Gab1 acts as an adapter molecule linking the cytokine receptor gp130 to ERK mitogen-activated protein kinase. *Mol Cell Biol* 18:4109–4117.
Tasken, K., and Aandahl, E. M. (2004). Localized effects of cAMP mediated by distinct routes of protein kinase A. *Physiol Rev* 84:137–167.
Tasken, K., and Stokka, A. J. (2006). The molecular machinery for cAMP-dependent immunomodulation in T-cells. *Biochem Soc Trans* 34:476–479.
Tedoldi, S., Paterson, J. C., Hansmann, M. L., et al. (2006). Transmembrane adaptor molecules: a new category of lymphoid-cell markers. *Blood* 107:213–221.
Thornton, M. V., Kudo, D., Rayman, P., et al. (2004). Degradation of NF-kappa B in T cells by gangliosides expressed on renal cell carcinomas. *J Immunol* 172:3480–3490.
Torgersen, K. M., Vaage, J. T., Levy, F. O., Hansson, V., Rolstad, B., and Tasken, K. (1997). Selective activation of cAMP-dependent protein kinase type I inhibits rat natural killer cell cytotoxicity. *J Biol Chem* 272:5495–5500.
Torgersen, K. M., Vang, T., Abrahamsen, H., et al. (2001). Release from tonic inhibition of T cell activation through transient displacement of C-terminal Src kinase (Csk) from lipid rafts. *J Biol Chem* 276:29313–29318.
Torgersen, K. M., Vang, T., Abrahamsen, H., Yaqub, S., and Tasken, K. (2002). Molecular mechanisms for protein kinase A-mediated modulation of immune function. *Cell Signal* 14:1–9.
Tran, D. D., Edgar, C. E., Heckman, K. L., et al. (2005). CAML is a p56Lck-interacting protein that is required for thymocyte development. *Immunity* 23:139–152.
Uzzo, R. G., Clark, P. E., Rayman, P., et al. (1999a). Alterations in NFkappaB activation in T lymphocytes of patients with renal cell carcinoma. *J Natl Cancer Inst* 91:718–721.
Uzzo, R. G., Rayman, P., Kolenko, V., et al. (1999b). Mechanisms of apoptosis in T cells from patients with renal cell carcinoma. *Clin Cancer Res* 5:1219–1229.
Valitutti, S., Muller, S., Salio, M., and Lanzavecchia, A. (1997). Degradation of T cell receptor (TCR)-CD3-zeta complexes after antigenic stimulation. *J Exp Med* 185:1859–1864.
Vang, T., Torgersen, K. M., Sundvold, V., et al. (2001). Activation of the COOH-terminal Src kinase (Csk) by cAMP-dependent protein kinase inhibits signaling through the T cell receptor. *J Exp Med* 193:497–507.
Varki, A., and Angata, T. (2006). Siglecs—the major subfamily of I-type lectins. *Glycobiology* 16:1R–27R.
Veillette, A., Horak, I. D., Horak, E. M., Bookman, M. A., and Bolen, J. B. (1988). Alterations of the lymphocyte-specific protein tyrosine kinase (p56lck) during T-cell activation. *Mol Cell Biol* 8:4353–4361.
Vivier, E., and Anfossi, N. (2004). Inhibitory NK-cell receptors on T cells: witness of the past, actors of the future. *Nat Rev Immunol* 4:190–198.
Vivier, E., and Daeron, M. (1997). Immunoreceptor tyrosine-based inhibition motifs. *Immunol Today* 18:286–291.
Wang, Q., Stanley, J., Kudoh, S., et al. (1995). T cells infiltrating non-Hodgkin's B cell lymphomas show altered tyrosine phosphorylation pattern even though T cell receptor/CD3-associated kinases are present. *J Immunol* 155:1382–1392.
Watts, J. D., Sanghera, J. S., Pelech, S. L., and Aebersold, R. (1993). Phosphorylation of serine 59 of p56lck in activated T cells. *J Biol Chem* 268:23275–23282.
Weiss, A., and Littman, D. R. (1994). Signal transduction by lymphocyte antigen receptors. *Cell* 76:263–274.
Whiteside, T. L. (1998). Immune cells in the tumor microenvironment. Mechanisms responsible for functional and signaling defects. *Adv Exp Med Biol* 451:167–171.
Whiteside, T. L. (1999). Signaling defects in T lymphocytes of patients with malignancy. *Cancer Immunol Immunother* 48:346–352.
Whiteside, T. L. (2004). Down-regulation of zeta-chain expression in T cells: a biomarker of prognosis in cancer? *Cancer Immunol Immunother* 53:865–878.

Whiteside, T. L., and Parmiani, G. (1994). Tumor-infiltrating lymphocytes: their phenotype, functions and clinical use. *Cancer Immunol Immunother* 39:15–21.

Wick, M., Dubey, P., Koeppen, H., et al. (1997). Antigenic cancer cells grow progressively in immune hosts without evidence for T cell exhaustion or systemic anergy. *J Exp Med* 186: 229–238.

Willimsky, G., and Blankenstein, T. (2005). Sporadic immunogenic tumours avoid destruction by inducing T-cell tolerance. *Nature* 437:141–146.

Winkler, D. G., Park, I., Kim, T., et al. (1993). Phosphorylation of Ser-42 and Ser-59 in the N-terminal region of the tyrosine kinase p56lck. *Proc Natl Acad Sci USA* 90:5176–5180.

Wolfl, M., Batten, W. Y., Posovszky, C., Bernhard, H., and Berthold, F. (2002). Gangliosides inhibit the development from monocytes to dendritic cells. *Clin Exp Immunol* 130:441–448.

Zamoyska, R. (1998). CD4 and CD8: modulators of T-cell receptor recognition of antigen and of immune responses? *Curr Opin Immunol* 10:82–87.

Zhang, S. Q., Yang, W., Kontaridis, M. I., et al. (2004). Shp2 regulates SRC family kinase activity and Ras/Erk activation by controlling Csk recruitment. *Mol Cell* 13:341–355.

Zhang, Z., Shen, K., Lu, W., and Cole, P. A. (2003). The role of C-terminal tyrosine phosphorylation in the regulation of SHP-1 explored via expressed protein ligation. *J Biol Chem* 278: 4668–4674.

Zhou, G., Drake, C. G., and Levitsky, H. I. (2006). Amplification of tumor-specific regulatory T cells following therapeutic cancer vaccines. *Blood* 107:628–636.

Zou, W. (2005). Immunosuppressive networks in the tumour environment and their therapeutic relevance. *Nat Rev Cancer* 5:263–274.

Immunobiology of Dendritic Cells in Cancer

Michael R. Shurin and Gurkamal S. Chatta

1 Introduction

Tumor progression is often associated with suppression or malfunction of the immune system, of which dendritic cells (DC) possess many key regulatory functions, especially those related to cytokine production, antigen presentation to naïve T cells and polarization of Th1/Th2/Th3/Treg subsets and their balance. DC are professional antigen-presenting cells, strategically positioned for bridging innate and adaptive immunity. DC can initiate T-cell responses against tumors due to their capacity to process and present tumor antigens and stimulate naïve T cells. However, little is known about DC behavior in vivo in tumor-bearing hosts. Although neglected for many years, the importance of the tumor microenvironment in regulating immunology of DC is becoming more defined.

Even though alterations in DC in the setting of cancer were described more than a decade ago, characterization of tumor-derived factors responsible for DC dysfunction and the molecular mechanisms of abnormal DC differentiation and function are still largely unknown. An understanding of how the tumor environment regulates the DC system and how it impacts the efficacy of DC vaccines and other immunotherapeutic approaches is far from complete and clinical trials focusing on the protection of DC from the detrimental effects of the tumor microenvironment are currently being tested. In addition to the tumor/stromal cell-derived factors and their interactions, the other agents impacting DC and, thus, vaccine efficacy in cancer include: (1) the psychological stress of both a potentially fatal disease as well as the psychological and physical stress associated with the treatment of cancer (radiation, surgery, hormones, chemotherapeutic agents) and (2) ageing immune system, since over 50% of cancer arises in people older than 65 years of age (Fig. 1). The deterioration of the immune system with progressive aging is also coupled with the increased incidence and severity of infections and autoimmune disorders. Thus, in patients with cancer, the DC system is functioning under the multidirectorial

---

M.R. Shurin
Clinical Immunopathology, 5725 CHP-MT, 200 Lothrop Street, Pittsburgh, PA 15213, USA.
e-mail: shurinmr@upmc.edu

influences of various local and systemic tumor-derived and tumor stroma-derived factors, acute and chronic stress hormones, therapeutic agents and factors, as well as multifaceted conditions associated with aging, infections, autoimmune diseases and other disorders (Fig. 1).

Modulation of DC generation and function by some of the above-mentioned factors or conditions has been partly described. However, a comprehensive and systematic analysis of the DC system in the tumor environment has not been reported. For instance, both tumor-derived (reviewed in Fricke and Gabrilovich, 2006; Shurin and Gabrilovich, 2001) factors, as well as non-tumor cells in the tumor milieu (reviewed in Shurin et al., 2006), have been reported to suppress DC maturation, function and longevity. Psychological and physical stressors may affect the functional activity of

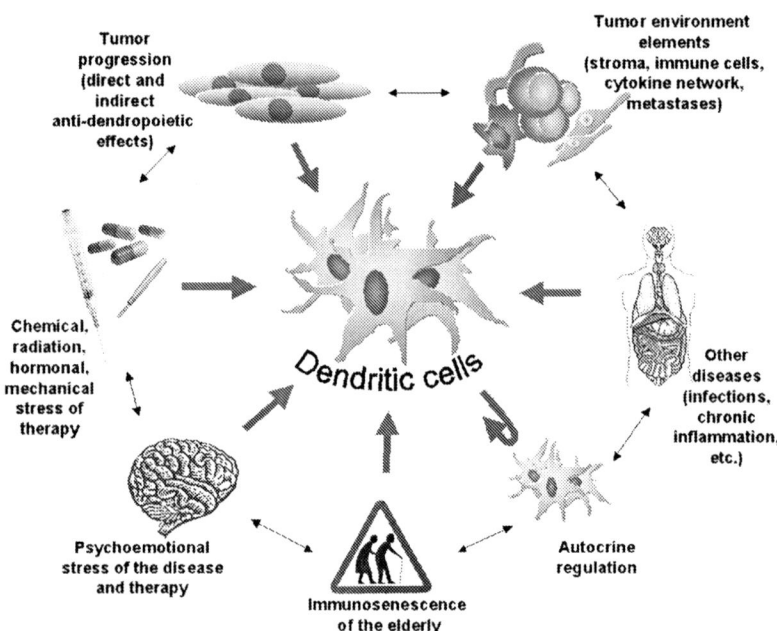

Fig. 1 The dendritic cell system in cancer patients. DC production, differentiation and function are under different influences and regulatory pathways operating in patients with cancer. This includes numerous tumor/stroma-derived factors that affect all stages of DC immunobiology and may be associated with cytokines, chemokines, growth factors, prostaglandins, gangliosides and many other soluble and membrane-bound molecules on different cell types in the tumor microenvironment. Additional modulation of DC function in cancer is a result of psychological and physical stressors associated with the diagnosis, as well as the effects of the treatment of a potentially fatal illness. Aging and immune-mediated diseases might also change DC differentiation and function, and, thus, DC-mediated immune responses. Finally, DC produce different factors in health and diseases, which may modulate DC by an autocrine and paracrine manner, and may also change DC responses to other molecules in the local environment. Thus, an understanding of the complex environmental conditions associated with DC function in cancer is necessary for harnessing the antitumor potential of these unique immunoregulatory cells

DC through a variety of hormones, neuromediators and neuropeptides (Maestroni, 2005; Saint-Mezard et al., 2003; Seiffert and Granstein, 2006). Indeed, modulation of DC maturation and function by glucocorticoids, neuropeptides and biogenic amines has been described. Glucocorticoid-treated DC showed a higher endocytic activity, a lower antigen-presenting function and a lower capacity to secrete cytokines (Piemonti et al., 1999). Norepinephrine can impede IL-12 and stimulate IL-10 production in DC, as well as inhibit their antigen-presenting capability and hamper their motility and chemotaxis (Maestroni, 2000; Maestroni and Mazzola, 2003). DC also express receptors for and respond to calcitonin gene-related peptide, neuropeptide Y, opioid peptides, prolactin, bombesin-like peptides, substance P and other neuropeptides, which might all be involved in stress-related modulation (Bedoui et al., 2007; Lambert and Granstein, 1998; Makarenkova et al., 2001, 2003; Marriott and Bost, 2001; Matera et al., 2001).

As shown in Fig. 1, surgery, radiation, chemotherapeutic agents and hormonal therapy might alter DC function and longevity (Bellik et al., 2006; Cao et al., 2004; Corrales et al., 2006; Schmidt et al., 2007). Furthermore, age-related alterations of DC maturity, function, longevity and subpopulation composition also play a significant role in the ability of the DC system to recognize tumor cells and T cells and induce and maintain an antitumor immune response in patients with cancer (reviewed in Shurin et al., 2007). For instance, increased levels of IL-6 and IL-10 repeatedly reported in old individuals might have a direct effect on dendropoiesis and/or maturation of DC and, thus, on their motility and ability to process and present tumor antigens. Finally, exposure to different stimuli induces DC to produce various endogenous mediators, including arachidonic acid-derived eicosanoids, cytokines, regulatory peptides and small molecules like nitric oxide (NO). Many secreted products of DC can act in an autocrine manner and modulate cell function; for instance, autocrine IL-10 can prevent spontaneous maturation of DC (Corinti et al., 2001). Interestingly, ageing has been associated with immunological changes (immunosenescence) that mimic changes observed in the setting of chronic stress as well as changes seen with cancer (Bauer, 2005; Tarazona et al., 2002). Thus there may be common mechanisms of immune alterations in the DC system in cancer, aging and chronic stress.

2 Alterations of the Dendritic Cell System in Cancer

Many mechanisms of DC alterations in the tumor microenvironment have been described during the last 10–15 years, and most of them relate to the inhibition of DC production/differentiation, functional deficiency and accelerated cell death. In Table 1, we summarize and group these mechanisms in four categories, which are briefly described here. This includes (1) *elimination* of functional DC by blocking their production/differentiation/maturation or inducing apoptosis in DC or their precursors; (2) *inhibition* of critical function of DC; (3) *polarization* of DC subpopulations toward immunosuppressive and tolerogenic DC subsets; and (4)

Table 1 Abnormalities of the DC system in cancer

	DC characteristics	Notes	References
1. Elimination	Quantitative inhibition of DC generation	From both CD34+ hematopoietic precursors and CD14+ macrophages in humans and bone marrow-derived precursors in mice	Ishida et al. (1998); Aalamian (2001); Shurin et al. (2001a,b); Tourkova et al. (2004); Ogden et al. (2006)
	Immaturity of DC at the tumor site	Low CD83, CD80 and CD86 expression	Melichar et al. (1998); Bell et al. (1999); Schwaab et al. (1999)
	Decreased number of circulating DC	Low levels of DC1 in blood	Savary et al. (1998); Hoffmann et al. (2002); Sakakura et al. (2006)
	Low expression of CD40	In vitro and in vivo data	Hasebe et al. (2000); Shurin et al. (2002)
	Downregulation of CCR7 on DC	DC prepared from tumor-bearing mice express low levels of CCR7 protein and mRNA	Walker et al. (2005)
	Induction of DC apoptosis and acceleration of DC rate	Tumor-derived factors induced apoptotic death of DC and increased rate of spontaneous apoptosis both in vitro and in vivo	Esche et al. (1999); Kiertscher et al. (2000); Pirtskhalaishvili et al. (2000a,b); Pinzon-Charry et al. (2006)
	Apoptotic death of DC precursors	Tumor cells may kill DC precursors and decrease numbers of CFU-DC	Katsenelson et al. (2001)
	Retention of DC inside the tumor lesions	Tumor-derived IL-8 might prevent DC emigration	Feijoo et al. (2005)
2. Inhibition	Inhibition of DC motility	IL-8 and TGF-β1, as well as low CCR7, expression might prevent DC emigration from tumor lesions	Feijoo et al. (2005); Weber et al. (2005); Walker et al. (2006)
	Inhibition of endocytic activity of DC	Phagocytosis and receptor-mediated endocytosis were inhibited in DC by different tumor cell lines	Tourkova et al. (2005, 2007)
	Inhibition of antigen processing in DC	Tumor suppress expression of MHC class I antigen-processing machinery proteins in DC	Tourkova et al. (2005)
	Inhibition of antigen presentation by DC	Due to low MHC class I and II and co-stimulatory molecules expression	Whiteside et al. (2004); Tourkova et al. (2005)

Table 1 (continued)

DC characteristics		Notes	References
	Suppression of DC–T cell contact interactions	Decreased ability of DC to form clusters (rosettes) with T cells	Tas et al. (1993); Shurin (unpublished data)
	Low expression of co-stimulatory molecules on DC	Downregulation of expression of B7 molecules	Nestle et al. (1997); Wang et al. (2004)
	Suppression of DC adhesiveness	Decreased adhesion of DC to matrix protein covered slides and tumor cell monolayers in vitro	Shurin (unpublished data)
	Downregulation of cytokine production in DC	E.g., low IL-12 production by blood DC in patients with breast cancer	Della Bella et al. (2003)
3. Polarization	Dysbalance between DC1/DC2/DCreg subpopulations	Decreased numbers of myeloid DC in blood; low mDC/pDC ratio in blood	Hoffmann et al. (2002); Della Bella et al. (2003); Vakkila et al. (2004); Ferrari et al. (2005); Bellone et al. (2006); Sakakura et al. [2006]; Takahashi et al. [2006]
	Dysbalance between DC, macrophages and MDSC	Tumor-associated increase in the levels of M2 macrophages and myeloid suppressor cells	Frey (2006); Huang et al. (2006); Ochoa et al. (2007)
	Transdifferentiation of bone marrow-derived DC into endothelial-like cells by tumor	Process of DC endothelialization is VEGF-dependent	Conejo-Garcia et al. (2004)
4. Avoidance	Loss of DC attracting chemokines at the tumor site	Loss or downregulation of CXCL14 expression in tumor cells	Shurin et al. (2005); Starnes et al. (2006)

avoidance of the contact with DC by downregulating expression of DC-attracting chemokines.

Stene et al. (1988) revealed that melanoma-associated skin DC (Langerhans cells) declined in number as melanoma progressed. In 1989, Alcalay et al. described a decreased number and altered morphology of Langerhans cells in squamous cell carcinomas of the skin (Alcalay et al., 1989) and showed later (1991) that antigen-presenting capacity of lymph node cells might be impaired during tumorigenesis (Alcalay and Kripke, 1991). Halliday et al. in 1991 demonstrated that tumor may regulate DC attraction and homing at the tumor site and suggested that yet unknown factors may inhibit function of DC and thus induction of antitumor immunity (Halliday et al., 1991, 1992). Becker speculated that outcome of a primary tumor in patients depends on the ability of DC to enter into tumors and that tumors might be different in their capacity to destroy or prevent DC from entering the tumor site (Becker, 1992). He also hypothesized that DC and tumor cells interacted and counteracted by releasing cytokines which abrogate tumor cells or DC, respectively (Becker, 1993). Tas et al. (1993) showed that DC are functionally abnormal in patients with cancer. Colasante et al. (1995) studied the role of cytokines in distribution and differentiation of DC lineage in primary lung carcinomas in humans and concluded on the potential role for GM-CSF, TNF-α, IL-1α and IL-1β in DC modulation. Gabrilovich et al. (1996a,b) reported that DC isolated from tumor-bearing mice showed a significantly reduced ability to induce syngeneic tumor-specific CTL and stimulate control allogeneic T cells and Chaux et al. (1996) revealed that tumor-associated DC express low levels of co-stimulatory molecules. Enk et al. (1997) showed that melanoma-derived factors converted DC antigen presenting cell function to tolerance induction against tumor tissue. Following these initial findings, other teams demonstrated suppression in preparation of human CD34-derived and CD14-derived DC, as well as murine bone marrow-derived DC by both identified and unidentified tumor-derived factors (Table 1). For example, Ninomiya et al. (1999) reported that DC propagated from patients with hepatocellular carcinoma expressed significantly lower levels of HLA-DR, had significantly lower capacity to stimulate allogeneic T cells and induced decreased amounts of IL-12. In vivo, Lissoni et al. (1999) revealed that the number of circulating DC in the peripheral blood of cancer patients was also significantly decreased, and these results were confirmed by others, e.g., in patients with squamous cell carcinoma of the head and neck (HNSCC) (Sakakura et al., 2006), leukemia (Maecker et al., 2006), hepatocellular carcinoma (Ormandy et al., 2006), lung cancer (Wojas et al., 2004) and invasive breast cancer (Della Bella et al., 2003). Metastasis development decreased the number of circulating DC even further (Bellik et al., 2006). Furthermore, blood monocytes isolated from both patients with glioblastoma and intracranial metastases had significantly reduced expression of GM-CSFR and showed a reduced capacity to differentiate into mature DC (Ogden et al., 2006). Similar data were reported for other cancers (Hasebe et al., 2000; Neves et al., 2005; Pedersen et al., 2005). Thus, local (at the tumor site) and systemic levels of DC might be markedly lower in cancer patients due to the inhibited or abnormal dendropoiesis (Shurin, 1999), i.e., DC generation and differentiation.

Elimination of functional DC in cancer may also be associated with the killing of DC or acceleration of their turnover. Induction of apoptosis in DC by tumor-derived factors was first reported by Esche et al. [1999] and confirmed by others (Kiertscher et al., 2000; Peguet-Navarro et al., 2003; Yang et al., 2002). Furthermore, the results were confirmed by documenting the presence of a significantly higher proportion of apoptotic blood DC in patients with early stage breast cancer compared to healthy volunteers (Pinzon-Charry et al., 2006). Similarly, tumor-mediated cell death of DC precursors (Katsenelson et al., 2001) and accelerated early apoptosis of DC (Kiertscher et al., 2000; Onishi et al., 2002) were also reported.

Second type of DC abnormalities in cancer includes their functional deficiency when compared to the cells derived from healthy age-matched controls (Table 1)

Decreased ability of DC obtained from cancer patients' blood or lymph nodes, or DC co-cultured with malignant cells, to stimulate allogeneic T cells, uptake, process and present antigen(s), provide co-stimulatory signal, migrate toward specific chemokines and produce IL-12 was repeatedly described for prostate, breast, renal, liver, lung cancer, HNSCC, melanoma, myeloma, leukemia, glioma, neuroblastoma and other tumor types (Aalamian, 2001; Brown et al., 2001; Katsenelson et al., 2001; Kichler-Lakomy et al., 2006; Onishi et al., 2002; Ratta et al., 2002; Satthaporn et al., 2004; Shurin et al., 2001a,b; Song et al., 2004). These and other results were also reviewed in Shurin and Gabrilovich (2001) Vicari et al. (2002), Shurin et al. (2003), Yang and Carbone (2004), Eisendle et al. (2005), Pinzon-Charry et al. (2005a,b) and Fricke and Gabrilovich (2006) and therefore are not detailed here.

Polarization of DC subtypes represents the third type of the DC system aberration in cancer (Table 1). For instance, there are substantial numbers of tumor-promoting functional plasmacytoid DC (pDC or DC2 by some classifications) (but not conventional DC or myeloid DC or DC1) accumulated in tumor ascites in patients with ovarian carcinomas (Zou et al., 2001). Similarly, estimating conventional and plasmacytoid subpopulations of DC in the peritoneal fluid of women with ovarian tumors, Wertel et al. (2006) reported that the percentage of pDC was higher in patients with ovarian cancer than in women with serous cystadenoma. Thus, decreased DC1/DC2 ratio at the tumor site in patients with cancer may favor Th2 lymphocyte differentiation and/or induction of immunological tolerance.

The levels of DC1 (i.e., myeloid or conventional DC subset) in circulation are also significantly lower, while the number of DC2 (lymphoid or plasmacytoid DC subset) might vary, as was repeatedly reported for patients with different tumor types compared to healthy donors (Hoffmann et al., 2002; Pinzon-Charry et al., 2005a; Wojas et al., 2004). Interestingly, these alterations were reverted by surgical resection of the tumor or chemoradiotherapy (Della Bella et al., 2003; Hoffmann et al., 2002; Takahashi et al., 2006; Yanagimoto et al., 2005) suggesting that tumor-derived factors are responsible for redirecting DC differentiation (dendropoiesis). Indeed, microvesicles isolated from plasma of advanced melanoma patients, but not from healthy donors, mediated the effect of tumor on $CD14^+$ monocytes and skewed their differentiation from DC toward $CD14^+HLA-DR^{low}$ cells with TGF-β-mediated suppressive activity on T-cell functions (Valenti et al., 2006). A subset of these TGF-β-secreting $CD14^+$ $HLA-DR^{low}$ cells was found to be significantly expanded in peripheral blood of melanoma patients compared with healthy donors.

Tumor-promoted redirection of dendropoiesis and its re-polarization are also associated with increased numbers of immature DC and appearance of other related immature cells of myeloid progeny. For example, in addition to having fewer levels of DC1 and DC2 in the peripheral blood, patients with breast and prostate cancer as well as patients with malignant glioma showed significant accumulation of abnormal population of $HLA-DR^+$ immature cells (DR^+IC), which in spite of HLA-DR, CD40 and CD86 expression had reduced capacity to capture antigens and elicited poor proliferation and IFN-γ secretion by T lymphocytes (Pinzon-Charry et al., 2005a,b). Immature DC fail to provide an appropriate costimulatory signal to T cells, might induce tolerance through abortive proliferation or anergy of antigen-specific $CD4^+$ and $CD8^+$ T cells or through the generation of regulatory T cells that suppress immune responses by producing IL-10 and TGF-β (Kim et al., 2006a,b). Immature DC were found at high levels within tumor-infiltrating leukocytes and increased circulating levels of immature DC have also been observed in the peripheral blood of patients with lung, breast, head and neck and esophageal cancer (Lizee et al., 2006). Whereas immature DC support Treg, immature myeloid precursors of DC suppress T-cell activation per se. Immature myeloid cells are a heterogeneous population of myeloid cells that comprises immature macrophages, granulocytes, DC and myeloid cells at early stages of differentiation (Gabrilovich, 2004). They have been reported in the spleen of mice with colon, prostate, lung, and breast cancer and in patients with cancer, i.e., RCC (Ochoa et al., 2007; Serafini et al., 2006). They suppress T-cell proliferation through a combination of nitric oxide and arginase production, as well as by reactive oxygen species, such as hydrogen peroxide (H_2O_2) (Frey, 2006). Arginase depletes arginine from the microenvironment, leading to distinct molecular changes in T cells, including the loss of T-cell receptor signaling and the inhibition of cell cycle in G_0 phase and peripheral T-cell tolerance. In addition, the production of nitric oxide may eventually lead to T-cell apoptosis in the tumor microenvironment (Ochoa et al., 2007). Recently, it has been reported that immature myeloid suppressor cells or myeloid-derived suppressor cells, probably the most appropriate term for this cell population (Gabrilovich et al., 2007), in addition to being able to suppress T-cell proliferation in vitro, can secrete IL-10 and TGF-β and induce the development of Foxp3+ Treg cells in vivo, which are anergic and suppressive (Huang et al., 2006). Similar to immature DC, circulating levels of immature myeloid cells have been well correlated with the stage of disease and poorer prognosis, and surgical resection of tumors has been shown to decrease the number of peripheral blood immature myeloid cells in both human and animal models (Lizee et al., 2006; Serafini et al., 2006).

Finally, the last mechanism of decreased number of DC associated with the tumor progression is the loss of expression of DC attracting chemokines at the tumor site (Table 1). For instance, it has been demonstrated that HNSCC cells do not express CXCL14 protein and mRNA, a potent DC-attracting chemokine (Shurin et al., 2005). This resulted in low chemoattraction of DC to the tumor bed, low numbers of tumor-associated DC and deficient induction of antitumor immunity; however, transduction of CXCL14-negative tumor cells with the CXCL14 gene was associated with increased DC infiltration, an antitumor immune response and inhibition of tumor growth in vivo. Interestingly, melanoma cells might utilize an opposite approach and can effectively chemoattract DC, modulate their phenotype and, eventually, severely damage DC mobility: melanoma-conditioned DC exhibited an increased adhesion capacity to a melanoma cell line in vitro and did not migrate in response to DC chemokines (Remmel et al., 2001). The explanation for abnormal DC retention inside of some human malignant lesions may come from another study where it was found that tumors from patients with hepatocellular carcinoma, colorectal or pancreatic cancer were producing IL-8 and that this chemokine attracted DC that uniformly express both IL-8 receptors CXCR1 and CXCR2 (Feijoo et al., 2005).

In summary, abnormal dendropoiesis, DC longevity and function and DC migration toward or from the tumor site are the key characteristics of the DC system dysfunction in tumor-bearing hosts that have a crucial role in immune non-responsiveness to tumors.

3 Mechanisms of Dendritic Cell Dysfunction in Cancer

3.1 Factors

Tumors exploit several strategies to evade immune recognition, including the production of a variety of immunosuppressive/immunomodulating factors, which might specifically block or redirect DC maturation, suppress DC survival and numbers and impair function of DC in the vicinity of tumors (Shurin et al., 2006) (Table 2). Historically, the first tumor-derived factor responsible for inhibiting DC differentiation in cancer was identified as vascular endothelial growth factor (VEGF) (Gabrilovich et al., 1996a,b). For instance, in patients affected by colorectal cancer, DC numbers inversely correlated with VEGF serum levels, suggesting a possible effect of this cytokine on DC compartment. In cultures, the exposure of monocyte-derived DC to VEGF produced a dramatic alteration of DC differentiation by induction of apoptosis, alteration of DC phenotypic profile and increased CXCR4 expression (Della Porta et al., 2005). VEGF blocks the functional maturation of DC from hematopoietic progenitor cells by blocking NF-κB transcription. The family of VEGF molecules also plays a key role in recruiting immature myeloid cells and immature DC from the bone marrow to enrich the tumor microenvironment (Kim et al., 2006b).

Tumor-derived TGF-β and IL-10 were shown to be responsible for downregulating CD80 expression on blood DC in myeloma patients (Brown et al., 2001). DC maturation, antigen presentation and IL-12 production induced by inflammatory cytokines IL-1 and TNF-α or by LPS might be inhibited by TGF-β (Geissmann et al., 1999). TGF-β might also induce apoptosis in DC (Ito et al., 2006). Increased levels of IL-10 in serum from patients with hepatocellular carcinoma and tumor progression were shown to correlate with profound numerical deficiencies and immature phenotype of circulating DC subsets (Beckebaum et al., 2004). Murine bone marrow-derived DC that were propagated in IL-10 and TGF-β (so-called "alternatively activated" DC) expressed low levels of TLR4, MHC class II, CD40, CD80, CD86, IL-12p70, programmed death-ligand 2 (B7-DC; CD273) and were resistant to maturation (Lan et al., 2006). They secreted much higher levels of IL-10 and efficiently expanded functional $CD4^+CD25^+Foxp3^+$Treg cells. We have shown earlier that murine colon adenocarcinoma cells produce IL-10 and that IL-10 causes downregulation of CD40 expression on DC and is responsible for inhibited CD40-dependent IL-12 production by DC (Shurin et al., 2002). These and other studies also revealed the tumor-associated in vivo effects of IL-10 on DC function in eliciting a type 1 immune response in both allogeneic and tumor-specific responses

Table 2 Mechanisms of DC dysfunction in cancer

Mechanisms and factors		Examples	References
Tumor/stroma-derived factors	Cytokines or their combinations	VEGF, M-CSF, IL-6, IL-10, TGF-β, IL-8	Gabrilovich et al. (1996a,b); Menetrier-Caux et al. (1998); Shurin et al. (2002); Yang et al. (2003); Bellone et al. (2006)
	CCL2	Mediates the migration of myeloid suppressors to tumors	Huang et al. (2007)
	CCL20/MIP3α	Involved in immature DC and their precursors attraction	Thomachot et al. (2004)
	Stromal-derived factor-1 and β-defensins	Attract proangiogenic DC subset	Zou et al. (2001); Conejo-Garcia et al. (2005)
	Prostanoids, prostaglandins	Regulate DC maturation	Sombroek et al. (2002)
	Gangliosides	Suppress dendropoiesis and DC longevity	Shurin et al. (2001a,b); Peguet-Navarro et al. (2003); Tourkova et al. (2005)
	Neuropeptides	Bombesin-like peptides (gastrin-releasing peptide (GRP) and neuromedin B (NMB))	Makarenkova et al. (2003)
	Tumor antigens	PSA is a serine protease MUC1 subverts DC function	Aalamian et al. (2003) Carlos et al. (2005)
	Other molecules	Lactic acid, Hyaluronan, NO spermine	Stanford et al. (2001); Yang et al. (2002); Della Bella et al. (2003); Gottfried et al. (2006)
	HLA-G	HLA-G might induce tolerogenic DC by disruption of the MHC class II presentation pathway	Ristich et al. (2005); Lemaoult et al. (2007)
	Reactive oxygen species	Hydrogen peroxide may activate p38 and JNK in DC and induce apoptosis	Handley et al. (2005)
	Tumor-derived microvesicles	Abnormal differentiation of monocytes into IL-6/TNF-α/TGF-β-producing cells	Valenti et al. (2006)
Affected signaling pathways in DC	Upregulation of Bax and downregulation of Bcl-2 and Bcl-X_L	Both extrinsic and intrinsic pathways are involved in tumor-induced apoptosis of DC	Esche et al. (1999, 2001); Pirtskhalaishvili et al. (2000a); Balkir et al. (2004)

Table 2 (continued)

Mechanisms and factors	Examples	References
Ceramide	Mediates tumor-induced DC apoptosis by downregulation of the PI3K pathway	Kanto et al. (2001)
Small Rho GTPases	Cdc42 and Rac 1 mediate tumor-induced dysfunction of DC	Tourkova et al. (2007)
STAT3	Tumor-mediated induction of STAT3 in DC results in reduced expression of IL-12, MHC class II and CD40	Gabrilovich (2004); Nefedova et al. (2004); Wang et al. (2004); Nefedova and Gabrilovich (2007)
SOCS1	SOCS1 functions as an antigen-presentation attenuator by controlling the tolerogenic state of DC and the magnitude of antigen presentation	Evel-Kabler et al. (2006)
p38 MAPK	Tumor activates p38 MAPK and JNK but inhibits ERK in DC	Wang et al. (2006)
H1^0 expression	Tumor-derived factors inhibit $h1°$ expression in DC precursors causing defective DC differentiation	Gabrilovich et al. (2002)

(Yang et al., 2003). Furthermore, analyzing pancreatic cancer-derived cytokines responsible for inhibition of DC differentiation, Bellone et al. (2006)reported that IL-10, TGF-β and IL-6, but not VEGF, cooperatively affect DC precursors in a manner consistent with ineffective antitumor immune responses. However, lung squamous cell carcinoma and adenocarcinoma cells have been shown to use different mediators to induce comparable phenotypic and functional changes in DC: IL-6 versus IL-10+IL-6+prostanoids, respectively (Avila-Moreno et al., 2006). RCC-derived IL-6 and VEGF were shown to block the ability of tumor antigen-loaded DC to induce CTL in the autologous system (Cabillic et al., 2006).

To define the pathways limiting DC function in the tumor microenvironment, Sharma et al. [2003] assessed the impact of tumor cyclooxygenase (COX)-2 expression on DC activities and reported that inhibition of tumor COX-2 expression or activity could prevent tumor-induced suppression of DC capacity to process and present antigens and secrete IL-12. COX-1- and COX-2-regulated prostanoids and IL-6 were found to be solely responsible for the hampered differentiation of monocyte-derived and CD34+ precursor-derived DC by freshly excised solid human tumors (colon, breast, renal cell carcinoma and melanoma) (Sombroek et al., 2002). An important role for the EP2 receptor in PGE_2-induced inhibition of DC differentiation and function and the diminished antitumor cellular immune responses in vivo has also been reported (Yang et al., 2003). Finally, PGE2 suppresses differentiation of DC and is a potent inducer of IL-10 in bone marrow-derived DC, and PGE2-induced IL-10 is a key regulator of the DC pro-inflammatory phenotype (Sombroek et al., 2002).

In addition to these "classic" tumor-derived anti-dendropoietic factors, other molecules were implemented to tumor-mediated dysfunction of the DC system (Table 2). Melanoma, neuroblastoma, RCC and lung cancer were shown to produce and shed various gangliosides, which may suppress dendropoiesis, inhibit DC function or induce apoptosis in DC (Peguet-Navarro et al., 2003; Shurin et al., 2001a,b; Tourkova et al., 2005). Tumor-derived lactic acid is also an important factor modulating the DC phenotype in the tumor environment, which may critically contribute to tumor escape mechanisms (Gottfried et al., 2006). Interestingly, several tumor antigens were recently found to display anti-dendropoietic properties. Prostate-specific antigen (PSA), which is a serine protease, was able to inhibit generation and maturation of DC from CD34+ hematopoietic precursors, assessed by the levels of expression of CD83, CD80, CD86 and HLA-DR, as well as the ability of DC to induce T-cell proliferation (Aalamian et al., 2003). When cultured with MUC1 glycoprotein, human monocyte-derived DC displayed decreased expression of CD86, CD40, CD1d, HLA-DR and CD83 and were defective in the ability to induce immune responses in both allogeneic and autologous settings. The modified phenotype of MUC1-treated DC corresponded to an altered balance in IL-12/IL-10 cytokine production with a failure to make IL-12 and induce Th1 responses (Carlos et al., 2005; Rughetti et al., 2005). Finally, human chorionic gonadotropin (hCG), which serves as an important tumor marker for trophoblastic disease, has been recently shown to upregulate expression of IDO in DC (Ueno et al., 2007).

Human leukocyte antigen G (HLA-G) molecules, which are normally expressed in cytotrophoblasts and play a key role in maintaining immune tolerance at the maternal–fetal interface, was also reported to be expressed on malignant cells and can be regulated by hypoxia (Mouillot et al., 2007; Wilczynski, 2006). As DC expressed immunoglobulin-like transcript 4 (ILT4), an inhibitory receptor capable of interacting with HLA-G, DC may be tolerized by HLA-G through inhibitory receptor interactions. Indeed, the HLA-G-ILT4 interaction leads to development of tolerogenic DC with the induction of anergic and immunosuppressive T cells (Ristich et al., 2005).

Finally, human tumors constitutively release endosome-derived microvesicles, transporting a broad array of biologically active molecules with potential modulatory effects on different immune cells. The first evidence that tumor-released microvesicles alter myeloid cell function by impairing monocyte differentiation into DC and promoting the generation of a myeloid immunosuppressive cell subset was recently published (Valenti et al., 2006, 2007).

3.2 Signaling Pathways

Many immunosuppressive factors produced by tumor cells induce STAT3 activation in DC, blocking their normal functioning. For instance, treatment of DC with melanoma-conditioned medium resulted in reduced expression of IL-12, MHC class II and CD40 due to the increased induction of STAT3 (Wang et al., 2004). The immunosuppressive effects of tumor-derived factors on DC differentiation were abrogated in cells from STAT3 knockout mice or by the treatment of DC precursors with a phosphopeptide that binds the STAT3 SH2 domain and blocks downstream STAT activation. Furthermore, IL-6-mediated suppression of DC maturation was also abrogated in STAT3-deficient DC precursors, indicating the significance of STAT3 in IL-6-mediated suppression of DC maturation and function (Wang et al., 2004). Furthermore, constitutive STAT3 activation in tumor cells was shown to inhibit DC function by the increased induction of STAT3 in immature DC. Thus, immunosuppression mediated by tumor cells results from a circuit of STAT3 signaling that begins in tumor cells and eventually activates inhibitory STAT3 signaling in DC in part due to the production of cytokines that increase STAT3 activation in DC (EGF, VEGF, IL-6, IL-10, G-CSF, M-CSF, GM-CSF) (Wang et al., 2004). In addition, STAT3 phosphorylation in DC was regulated by IL-6 in vivo, and STAT3 was necessary for the IL-6 suppression of DC activation/maturation (Park et al., 2004). Interestingly, CD4+CD25+FoxP3+ regulatory T cells from tumor-bearing animals may also impede DC function by activating STAT3 and inducing the Smad signaling pathway (Larmonier et al., 2007). The suppression mechanism was also associated with downregulation of activation of the transcription factor NF-κB, required TGF-β and IL-10 and resulted in strong inhibition of expression of the co-stimulatory molecules CD80, CD86 and CD40, the production of TNF-α, IL-12 and CCL5/RANTES by DC (Table 2).

Many STAT family members are developmentally regulated and play a role in DC differentiation and maturation. For instance, the STAT6 signaling pathway is constitutively activated in immature DC and declines as they differentiate into mature DC. Downregulation of STAT6 pathway is accompanied by dramatic induction of suppressors of cytokine signaling 1 (SOCS1), SOCS2, SOCS3 and cytokine-induced Src homology 2-containing protein expression (Jackson et al., 2004). In contrast, STAT1 signaling is most robust in mature DC. Thus, it is likely that cytokine-induced maturation of DC is under feedback regulation by SOCS proteins and that the switch from constitutive activation of the STAT6 pathway in immature DC to predominant use of STAT1 signals in mature DC is mediated in part by STAT1-induced SOCS expression (Jackson et al., 2004). Recent studies also demonstrate that SOCS1 functions as an antigen-presentation attenuator by controlling the tolerogenic state of DC and the magnitude of antigen presentation (Evel-Kabler et al., 2006). Since SOCS1 restricts DC ability to break self-tolerance and induce antitumor immunity by regulating IL-12 production and signaling, it is quite possible that some products of tumor cells or other cells within the tumor milieu might induce SOCS1 expression in DC. Although not proven experimentally, this pathway may operate in the tumor microenvironment limiting the ability of DC to process and present tumor antigens and secrete IL-12.

Recent data allowed Wang et al. [2006] to speculate that tumor-induced p38 MAPK activation and ERK inhibition in DC may be a new mechanism for tumor evasion. They showed that tumor supernatant-treated DC were inferior to normal

DC at priming tumor-specific immune responses, but inhibiting p38 mitogen-activated protein kinase (MAPK) restored the phenotype, cytokine secretion and function of tumor-treated DC. Tumor-derived factors activated p38 MAPK and Janus kinase (JNK) but inhibited extracellular regulated kinase (ERK) in DC (Table 2).

Since many functions of DC, such as endocytosis, exocytosis, adhesiveness and motility, depend on actin polymerization and membrane rearrangements, Tourkova et al. (2005) analyzed whether small Rho GTPases (Cdc42, RhoA and Rac1/2), which are primary involved in regulating these functions in DC, might be affected by tumor-derived factors. They found that impaired endocytic activity of DC co-cultured with tumor cells was associated with decreased levels of active Cdc42 and Rac1. Transduction of DC with the dominant negative Cdc42 and Rac1 genes also leads to reduced phagocytosis and receptor-mediated endocytosis, while transduction of DC with the constitutively active Cdc42 and Rac1 genes restored endocytic activity of DC that were inhibited by the tumors (Tourkova et al., 2007).

Less is known about signaling pathways that control DC longevity and DC sensitivity to tumor-induced cell death. Early studies showed that Bcl-X_L, Bcl-2 and mitochondrial cytochrome *c* release mediate resistance of DC to tumor-induced apoptosis (Esche et al., 2001; Pirtskhalaishvili et al., 2000a,b). Other data demonstrated that down-regulation of phosphoinositide 3-kinase (PI3K) is the major facet of tumor-induced DC apoptosis (Kanto et al., 2001). Interestingly, it is known that cancer cells have increased production of hydrogen peroxide (H_2O_2) (Lopez-Lazaro, 2007; Szatrowski and Nathan, 1991) and in DC hydrogen peroxide activates two key MAPK, p38 and JNK. Activation of JNK, which is associated with inhibition of tyrosine phosphatases in DC, is linked to the induction of DC apoptosis (Handley et al., 2005). By targeting different anti-apoptotic molecules, including FLIP, XIAP/hILP, procaspase-9 and HSP70, Balkir et al. [2004] demonstrated that anti-apoptotic molecules other than the Bcl-2 family of proteins were involved in tumor-induced apoptosis in DC. This suggests that tumor-induced apoptosis of DC is not limited to the mitochondrial pathway of cell death and that both extrinsic and intrinsic apoptotic pathways play a role in DC survival in the tumor microenvironment.

4 Role of Dendritic Cells in Tumor Escape Mechanisms

A growing body of evidence clearly demonstrates that the DC system is directly and indirectly involved in controlling tumor growth and progression and there are finally no doubts that DC modified in the tumor environment play an important role in tumor escape from immune recognition and elimination. However, with the realization that the DC lineage represents a varied collection of distinct populations, a question has arisen as to whether certain types of DC are dysregulated in tumor-bearing hosts, or whether the nature of immunological challenge and state of DC maturation define particular facets of innate/aquired/tolerogenic responses in the tumor environment. Numerous studies have revealed that specific DC subsets might be linked to immunological unresponsiveness and/or tolerance to tumor antigens. For instance, clinical outcome of the children with cancer has been shown to correlate with circulating plasmacytoid DC count: Children with high plasmacytoid DC counts at diagnosis survived significantly worse than those with low counts and the development of cancer was associated with low number of conventional DC (Vakkila et al., 2004). Thus, tumor-associated plasmacytoid DC contribute to the tumor environmental immunosuppressive network. Indeed, tumor ascites pDC induced IL-10+CCR7+CD45RO+CD8+ regulatory T cells, which significantly suppress myeloid DC-mediated tumor-associated antigen-specific T-cell effector functions through IL-10 (Wei et al., 2005). Plasmacytoid DC in tumor-draining lymph nodes might create a local microenvironment that is potently suppressive of host antitumor T-cell responses and this mechanism may be mediated by immunosuppressive indoleamine-2,3-dioxygenase (IDO) (Table 3).

IDO degrades tryptophan to kynurenine, which is further metabolized to 3-hydroxyanthranilic acid and thus initiates the immunosuppressive pathway of tryptophan catabolism. Emerging evidence suggests that Tregs may be generated de novo against specific tumor-derived antigens, and thus they arise as a direct consequence of antigen presentation in the tumor-draining lymph nodes (Munn, 2006). IDO can also be expressed within the tumor itself, by tumor cells or host stromal cells, where it can inhibit the effector phase of the immune response (Munn, 2006). Other data indicate that kynurenine pathway enzymes downstream of IDO can initiate tolerogenesis by DC independently of tryptophan deprivation, as tolerogenic DC can confer suppressive ability on otherwise immunogenic DC in an IDO-dependent fashion (Belladonna et al., 2006). Thus, the paracrine production of kynurenines might be one mechanism used by IDO-expressing cells in the tumor microenvironment to convert DC lacking functional IDO to a tolerogenic phenotype. IDO, i.e., tryptophan, kynurenine or 3-hydroxyanthranilic acid, could also induce expression of the tolerogenic molecule HLA-G in DC (Lopez et al., 2006). Thus, IDO and HLA-G can cooperate in the immune suppression, since HLA-G-expressing DC might suppress or alter effector T cells as well. Indeed, activated CD4+ and CD8+ T cells could efficiently acquire immunosuppressive HLA-G from antigen-presenting cells through membrane transfers (a process called trogocytosis) and acquisition of HLA-G immediately reversed T-cell function from effectors to regulatory cells. These regulatory T cells were able to inhibit proliferative responses through HLA-G that they acquired (Lemaoult et al., 2007).

In support of the concept that certain DC subpopulations play crucial roles in tumor escape, it was recently reported that tumor expansion could stimulate Treg cells via a specific DC subset: During tumor progression, a subset of DC exhibiting a myeloid immature phenotype may be recruited to draining lymph nodes and selectively promote proliferation of Treg cells in a TGF-β-dependent manner (Ghiringhelli et al., 2005). Importantly, tumor cells are necessary and sufficient to convert DC into regulatory cells that secrete TGF-β and stimulate Treg cell proliferation (Table 3).

Table 3 DC-mediated mechanisms of tumor escape

Mechanisms	Notes	References
Inability to present tumor antigens to T cells and induce tumor-specific CTL	Tumor-mediated suppression of critical functions of DC	Gabrilovich et al. (1996a,b); Shurin and Gabrilovich (2001); Cabillic et al. (2006)
Immaturity of DC in the tumor environment	High levels of immature DC can actively induce T-cell tolerance to tumor antigens and promote cancer progression	Ninomiya et al. (1999); Beckebaum et al. (2004); Thomachot et al. (2004); Bellone et al. (2006)
Expression of IL-10	IL-10-producing DC efficiently expand functional $CD4^+CD25^+Foxp3^+$Treg cells	Mahnke and Enk (2005); Lan et al. (2006)
Expression of TGF-β	DC exhibiting a myeloid immature phenotype may promote proliferation of Treg cells in a TGF-β-dependent manner	Ghiringhelli et al. (2005)
Low IL-12 production	DC isolated from tumor-bearing animals or cancer patients produce low levels of spontaneous and inducible IL-12	Shurin et al. (2002); Satthaporn et al. (2004); Bellone et al. (2006)
Low IL-18 expression	Melanoma might block IL-18 synthesis in DC and thus prevent activation of NK cells by DC	Capobianco et al. (2006)
IDO-mediated mechanisms	Tolerogenic DC can confer suppressive ability on otherwise immunogenic DC in an IDO-dependent fashion	Munn (2006); Hou et al. (2007)
Immunoglobulin-like transcript 3 (ILT3) and ILT4 (human) Paired immunoglobulin-like inhibitory receptor (PIR-B) (murine)	The HLA-G-ILT4 (immunoglobulin-like transcript) interaction leads to development of tolerogenic DC with the induction of anergic and immunosuppressive T cells	Chang et al. (2002); Suciu-Foca et al. (2005)
Expression of HLA-G molecules	HLA-G on DC induces immune suppression and tolerance	Lopez et al. (2006); Lemaoult et al. (2007)
Expression of B7-H1	Upregulation of B7-H1 on DC in the tumor microenvironment downregulates T-cell immunity	Curiel et al. (2003); Perrot et al. (2007)
Attraction and protection of pDC precursors	Accumulation of pDC in peritoneal fluid, ascites in ovarian cancer	Zou et al. (2001); Wertel et al. (2006)
Stimulation of angiogenesis by tumor-associated pDC and DC precursors	Tumor pDC produce high levels of TNF-α and IL-8 and induce neovascularization in vivo	Conejo-Garcia et al. (2004); Curiel et al. (2004)

Table 3 (continued)

Mechanisms	Notes	References
Induction of Treg	Tumor ascites pDC induce IL-10+CCR7+ CD45RO+CD8+ Treg	Wei et al. [2005]; Munn and Mellor (2006)
Induction of DCreg	Tumor might induce a subpopulation of DC which secrete TGF-β and support proliferation of Treg cells	Sato et al. [2003]; Ghiringhelli et al. [2005]; Mellor et al. [2005]
Inhibition of IKDC subset (?)	IFN-producing killer DC (IKDC) are $B220^+NK1.1^+Gr1^-$ DC that kill tumor cells through the TRAIL pathway	Taieb et al. (2006); Ghiringhelli et al. (2007)
Impairment of DC migration toward tumor site or lymphoid tissue	Decreased attraction of increased retention of DC at the tumor site	Feijoo et al. (2005); Shurin et al. (2005)
Anti-inflammatory and tolerogenic properties of tumor-induced immature DC	Decreased production of IL-1, IL-6 and TNF-α and increased expression of IL-10 and TGF-β after capturing apoptotic tumor cells	Kim et al. (2006a)

Another subset of DC might contribute to neovascularization at the tumor site. Recently, Conejo-Garcia et al. (2004) reported that within 3 weeks of culture with tumor cell-conditioned medium, bone marrow-derived DC could be transdifferentiated into endothelial-like cells in vitro. They also identified a novel leukocyte subset within ovarian carcinoma that co-expressed endothelial and DC markers which may play a role in the formation of blood vessels (Conejo-Garcia et al., 2005). Curiel et al. (2004) observed high numbers of plasmacytoid DC in malignant ascites of patients with untreated ovarian carcinoma and showed that tumor-associated pDC induced angiogenesis in vivo through production of TNF-α and IL-8. By contrast, conventional (or myeloid) DC, which might suppress angiogenesis in vivo through production of IL-12, were absent from malignant ascites. Thus, the tumor may attract pDC to augment neovascularization while excluding myeloid DC to prevent angiogenesis inhibition.

Thus, one mechanism contributing to immunologic unresponsiveness toward tumors may be presentation of tumor antigens by tolerogenic/regulatory host DC. Indeed, using bone marrow chimeras in transgenic mice, Mihalyo et al. (2007) have recently reported that DC, but not CD4+CD25+ T regulatory cells, play a critical role in programming CD4 cell responses to tumor antigens during tumorigenesis. Regulatory DC could be produced from bone marrow precursors in the presence of GM-CSF, IL-10, TGF-β1 and LPS or TNF-α and they retained their T-cell regulatory property in vitro and in vivo even under inflammatory conditions (Rutella et al., 2006; Sato et al., 2003). Another minor subpopulation of regulatory DC has been recently described in murine spleen. These splenic CD19+ DC that did not express the plasmacytoid DC marker acquired potent IDO-dependent T-cell-suppressive functions (Mellor et al., 2005).

However, proponents of the "maturation" hypothesis suggest that the maturation state of the DC in the premalignant/inflammatory milieu or in the newly formed tumor setting predicts the development of an antitumor immune response or tumor tolerance. It was also proven that DC that capture and present antigen under non-inflammatory conditions maintain an immature phenotype and acquire tolerogenic properties. These DC generate regulatory T lymphocytes that potentiate tolerogenic responses (Ureta et al., 2007). An increased proportion of immature DC with reduced expression of co-stimulatory molecules was seen or isolated from tumor mass of patients with RCC, prostate cancer, basal-cell carcinoma and melanoma or was found in the peripheral blood of patients with breast, head and neck, lung or esophageal cancer (Gabrilovich, 2004). Similar data have been obtained using several mouse tumor models. The maturation hypothesis was also bolstered by studies showing that in tumor tissues, immature DC resided within the tumor, whereas mature DC were located in peritumoral areas (Bell et al., 1999). Immature DC cannot induce antitumor immune responses and, most importantly, immature DC can induce T-cell tolerance or anergy. Thomachot et al. (2004) showed that breast carcinoma cells produce soluble factors (CCL20 and TGF-β), which attract DC precursors in vivo and promote their differentiation into immature DC with altered functional capacities, and that these altered DC may contribute to the impaired immune response against the tumor. Similarly, medium conditioned by human pancreatic carcinoma cells induced monocyte-derived immature DC with inhibited proliferation, expression of costimulatory molecules (CD80 and CD40) and HLA-DR and functional activity as assessed by T-cell activation and IL-12p70 production (Bellone et al., 2006). Immature DC generated from pancreatic carcinoma patients in advanced stages of the disease similarly showed decreased levels of HLA-DR expression and reduced ability to stimulate T cells. Direct ex vivo flow cytometric analysis of various DC subpopulations in peripheral blood from hepatocellular carcinoma patients revealed an immature phenotype of circulating DC that was associated with increased IL-10 concentrations in serum and with tumor progression (Beckebaum et al., 2004; Ninomiya et al., 1999).

To evaluate whether and to what extent the capacity of tumor-infiltrating DC to drive immunization can be turned off by tumor cells, leading to tumor-specific tolerance rather than immunization, Perrot et al. have characterized the DC isolated from human non-small cell lung cancer based on the expression of CD11c. All isolated DC, including CD11c^{high} myeloid DC, CD11c$^-$ plasmacytoid DC and a third DC subset expressing intermediate level of CD11c, were immature and display the poor antigen-presenting function even after TLR stimulation and the reduced migratory response toward CCL21 and SDF-1 (Perrot et al., 2007). Interestingly, CD11c^{int} myeloid DC, which represented approximately 25% of total DC in tumoral and peritumoral tissues, expressed low levels of costimulatory molecules contrasting with high levels of the immunoinhibitory molecule B7-H1. These data suggest that immature tumor-associated DC have an ability to compromise the tumor-specific immune response in draining lymph nodes in vivo (Table 3).

In spite of multiple evidence supporting both "subpopulation"-based and "maturation"-based explanations of how the DC system is involved in tumor escape, additional data suggest that the real situation might be significantly more complex. The first layer of complexity arrives from the results showing that DC subsets may induce both tolerogenic and immunogenic responses depending on the environmental stimuli. For example, although the general thought is that pDC are commonly tolerogenic, it appears that the functional role of pDC in cancer immunity depends on cytokines that affect the balance between immunity and tolerance in the tumor and lymphoid organ microenvironment: In an analysis of draining lymph nodes in breast cancer, pDC with a relative increase in IL-12 and IFN-γ were associated with a good prognosis, whereas pDC with a relative increase in IL-10 and IL-4 were associated with a poor prognosis (Cox et al., 2005). In confirmation of this conclusion, Kim et al. [2007] have reported that although pDC recruited to the tumor site are implicated in facilitating tumor growth via immune suppression, they can be released from the tumor as a result of cell death caused by primary systemic chemotherapy and can then be activated through TLR9. Thus, synergistically with conventional DC, pDC may also play a crucial role in mediating cancer immunity.

Thus, we can conclude that pDC, as well as myeloid DC, have a dual role not only in initiating immune responses but also in inducing tolerance to tumor antigens.

Additional layer of complexity of DC subset versus DC maturation problem in cancer comes from the data revealing different maturation patterns of different DC subsets and its differential regulation by other immune cells. For example, analysis of the maturation of human blood-derived conventional myeloid DC and plasmacytoid DC activated with TLR ligands in the presence of Treg revealed that preactivated Treg suppressed strongly TLR-triggered myeloid DC maturation, as judged by the blocking of costimulatory molecule upregulation and the inhibition of proinflammatory cytokines secretion that resulted in poor antigen presentation capacity. Although IL-10 played a prominent role in inhibiting cytokines secretion, suppression of phenotypic maturation required cell–cell contact and was independent of TGF-β and CTLA-4. In contrast, the acquisition of maturation markers and production of cytokines by plasmacytoid DC triggered with TLR ligands were insensitive to regulatory T cells (Houot et al., 2006). Therefore, human Treg may enlist conventional, but not plasmacytoid DC for the initiation and amplification of tolerance in vivo by restraining their maturation after TLR stimulation.

In another study, evidence was provided that maturing conventional DC and plasmacytoid DC express different sets of molecules that drive distinct types of T-cell responses (Ito et al., 2007). Although both maturing myeloid DC and pDC upregulate the expression of CD80 and CD86, only pDC upregulate the expression of inducible costimulator ligand (ICOS-L) and maintain high expression levels upon differentiation into mature DC. High ICOS-L expression endows maturing pDC with the ability to induce the differentiation of naïve CD4 T cells to produce IL-10 but not the Th2 cytokines IL-4, IL-5 and IL-13. These IL-10-producing T cells are T-regulatory cells, and their generation by ICOS-L is independent of pDC-driven Th1 and Th2 differentiation. Thus, in contrast to myeloid DC, pDC are poised to express ICOS-L upon maturation, which leads to the generation of IL-10-producing Treg cells (Ito et al., 2007).

As such there are still many confusions in the field as to whether certain DC subpopulations have evolved to fulfill unique immunological roles in cancer (Th1/Th2/Th3/Th17 polarization, Treg induction, tolerance, etc.) or whether distinct DC subsets exist to uniquely respond to tumor-derived stimuli but each participate in maintaining tolerance or immunity in immature or mature state. It is also somewhat undecided whether some of the diversity in the DC lineage as determined by cell surface molecule expression represents genuine distinct DC subsets or particular developmental/activation states of the same DC subtype. However, collectively, an emerging view in the field is that DC control the course of tumor immunity/tolerance on at least three levels: (1) the developmental repertoire of DC lineage populations which can dictate the nature of DC response to a particular stimulus in the tumor microenvironment, (2) the maturation stage of DC when cells interact with other immune cells or respond to immunological signals (i.e., cytokines, chemokines and TLR ligands) and (3) the environment within which DC encounter tumor antigens, as defined by the tissue type, infiltrating leukocytes and inflammatory cytokine milieu (Table 3).

The importance of these issues and mechanisms controlling them are significant, as efforts to harness the power of DC in vaccination strategies against tumors would ultimately aim to identify the correct type of DC for a particular approach and insure that these cells are appropriately activated or protected from tumor influence to elicit the desired response.

5 Concluding Remarks

Numerous experimental and clinical observations discussed above suggest that tumor-induced apoptosis or altered differentiation and function of DC as well as accumulation of immature DC or DC precursors with inhibitory and tolerogenic function could impair antitumor immune responses. For patients with cancer, the resulting dysfunction of the DC system would result in marked deficiency in the induction of antitumor immunity, tumor progression and, probably, low response to immunotherapy. This is really important for understanding tumor immunopathology as well as re-evaluating tumor immunotherapeutic strategies since DC prepared from patients with cancer are being evaluated as cellular vaccine in multiple clinical trials worldwide. However, to date, DC-based immunotherapies have met limited success for several reasons, including the restricted longevity and efficacy of administered DC in suppressive tumor environment. Therefore, alternative approaches, including protection of DC longevity, blockade of tumor-mediated inhibitory pathways and prevention of DC dysfunction/polarization ex vivo, should be evaluated to potentiate the efficacy of DC-based cancer vaccines. Given that endogenous DC might be important for fulfilling the potential of various cellular vaccines, gained knowledge in the area of DC immunobiology in cancer may help to find new drugs to selectively block suppressive pathways and restore the original function of DC.

References

Aalamian, M., Pirtskhalaishvili, G., Nunez, A., Esche, C., Shurin, G. V., Huland, E., Huland, H., and Shurin, M. R. (2001). Human prostate cancer regulates generation and maturation of monocyte-derived dendritic cells. *Prostate* 46:68–75.

Aalamian, M., Tourkova, I. L., Chatta, G. S., Lilja, H., Huland, E., Huland, H., Shurin, G. V., and Shurin, M. R. (2003). Inhibition of dendropoiesis by tumor derived and purified prostate specific antigen. *J Urol* 170:2026–2030.

Alcalay, J., Goldberg, L. H., Wolf, J. E., Jr, and Kripke, M. L. (1989). Variations in the number and morphology of Langerhans' cells in the epidermal component of squamous cell carcinomas. *Arch Dermatol* 125:917–920.

Alcalay, J., and Kripke, M. L. (1991). Antigen-presenting activity of draining lymph node cells from mice painted with a contact allergen during ultraviolet carcinogenesis. *J Immunol* 146:1717–1721.

Avila-Moreno, F., Lopez-Gonzalez, J. S., Galindo-Rodriguez, G., Prado-Garcia, H., Bajana, S., and Sanchez-Torres, C. (2006). Lung squamous cell carcinoma and adenocarcinoma cell lines use different mediators to induce comparable phenotypic and functional changes in human monocyte-derived dendritic cells. *Cancer Immunol Immunother* 55:598–611.

Balkir, L., Tourkova, I. L., Makarenkova, V. P., Shurin, G. V., Robbins, P. D., Yin, X. M., Chatta, G., and Shurin, M. R. (2004). Comparative analysis of dendritic cells transduced with different anti-apoptotic molecules: sensitivity to tumor-induced apoptosis. *J Gene Med* 6:537–544.

Bauer, M. E. (2005). Stress, glucocorticoids and ageing of the immune system. *Stress* 8: 69–83.

Beckebaum, S., Zhang, X., Chen, X., Yu, Z., Frilling, A., Dworacki, G., Grosse-Wilde, H., Broelsch, C. E., Gerken, G., and Cicinnati, V. R. (2004). Increased levels of interleukin-10 in serum from patients with hepatocellular carcinoma correlate with profound numerical deficiencies and immature phenotype of circulating dendritic cell subsets. *Clin Cancer Res* 10: 7260–7269.

Becker, Y. (1992). Anticancer role of dendritic cells (DC) in human and experimental cancers—a review. *Anticancer Res* 12:511–520.

Becker, Y. (1993). Dendritic cell activity against primary tumors: an overview. *In Vivo* 7:187–191.

Bedoui, S., von Horsten, S., and Gebhardt, T. (2007). A role for neuropeptide Y (NPY) in phagocytosis: implications for innate and adaptive immunity. *Peptides* 28:373–376.

Bell, D., Chomarat, P., Broyles, D., Netto, G., Harb, G. M., Lebecque, S., Valladeau, J., Davoust, J., Palucka, K. A., and Banchereau, J. (1999). In breast carcinoma tissue, immature dendritic cells reside within the tumor, whereas mature dendritic cells are located in peritumoral areas. *J Exp Med* 190:1417–1426.

Belladonna, M. L., Grohmann, U., Guidetti, P., Volpi, C., Bianchi, R., Fioretti, M. C., Schwarcz, R., Fallarino, F., and Puccetti, P. (2006). Kynurenine pathway enzymes in dendritic cells initiate tolerogenesis in the absence of functional IDO. *J Immunol* 177:130–137.

Bellik, L., Gerlini, G., Parenti, A., Ledda, F., Pimpinelli, N., Neri, B., and Pantalone, D. (2006). Role of conventional treatments on circulating and monocyte-derived dendritic cells in colorectal cancer. *Clin Immunol* 121:74–80.

Bellone, G., Carbone, A., Smirne, C., Scirelli, T., Buffolino, A., Novarino, A., Stacchini, A., Bertetto, O., Palestro, G., Sorio, C., Scarpa, A., Emanuelli, G., and Rodeck, U. (2006). Cooperative induction of a tolerogenic dendritic cell phenotype by cytokines secreted by pancreatic carcinoma cells. *J Immunol* 177:3448–3460.

Brown, R. D., Pope, B., Murray, A., Esdale, W., Sze, D. M., Gibson, J., Ho, P. J., Hart, D., and Joshua, D. (2001). Dendritic cells from patients with myeloma are numerically normal but functionally defective as they fail to up-regulate CD80 (B7-1) expression after huCD40LT stimulation because of inhibition by transforming growth factor-beta1 and interleukin-10. *Blood* 98:2992–2998.

Cabillic, F., Bouet-Toussaint, F., Toutirais, O., Rioux-Leclercq, N., Fergelot, P., de la Pintiere, C. T., Genetet, N., Patard, J. J., and Catros-Quemener, V. (2006). Interleukin-6 and vascular endothe-

lial growth factor release by renal cell carcinoma cells impedes lymphocyte-dendritic cell crosstalk. *Clin Exp Immunol* 146:518–523.
Cao, M. D., Chen, Z. D., and Xing, Y. (2004). Gamma irradiation of human dendritic cells influences proliferation and cytokine profile of T cells in autologous mixed lymphocyte reaction. *Cell Biol Int* 28:223–228.
Capobianco, A., Rovere-Querini, P., Rugarli, C., and Manfredi, A. A. (2006). Melanoma cells interfere with the interaction of dendritic cells with NK/LAK cells. *Int J Cancer* 119: 2861–2869.
Carlos, C. A., Dong, H. F., Howard, O. M., Oppenheim, J. J., Hanisch, F. G., and Finn, O. J. (2005). Human tumor antigen MUC1 is chemotactic for immature dendritic cells and elicits maturation but does not promote Th1 type immunity. *J Immunol* 175:1628–1635.
Chang, C. C., Ciubotariu, R., Manavalan, J. S., Yuan, J., Colovai, A. I., Piazza, F., Lederman, S., Colonna, M., Cortesini, R., Dalla-Favera, R., and Suciu-Foca, N. (2002). Tolerization of dendritic cells by T(S) cells: the crucial role of inhibitory receptors ILT3 and ILT4. *Nat Immunol* 3:237–243.
Chaux, P., Moutet, M., Faivre, J., Martin, F., and Martin, M. (1996). Inflammatory cells infiltrating human colorectal carcinomas express HLA class II but not B7-1 and B7-2 costimulatory molecules of the T-cell activation. *Lab Invest* 74:975–983.
Colasante, A., Castrilli, G., Aiello, F. B., Brunetti, M., and Musiani, P. (1995). Role of cytokines in distribution and differentiation of dendritic cell/Langerhans' cell lineage in human primary carcinomas of the lung. *Hum Pathol* 26:866–872.
Conejo-Garcia, J. R., Benencia, F., Courreges, M. C., Kang, E., Mohamed-Hadley, A., Buckanovich, R. J., Holtz, D. O., Jenkins, A., Na, H., Zhang, L., Wagner, D. S., Katsaros, D., Caroll, R., and Coukos, G. (2004). Tumor-infiltrating dendritic cell precursors recruited by a beta-defensin contribute to vasculogenesis under the influence of Vegf-A. *Nat Med* 10:950–958.
Conejo-Garcia, J. R., Buckanovich, R. J., Benencia, F., Courreges, M. C., Rubin, S. C., Carroll, R. G., and Coukos, G. (2005). Vascular leukocytes contribute to tumor vascularization. *Blood* 105:679–681.
Corinti, S., Albanesi, C., la Sala, A., Pastore, S., and Girolomoni, G. (2001). Regulatory activity of autocrine IL-10 on dendritic cell functions. *J Immunol* 166:4312–4318.
Corrales, J. J., Almeida, M., Burgo, R., Mories, M. T., Miralles, J. M., and Orfao, A. (2006). Androgen-replacement therapy depresses the ex vivo production of inflammatory cytokines by circulating antigen-presenting cells in aging type-2 diabetic men with partial androgen deficiency. *J Endocrinol* 189:595–604.
Cox, K., North, M., Burke, M., Singhal, H., Renton, S., Aqel, N., Islam, S., and Knight, S. C. (2005). Plasmacytoid dendritic cells (PDC) are the major DC subset innately producing cytokines in human lymph nodes. *J Leukoc Biol* 78:1142–1152.
Curiel, T. J., Cheng, P., Mottram, P., Alvarez, X., Moons, L., Evdemon-Hogan, M., Wei, S., Zou, L., Kryczek, I., Hoyle, G., Lackner, A., Carmeliet, P., and Zou, W. (2004). Dendritic cell subsets differentially regulate angiogenesis in human ovarian cancer. *Cancer Res* 64:5535–5538.
Curiel, T. J., Wei, S., Dong, H., Alvarez, X., Cheng, P., Mottram, P., Krzysiek, R., Knutson, K. L., Daniel, B., Zimmermann, M. C., David, O., Burow, M., Gordon, A., Dhurandhar, N., Myers, L., Berggren, R., Hemminki, A., Alvarez, R. D., Emilie, D., Curiel, D. T., Chen, L., and Zou, W. (2003). Blockade of B7-H1 improves myeloid dendritic cell-mediated antitumor immunity. *Nat Med* 9:562–567.
Della Bella, S., Gennaro, M., Vaccari, M., Ferraris, C., Nicola, S., Riva, A., Clerici, M., Greco, M., and Villa, M. L. (2003). Altered maturation of peripheral blood dendritic cells in patients with breast cancer. *Br J Cancer* 89:1463–1472.
Della Porta, M., Danova, M., Rigolin, G. M., Brugnatelli, S., Rovati, B., Tronconi, C., Fraulini, C., Russo Rossi, A., Riccardi, A., and Castoldi, G. (2005). Dendritic cells and vascular endothelial growth factor in colorectal cancer: correlations with clinicobiological findings. *Oncology* 68:276–284.
Eisendle, K., Wolf, D., Gastl, G., and Kircher-Eibl, B. (2005). Dendritic cells from patients with chronic myeloid leukemia: functional and phenotypic features. *Leuk Lymphoma* 46:663–670.

Enk, A. H., Jonuleit, H., Saloga, J., and Knop, J. (1997). Dendritic cells as mediators of tumor-induced tolerance in metastatic melanoma. *Int J Cancer* 73:309–316.

Esche, C., Lokshin, A., Shurin, G. V., Gastman, B. R., Rabinowich, H., Watkins, S. C., Lotze, M. T., and Shurin, M. R. (1999). Tumor's other immune targets: dendritic cells. *J Leukoc Biol* 66: 336–344.

Esche, C., Shurin, G. V., Kirkwood, J. M., Wang, G. Q., Rabinowich, H., Pirtskhalaishvili, G., and Shurin, M. R. (2001). Tumor necrosis factor-alpha-promoted expression of Bcl-2 and inhibition of mitochondrial cytochrome c release mediate resistance of mature dendritic cells to melanoma-induced apoptosis. *Clin Cancer Res* 7:974s–979s.

Evel-Kabler, K., Song, X. T., Aldrich, M., Huang, X. F., and Chen, S. Y. (2006). SOCS1 restricts dendritic cells' ability to break self tolerance and induce antitumor immunity by regulating IL-12 production and signaling. *J Clin Invest* 116:90–100.

Feijoo, E., Alfaro, C., Mazzolini, G., Serra, P., Penuelas, I., Arina, A., Huarte, E., Tirapu, I., Palencia, B., Murillo, O., Ruiz, J., Sangro, B., Richter, J. A., Prieto, J., and Melero, I. (2005). Dendritic cells delivered inside human carcinomas are sequestered by interleukin-8. *Int J Cancer* 116:275–281.

Ferrari, S., Malugani, F., Rovati, B., Porta, C., Riccardi, A., and Danova, M. (2005). Flow cytometric analysis of circulating dendritic cell subsets and intracellular cytokine production in advanced breast cancer patients. *Oncol Rep* 14:113–120.

Frey, A. B. (2006). Myeloid suppressor cells regulate the adaptive immune response to cancer. *J Clin Invest* 116:2587–2590.

Fricke, I., and Gabrilovich, D. I. (2006). Dendritic cells and tumor microenvironment: a dangerous liaison. *Immunol Investig* 35:459–483.

Gabrilovich, D. (2004). Mechanisms and functional significance of tumour-induced dendritic-cell defects. *Nat Rev Immunol* 4:941–952.

Gabrilovich, D. I., Bronte, V., Chen, S. H., Colombo, M. P., Ochoa, A., Ostrand-Rosenberg, S., and Schreiber, H. (2007). The terminology issue for myeloid-derived suppressor cells. *Cancer Res* 67:425; author reply 426.

Gabrilovich, D. I., Chen, H. L., Girgis, K. R., Cunningham, H. T., Meny, G. M., Nadaf, S., Kavanaugh, D., and Carbone, D. P. (1996a). Production of vascular endothelial growth factor by human tumors inhibits the functional maturation of dendritic cells. *Nat Med* 2: 1096–1103.

Gabrilovich, D. I., Cheng, P., Fan, Y., Yu, B., Nikitina, E., Sirotkin, A., Shurin, M., Oyama, T., Adachi, Y., Nadaf, S., Carbone, D. P., and Skoultchi, A. I. (2002). H1(0) histone and differentiation of dendritic cells. A molecular target for tumor-derived factors. *J Leukoc Biol* 72: 285–296.

Gabrilovich, D. I., Nadaf, S., Corak, J., Berzofsky, J. A., and Carbone, D. P. (1996b). Dendritic cells in antitumor immune responses. II. Dendritic cells grown from bone marrow precursors, but not mature DC from tumor bearing mice, are effective antigen carriers in the therapy of established tumors. *Cell Immunol* 170:111–119.

Geissmann, F., Revy, P., Regnault, A., Lepelletier, Y., Dy, M., Brousse, N., Amigorena, S., Hermine, O., and Durandy, A. (1999). TGF-beta 1 prevents the noncognate maturation of human dendritic Langerhans cells. *J Immunol* 162:4567–4575.

Ghiringhelli, F., Apetoh, L., Housseau, F., Kroemer, G., and Zitvogel, L. (2007). Links between innate and cognate tumor immunity. *Curr Opin Immunol* 19:224–231.

Ghiringhelli, F., Puig, P. E., Roux, S., Parcellier, A., Schmitt, E., Solary, E., Kroemer, G., Martin, F., Chauffert, B., and Zitvogel, L. (2005). Tumor cells convert immature myeloid dendritic cells into TGF-beta-secreting cells inducing CD4+CD25+ regulatory T cell proliferation. *J Exp Med* 202:919–929.

Gottfried, E., Kunz-Schughart, L. A., Ebner, S., Mueller-Klieser, W., Hoves, S., Andreesen, R., Mackensen, A., and Kreutz, M. (2006). Tumor-derived lactic acid modulates dendritic cell activation and antigen expression. *Blood* 107:2013–2021.

Halliday, G. M., Lucas, A. D., and Barnetson, R. S. (1992). Control of Langerhans' cell density by a skin tumour-derived cytokine. *Immunology* 77:13–18.

Halliday, G. M., Reeve, V. E., and Barnetson, R. S. (1991). Langerhans cell migration into ultraviolet light-induced squamous skin tumors is unrelated to anti-tumor immunity. *J Invest Dermatol* 97:830–834.

Handley, M. E., Thakker, M., Pollara, G., Chain, B. M., and Katz, D. R. (2005). JNK activation limits dendritic cell maturation in response to reactive oxygen species by the induction of apoptosis. *Free Radic Biol Med* 38:1637–1652.

Hasebe, H., Nagayama, H., Sato, K., Enomoto, M., Takeda, Y., Takahashi, T. A., Hasumi, K., and Eriguchi, M. (2000). Dysfunctional regulation of the development of monocyte-derived dendritic cells in cancer patients. *Biomed Pharmacother* 54:291–298.

Hoffmann, T. K., Muller-Berghaus, J., Ferris, R. L., Johnson, J. T., Storkus, W. J., and Whiteside, T. L. (2002). Alterations in the frequency of dendritic cell subsets in the peripheral circulation of patients with squamous cell carcinomas of the head and neck. *Clin Cancer Res* 8:1787–1793.

Hou, D. Y., Muller, A. J., Sharma, M. D., DuHadaway, J., Banerjee, T., Johnson, M., Mellor, A. L., Prendergast, G. C., and Munn, D. H. (2007). Inhibition of indoleamine 2,3-dioxygenase in dendritic cells by stereoisomers of 1-methyl-tryptophan correlates with antitumor responses. *Cancer Res* 67:792–801.

Houot, R., Perrot, I., Garcia, E., Durand, I., and Lebecque, S. (2006). Human CD4+CD25high regulatory T cells modulate myeloid but not plasmacytoid dendritic cells activation. *J Immunol* 176:5293–5298.

Huang, B., Lei, Z., Zhao, J., Gong, W., Liu, J., Chen, Z., Liu, Y., Li, D., Yuan, Y., Zhang, G. M., and Feng, Z. H. (2007). CCL2/CCR2 pathway mediates recruitment of myeloid suppressor cells to cancers. *Cancer Lett* 252:86–92.

Huang, B., Pan, P. Y., Li, Q., Sato, A. I., Levy, D. E., Bromberg, J., Divino, C. M., and Chen, S. H. (2006). Gr-1+CD115+ immature myeloid suppressor cells mediate the development of tumor-induced T regulatory cells and T-cell anergy in tumor-bearing host. *Cancer Res* 66:1123–1131.

Ishida, T., Oyama, T., Carbone, D. P., and Gabrilovich, D. I. (1998). Defective function of Langerhans cells in tumor-bearing animals is the result of defective maturation from hemopoietic progenitors. *J Immunol* 161:4842–4851.

Ito, M., Minamiya, Y., Kawai, H., Saito, S., Saito, H., Nakagawa, T., Imai, K., Hirokawa, M., and Ogawa, J. (2006). Tumor-derived TGFbeta-1 induces dendritic cell apoptosis in the sentinel lymph node. *J Immunol* 176:5637–5643.

Ito, T., Yang, M., Wang, Y. H., Lande, R., Gregorio, J., Perng, O. A., Qin, X. F., Liu, Y. J., and Gilliet, M. (2007). Plasmacytoid dendritic cells prime IL-10-producing T regulatory cells by inducible costimulator ligand. *J Exp Med* 204:105–115.

Jackson, S. H., Yu, C. R., Mahdi, R. M., Ebong, S., and Egwuagu, C. E. (2004). Dendritic cell maturation requires STAT1 and is under feedback regulation by suppressors of cytokine signaling. *J Immunol* 172:2307–2315.

Kanto, T., Kalinski, P., Hunter, O. C., Lotze, M. T., and Amoscato, A. A. (2001). Ceramide mediates tumor-induced dendritic cell apoptosis. *J Immunol* 167:3773–3784.

Katsenelson, N. S., Shurin, G. V., Bykovskaia, S. N., Shogan, J., and Shurin, M. R. (2001). Human small cell lung carcinoma and carcinoid tumor regulate dendritic cell maturation and function. *Mod Pathol* 14:40–45.

Kichler-Lakomy, C., Budinsky, A. C., Wolfram, R., Hellan, M., Wiltschke, C., Brodowicz, T., Viernstein, H., and Zielinski, C. C. (2006). Deficiences in phenotype expression and function of dendritic cells from patients with early breast cancer. *Eur J Med Res* 11:7–12.

Kiertscher, S. M., Luo, J., Dubinett, S. M., and Roth, M. D. (2000). Tumors promote altered maturation and early apoptosis of monocyte-derived dendritic cells. *J Immunol* 164:1269–1276.

Kim, R., Emi, M., and Tanabe, K. (2006a). Functional roles of immature dendritic cells in impaired immunity of solid tumour and their targeted strategies for provoking tumour immunity. *Clin Exp Immunol* 146:189–196.

Kim, R., Emi, M., Tanabe, K., and Arihiro, K. (2006b). Tumor-driven evolution of immunosuppressive networks during malignant progression. *Cancer Res* 66:5527–5536.

Kim, R., Emi, M., Tanabe, K., and Arihiro, K. (2007). Potential functional role of plasmacytoid dendritic cells in cancer immunity. *Immunology* 121:149–157.
Lambert, R. W., and Granstein, R. D. (1998). Neuropeptides and Langerhans cells. *Exp Dermatol* 7:73–80.
Lan, Y. Y., Wang, Z., Raimondi, G., Wu, W., Colvin, B. L., de Creus, A., and Thomson, A. W. (2006). "Alternatively activated" dendritic cells preferentially secrete IL-10, expand Foxp3+CD4+ T cells, and induce long-term organ allograft survival in combination with CTLA4-Ig. *J Immunol* 177:5868–5877.
Larmonier, N., Marron, M., Zeng, Y., Cantrell, J., Romanoski, A., Sepassi, M., Thompson, S., Chen, X., Andreansky, S., and Katsanis, E. (2007). Tumor-derived CD4(+)CD25(+) regulatory T cell suppression of dendritic cell function involves TGF-beta and IL-10. *Cancer Immunol Immunother* 56:48–59.
Lemaoult, J., Caumartin, J., Daouya, M., Favier, B., Rond, S. L., Gonzalez, A., and Carosella, E. D. (2007). Immune regulation by pretenders: cell-to-cell transfers of HLA-G make effector T cells act as regulatory cells. *Blood* 109:2040–2048.
Lissoni, P., Bolis, S., Mandala, M., Viviani, S., Pogliani, E., and Barni, S. (1999). Blood concentrations of tumor necrosis factor-alpha in malignant lymphomas and their decrease as a predictor of disease control in response to low-dose subcutaneous immunotherapy with interleukin-2. *Int J Biol Markers* 14:167–171.
Lizee, G., Radvanyi, L. G., Overwijk, W. W., and Hwu, P. (2006). Improving antitumor immune responses by circumventing immunoregulatory cells and mechanisms. *Clin Cancer Res* 12:4794–4803.
Lopez, A. S., Alegre, E., LeMaoult, J., Carosella, E., and Gonzalez, A. (2006). Regulatory role of tryptophan degradation pathway in HLA-G expression by human monocyte-derived dendritic cells. *Mol Immunol* 43:2151–2160.
Lopez-Lazaro, M. (2007). Excessive superoxide anion generation plays a key role in carcinogenesis. *Int J Cancer* 120:1378–1380.
Maecker, B., Mougiakakos, D., Zimmermann, M., Behrens, M., Hollander, S., Schrauder, A., Schrappe, M., Welte, K., and Klein, C. (2006). Dendritic cell deficiencies in pediatric acute lymphoblastic leukemia patients. *Leukemia* 20:645–649.
Maestroni, G. J. (2000). Dendritic cell migration controlled by alpha 1b-adrenergic receptors. *J Immunol* 165:6743–6747.
Maestroni, G. J. (2005). Adrenergic modulation of dendritic cells function: relevance for the immune homeostasis. *Curr Neurovasc Res* 2:169–173.
Maestroni, G. J., and Mazzola, P. (2003). Langerhans cells beta 2-adrenoceptors: role in migration, cytokine production, Th priming and contact hypersensitivity. *J Neuroimmunol* 144:91–99.
Mahnke, K., and Enk, A. H. (2005). Dendritic cells: key cells for the induction of regulatory T cells? *Curr Top Microbiol Immunol* 293:133–150.
Makarenkova, V. P., Esche, C., Kost, N. V., Shurin, G. V., Rabin, B. S., Zozulya, A. A., and Shurin, M. R. (2001). Identification of delta- and mu-type opioid receptors on human and murine dendritic cells. *J Neuroimmunol* 117:68–77.
Makarenkova, V. P., Shurin, G. V., Tourkova, I. L., Balkir, L., Pirtskhalaishvili, G., Perez, L., Gerein, V., Siegfried, J. M., and Shurin, M. R. (2003). Lung cancer-derived bombesin-like peptides down-regulate the generation and function of human dendritic cells. *J Neuroimmunol* 145:55–67.
Marriott, I., and Bost, K. L. (2001). Expression of authentic substance P receptors in murine and human dendritic cells. *J Neuroimmunol* 114:131–141.
Matera, L., Mori, M., and Galetto, A. (2001). Effect of prolactin on the antigen presenting function of monocyte-derived dendritic cells. *Lupus* 10:728–734.
Melichar, B., Savary, C., Kudelka, A. P., Verschraegen, C., Kavanagh, J. J., Edwards, C. L., Platsoucas, C. D., and Freedman, R. S. (1998). Lineage-negative human leukocyte antigen-DR+ cells with the phenotype of undifferentiated dendritic cells in patients with carcinoma of the abdomen and pelvis. *Clin Cancer Res* 4:799–809.

Mellor, A. L., Baban, B., Chandler, P. R., Manlapat, A., Kahler, D. J., and Munn, D. H. (2005). Cutting edge: CpG oligonucleotides induce splenic CD19+ dendritic cells to acquire potent indoleamine 2,3-dioxygenase-dependent T cell regulatory functions via IFN Type 1 signaling. *J Immunol* 175:5601–5605.

Menetrier-Caux, C., Montmain, G., Dieu, M. C., Bain, C., Favrot, M. C., Caux, C., and Blay, J. Y. (1998). Inhibition of the differentiation of dendritic cells from CD34(+) progenitors by tumor cells: role of interleukin-6 and macrophage colony-stimulating factor. *Blood* 92: 4778–4791.

Mihalyo, M. A., Hagymasi, A. T., Slaiby, A. M., Nevius, E. E., and Adler, A. J. (2007). Dendritic cells program non-immunogenic prostate-specific T cell responses beginning at early stages of prostate tumorigenesis. *Prostate* 67:536–546.

Mouillot, G., Marcou, C., Zidi, I., Guillard, C., Sangrouber, D., Carosella, E. D., and Moreau, P. (2007). Hypoxia modulates HLA-G gene expression in tumor cells. *Hum Immunol* 68:277–285.

Munn, D. H. (2006). Indoleamine 2,3-dioxygenase, tumor-induced tolerance and counter-regulation. *Curr Opin Immunol* 18:220–225.

Munn, D. H., and Mellor, A. L. (2006). The tumor-draining lymph node as an immune-privileged site. *Immunol Rev* 213:146–158.

Nefedova, Y., and Gabrilovich, D. I. (2007). Targeting of Jak/STAT pathway in antigen presenting cells in cancer. *Curr Cancer Drug Targets* 7:71–77.

Nefedova, Y., Huang, M., Kusmartsev, S., Bhattacharya, R., Cheng, P., Salup, R., Jove, R., and Gabrilovich, D. (2004). Hyperactivation of STAT3 is involved in abnormal differentiation of dendritic cells in cancer. *J Immunol* 172:464–474.

Nestle, F. O., Burg, G., Fah, J., Wrone-Smith, T., and Nickoloff, B. J. (1997). Human sunlight-induced basal-cell-carcinoma-associated dendritic cells are deficient in T cell co-stimulatory molecules and are impaired as antigen-presenting cells. *Am J Pathol* 150:641–651.

Neves, A. R., Ensina, L. F., Anselmo, L. B., Leite, K. R., Buzaid, A. C., Camara-Lopes, L. H., and Barbuto, J. A. (2005). Dendritic cells derived from metastatic cancer patients vaccinated with allogeneic dendritic cell-autologous tumor cell hybrids express more CD86 and induce higher levels of interferon-gamma in mixed lymphocyte reactions. *Cancer Immunol Immunother* 54:61–66.

Ninomiya, T., Akbar, S. M., Masumoto, T., Horiike, N., and Onji, M. (1999). Dendritic cells with immature phenotype and defective function in the peripheral blood from patients with hepatocellular carcinoma. *J Hepatol* 31:323–331.

Ochoa, A. C., Zea, A. H., Hernandez, C., and Rodriguez, P. C. (2007). Arginase, prostaglandins, and myeloid-derived suppressor cells in renal cell carcinoma. *Clin Cancer Res* 13:721s–726s.

Ogden, A. T., Horgan, D., Waziri, A., Anderson, D., Louca, J., McKhann, G. M., Sisti, M. B., Parsa, A. T., and Bruce, J. N. (2006). Defective receptor expression and dendritic cell differentiation of monocytes in glioblastomas. *Neurosurgery* 59:902–909; discussion 909–910.

Onishi, H., Morisaki, T., Baba, E., Kuga, H., Kuroki, H., Matsumoto, K., Tanaka, M., and Katano, M. (2002). Dysfunctional and short-lived subsets in monocyte-derived dendritic cells from patients with advanced cancer. *Clin Immunol* 105:286–295.

Ormandy, L. A., Farber, A., Cantz, T., Petrykowska, S., Wedemeyer, H., Horning, M., Lehner, F., Manns, M. P., Korangy, F., and Greten, T. F. (2006). Direct ex vivo analysis of dendritic cells in patients with hepatocellular carcinoma. *World J Gastroenterol* 12:3275–3282.

Park, S. J., Nakagawa, T., Kitamura, H., Atsumi, T., Kamon, H., Sawa, S., Kamimura, D., Ueda, N., Iwakura, Y., Ishihara, K., Murakami, M., and Hirano, T. (2004). IL-6 regulates in vivo dendritic cell differentiation through STAT3 activation. *J Immunol* 173:3844–3854.

Pedersen, A. E., Thorn, M., Gad, M., Walter, M. R., Johnsen, H. E., Gaarsdal, E., Nikolajsen, K., Buus, S., Claesson, M. H., and Svane, I. M. (2005). Phenotypic and functional characterization of clinical grade dendritic cells generated from patients with advanced breast cancer for therapeutic vaccination. *Scand J Immunol* 61:147–156.

Peguet-Navarro, J., Sportouch, M., Popa, I., Berthier, O., Schmitt, D., and Portoukalian, J. (2003). Gangliosides from human melanoma tumors impair dendritic cell differentiation from monocytes and induce their apoptosis. *J Immunol* 170:3188–3194.

Perrot, I., Blanchard, D., Freymond, N., Isaac, S., Guibert, B., Pacheco, Y., and Lebecque, S. (2007). Dendritic cells infiltrating human non-small cell lung cancer are blocked at immature stage. *J Immunol* 178:2763–2769.
Piemonti, L., Monti, P., Allavena, P., Sironi, M., Soldini, L., Leone, B. E., Socci, C., and Di Carlo, V. (1999). Glucocorticoids affect human dendritic cell differentiation and maturation. *J Immunol* 162:6473–6481.
Pinzon-Charry, A., Ho, C. S., Laherty, R., Maxwell, T., Walker, D., Gardiner, R. A., O'Connor, L., Pyke, C., Schmidt, C., Furnival, C., and Lopez, J. A. (2005a). A population of HLA-DR+ immature cells accumulates in the blood dendritic cell compartment of patients with different types of cancer. *Neoplasia* 7:1112–1122.
Pinzon-Charry, A., Maxwell, T., and Lopez, J. A. (2005b). Dendritic cell dysfunction in cancer: a mechanism for immunosuppression. *Immunol Cell Biol* 83:451–461.
Pinzon-Charry, A., Maxwell, T., McGuckin, M. A., Schmidt, C., Furnival, C., and Lopez, J. A. (2006). Spontaneous apoptosis of blood dendritic cells in patients with breast cancer. *Breast Cancer Res* 8:R5.
Pirtskhalaishvili, G., Shurin, G. V., Esche, C., Cai, Q., Salup, R. R., Bykovskaia, S. N., Lotze, M. T., and Shurin, M. R. (2000a). Cytokine-mediated protection of human dendritic cells from prostate cancer-induced apoptosis is regulated by the Bcl-2 family of proteins. *Br J Cancer* 83:506–513.
Pirtskhalaishvili, G., Shurin, G. V., Gambotto, A., Esche, C., Wahl, M., Yurkovetsky, Z. R., Robbins, P. D., and Shurin, M. R. (2000b). Transduction of dendritic cells with Bcl-xL increases their resistance to prostate cancer-induced apoptosis and antitumor effect in mice. *J Immunol* 165:1956–1964.
Ratta, M., Fagnoni, F., Curti, A., Vescovini, R., Sansoni, P., Oliviero, B., Fogli, M., Ferri, E., Della Cuna, G. R., Tura, S., Baccarani, M., and Lemoli, R. M. (2002). Dendritic cells are functionally defective in multiple myeloma: the role of interleukin-6. *Blood* 100:230–237.
Remmel, E., Terracciano, L., Noppen, C., Zajac, P., Heberer, M., Spagnoli, G. C., and Padovan, E. (2001). Modulation of dendritic cell phenotype and mobility by tumor cells in vitro. *Hum Immunol* 62:39–49.
Ristich, V., Liang, S., Zhang, W., Wu, J., and Horuzsko, A. (2005). Tolerization of dendritic cells by HLA-G. *Eur J Immunol* 35:1133–1142.
Rughetti, A., Pellicciotta, I., Biffoni, M., Backstrom, M., Link, T., Bennet, E. P., Clausen, H., Noll, T., Hansson, G. C., Burchell, J. M., Frati, L., Taylor-Papadimitriou, J., and Nuti, M. (2005). Recombinant tumor-associated MUC1 glycoprotein impairs the differentiation and function of dendritic cells. *J Immunol* 174:7764–7772.
Rutella, S., Danese, S., and Leone, G. (2006). Tolerogenic dendritic cells: cytokine modulation comes of age. *Blood* 108:1435–1440.
Saint-Mezard, P., Chavagnac, C., Bosset, S., Ionescu, M., Peyron, E., Kaiserlian, D., Nicolas, J. F., and Berard, F. (2003). Psychological stress exerts an adjuvant effect on skin dendritic cell functions in vivo. *J Immunol* 171:4073–4080.
Sakakura, K., Chikamatsu, K., Takahashi, K., Whiteside, T. L., and Furuya, N. (2006). Maturation of circulating dendritic cells and imbalance of T-cell subsets in patients with squamous cell carcinoma of the head and neck. *Cancer Immunol Immunother* 55:151–159.
Sato, K., Yamashita, N., Baba, M., and Matsuyama, T. (2003). Regulatory dendritic cells protect mice from murine acute graft-versus-host disease and leukemia relapse. *Immunity* 18:367–379.
Satthaporn, S., Robins, A., Vassanasiri, W., El-Sheemy, M., Jibril, J. A., Clark, D., Valerio, D., and Eremin, O. (2004). Dendritic cells are dysfunctional in patients with operable breast cancer. *Cancer Immunol Immunother* 53:510–518.
Savary, C. A., Grazziutti, M. L., Melichar, B., Przepiorka, D., Freedman, R. S., Cowart, R. E., Cohen, D. M., Anaissie, E. J., Woodside, D. G., McIntyre, B. W., Pierson, D. L., Pellis, N. R., and Rex, J. H. (1998). Multidimensional flow-cytometric analysis of dendritic cells in peripheral blood of normal donors and cancer patients. *Cancer Immunol Immunother* 45:234–240.

Schmidt, J., Jager, D., Hoffmann, K., Buchler, M. W., and Marten, A. (2007). Impact of interferon-alpha in combined chemoradioimmunotherapy for pancreatic adenocarcinoma (CapRI): first data from the immunomonitoring. *J Immunother* 30:108–115.

Schwaab, T., Schned, A. R., Heaney, J. A., Cole, B. F., Atzpodien, J., Wittke, F., and Ernstoff, M. S. (1999). In vivo description of dendritic cells in human renal cell carcinoma. *J Urol* 162: 567–573.

Seiffert, K., and Granstein, R. D. (2006). Neuroendocrine regulation of skin dendritic cells. *Ann N Y Acad Sci* 1088:195–206.

Serafini, P., Borrello, I., and Bronte, V. (2006). Myeloid suppressor cells in cancer: recruitment, phenotype, properties, and mechanisms of immune suppression. *Semin Cancer Biol* 16: 53–65.

Sharma, S., Stolina, M., Yang, S. C., Baratelli, F., Lin, J. F., Atianzar, K., Luo, J., Zhu, L., Lin, Y., Huang, M., Dohadwala, M., Batra, R. K., and Dubinett, S. M. (2003). Tumor cyclooxygenase 2-dependent suppression of dendritic cell function. *Clin Cancer Res* 9:961–968.

Shurin, G. V., Aalamian, M., Pirtskhalaishvili, G., Bykovskaia, S., Huland, E., Huland, H., and Shurin, M. R. (2001a). Human prostate cancer blocks the generation of dendritic cells from CD34+ hematopoietic progenitors. *Eur Urol* 39(Suppl 4):37–40.

Shurin, G. V., Ferris, R. L., Tourkova, I. L., Perez, L., Lokshin, A., Balkir, L., Collins, B., Chatta, G. S., and Shurin, M. R. (2005). Loss of new chemokine CXCL14 in tumor tissue is associated with low infiltration by dendritic cells (DC), while restoration of human CXCL14 expression in tumor cells causes attraction of DC both in vitro and in vivo. *J Immunol* 174:5490–5498.

Shurin, G. V., Shurin, M. R., Bykovskaia, S., Shogan, J., Lotze, M. T., and Barksdale, E. M., Jr (2001b). Neuroblastoma-derived gangliosides inhibit dendritic cell generation and function. *Cancer Res* 61:363–369.

Shurin, G. V., Yurkovetsky, Z. R., and Shurin, M. R. (2003). Tumor-induced dendritic cell dysfunction. In: A. Ochoa (ed.) *Mechanisms of Tumor Escape from the Immune Response*. London: Taylor and Francis, pp. 112–138.

Shurin, M. R. (1999). Regulation of dendropoiesis in cancer. *Clin Immunol Newslett* 19:135–139.

Shurin, M. R., Chatta, G. S., and Shurin, G. V. (2007). Aging and the dendritic cell system: implications for cancer. *Crit Rev Oncol Hematol.* 64:90–105.

Shurin, M. R., and Gabrilovich, D. I. (2001). Regulation of dendritic cell system by tumor. *Cancer Res Ther Control* 11:65–78.

Shurin, M. R., Shurin, G. V., Lokshin, A., Yurkovetsky, Z. R., Gutkin, D. W., Chatta, G., Zhong, H., Han, B., and Ferris, R. L. (2006). Intratumoral cytokines/chemokines/growth factors and tumor infiltrating dendritic cells: friends or enemies? *Cancer Metastasis Rev* 25:333–356.

Shurin, M. R., Yurkovetsky, Z. R., Tourkova, I. L., Balkir, L., and Shurin, G. V. (2002). Inhibition of CD40 expression and CD40-mediated dendritic cell function by tumor-derived IL-10. *Int J Cancer* 101:61–68.

Sombroek, C. C., Stam, A. G., Masterson, A. J., Lougheed, S. M., Schakel, M. J., Meijer, C. J., Pinedo, H. M., van den Eertwegh, A. J., Scheper, R. J., and de Gruijl, T. D. (2002). Prostanoids play a major role in the primary tumor-induced inhibition of dendritic cell differentiation. *J Immunol* 168:4333–4343.

Song, E. Y., Shurin, M. R., Tourkova, I. L., Chatta, G., and Shurin, G. V. (2004). Human renal cell carcinoma inhibits dendritic cell maturation and functions. *Urologe A* 43(Suppl 3):128–130.

Stanford, A., Chen, Y., Zhang, X. R., Hoffman, R., Zamora, R., and Ford, H. R. (2001). Nitric oxide mediates dendritic cell apoptosis by downregulating inhibitors of apoptosis proteins and upregulating effector caspase activity. *Surgery* 130:326–332.

Starnes, T., Rasila, K. K., Robertson, M. J., Brahmi, Z., Dahl, R., Christopherson, K., and Hromas, R. (2006). The chemokine CXCL14 (BRAK) stimulates activated NK cell migration: implications for the downregulation of CXCL14 in malignancy. *Exp Hematol* 34:1101–1105.

Stene, M. A., Babajanians, M., Bhuta, S., and Cochran, A. J. (1988). Quantitative alterations in cutaneous Langerhans cells during the evolution of malignant melanoma of the skin. *J Invest Dermatol* 91:125–128.

Suciu-Foca, N., Manavalan, J. S., Scotto, L., Kim-Schulze, S., Galluzzo, S., Naiyer, A. J., Fan, J., Vlad, G., and Cortesini, R. (2005). Molecular characterization of allospecific T suppressor and tolerogenic dendritic cells: review. *Int Immunopharmacol* 5:7–11.

Szatrowski, T. P., and Nathan, C. F. (1991). Production of large amounts of hydrogen peroxide by human tumor cells. *Cancer Res* 51:794–798.

Taieb, J., Chaput, N., Menard, C., Apetoh, L., Ullrich, E., Bonmort, M., Pequignot, M., Casares, N., Terme, M., Flament, C., Opolon, P., Lecluse, Y., Metivier, D., Tomasello, E., Vivier, E., Ghiringhelli, F., Martin, F., Klatzmann, D., Poynard, T., Tursz, T., Raposo, G., Yagita, H., Ryffel, B., Kroemer, G., and Zitvogel, L. (2006). A novel dendritic cell subset involved in tumor immunosurveillance. *Nat Med* 12:214–219.

Takahashi, K., Toyokawa, H., Takai, S., Satoi, S., Yanagimoto, H., Terakawa, N., Araki, H., Kwon, A. H., and Kamiyama, Y. (2006). Surgical influence of pancreatectomy on the function and count of circulating dendritic cells in patients with pancreatic cancer. *Cancer Immunol Immunother* 55:775–784.

Tarazona, R., Solana, R., Ouyang, Q., and Pawelec, G. (2002). Basic biology and clinical impact of immunosenescence. *Exp. Gerontol* 37:183–189.

Tas, M. P., Simons, P. J., Balm, F. J., and Drexhage, H. A. (1993). Depressed monocyte polarization and clustering of dendritic cells in patients with head and neck cancer: in vitro restoration of this immunosuppression by thymic hormones. *Cancer Immunol Immunother* 36:108–114.

Thomachot, M. C., Bendriss-Vermare, N., Massacrier, C., Biota, C., Treilleux, I., Goddard, S., Caux, C., Bachelot, T., Blay, J. Y., and Menetrier-Caux, C. (2004). Breast carcinoma cells promote the differentiation of CD34+ progenitors towards 2 different subpopulations of dendritic cells with CD1a(high)CD86(−)Langerin− and CD1a(+)CD86(+)Langerin+ phenotypes. *Int J Cancer* 110:710–720.

Tourkova, I. L., Shurin, G. V., Chatta, G. S., Perez, L., Finke, J., Whiteside, T. L., Ferrone, S., and Shurin, M. R. (2005). Restoration by IL-15 of MHC class I antigen-processing machinery in human dendritic cells inhibited by tumor-derived gangliosides. *J Immunol* 175:3045–3052.

Tourkova, I. L., Shurin, G. V., Wei, S., and Shurin, M. R. (2007). Small Rho GTPases mediate tumor-induced Inhibition of endocytic activity of dendritic cells. *J Immunol* 178:7787–7793.

Tourkova, I. L., Yamabe, K., Foster, B., Chatta, G., Perez, L., Shurin, G. V., and Shurin, M. R. (2004). Murine prostate cancer inhibits both in vivo and in vitro generation of dendritic cells from bone marrow precursors. *Prostate* 59:203–213.

Ueno, A., Cho, S., Cheng, L., Wang, J., Hou, S., Nakano, H., Santamaria, P., and Yang, Y. (2007). Transient upregulation of IDO in dendritic cells by human chorionic gonadotropin downregulates autoimmune diabetes. *Diabetes* 56:1686–1693.

Ureta, G., Osorio, F., Morales, J., Rosemblatt, M., Bono, M. R., and Fierro, J. A. (2007). Generation of dendritic cells with regulatory properties. *Transplant Proc* 39:633–637.

Vakkila, J., Thomson, A. W., Vettenranta, K., Sariola, H., and Saarinen-Pihkala, U. M. (2004). Dendritic cell subsets in childhood and in children with cancer: relation to age and disease prognosis. *Clin Exp Immunol* 135:455–461.

Valenti, R., Huber, V., Filipazzi, P., Pilla, L., Sovena, G., Villa, A., Corbelli, A., Fais, S., Parmiani, G., and Rivoltini, L. (2006). Human tumor-released microvesicles promote the differentiation of myeloid cells with transforming growth factor-beta-mediated suppressive activity on T lymphocytes. *Cancer Res* 66:9290–9298.

Valenti, R., Huber, V., Iero, M., Filipazzi, P., Parmiani, G., and Rivoltini, L. (2007). Tumor-released microvesicles as vehicles of immunosuppression. *Cancer Res* 67:2912–2915.

Vicari, A. P., Caux, C., and Trinchieri, G. (2002). Tumour escape from immune surveillance through dendritic cell inactivation. *Semin Cancer Biol* 12:33–42.

Walker, S. R., Aboka, A., Ogagan, P. D., and Barksdale, E. M., Jr (2005). Murine neuroblastoma attenuates dendritic cell cysteine cysteine receptor 7 (CCR7) expression. *J Pediatr Surg* 40: 983–987.

Walker, S. R., Ogagan, P. D., DeAlmeida, D., Aboka, A. M., and Barksdale, E. M., Jr (2006). Neuroblastoma impairs chemokine-mediated dendritic cell migration in vitro. *J Pediatr Surg* 41:260–265.

Wang, S., Yang, J., Qian, J., Wezeman, M., Kwak, L. W., and Yi, Q. (2006). Tumor evasion of the immune system: inhibiting p38 MAPK signaling restores the function of dendritic cells in multiple myeloma. *Blood* 107:2432–2439.

Wang, T., Niu, G., Kortylewski, M., Burdelya, L., Shain, K., Zhang, S., Bhattacharya, R., Gabrilovich, D., Heller, R., Coppola, D., Dalton, W., Jove, R., Pardoll, D., and Yu, H. (2004). Regulation of the innate and adaptive immune responses by Stat-3 signaling in tumor cells. *Nat Med* 10:48–54.

Weber, F., Byrne, S. N., Le, S., Brown, D. A., Breit, S. N., Scolyer, R. A., and Halliday, G. M. (2005). Transforming growth factor-beta1 immobilises dendritic cells within skin tumours and facilitates tumour escape from the immune system. *Cancer Immunol Immunother* 54: 898–906.

Wei, S., Kryczek, I., Zou, L., Daniel, B., Cheng, P., Mottram, P., Curiel, T., Lange, A., and Zou, W. (2005). Plasmacytoid dendritic cells induce CD8+ regulatory T cells in human ovarian carcinoma. *Cancer Res* 65:5020–5026.

Wertel, F., Polak, G., Rolinski, J., Barczynski, B., and Kotarski, J. (2006). Myeloid and lymphoid dendritic cells in the peritoneal fluid of women with ovarian cancer. *Adv Med Sci* 51:174–177.

Whiteside, T. L., Stanson, J., Shurin, M. R., and Ferrone, S. (2004). Antigen-processing machinery in human dendritic cells: up-regulation by maturation and down-regulation by tumor cells. *J Immunol* 173:1526–1534.

Wilczynski, J. R. (2006). Cancer and pregnancy share similar mechanisms of immunological escape. *Chemotherapy* 52:107–110.

Wojas, K., Tabarkiewicz, J., Jankiewicz, M., and Rolinski, J. (2004). Dendritic cells in peripheral blood of patients with breast and lung cancer—a pilot study. *Folia Histochem Cytobiol* 42: 45–48.

Yanagimoto, H., Takai, S., Satoi, S., Toyokawa, H., Takahashi, K., Terakawa, N., Kwon, A. H., and Kamiyama, Y. (2005). Impaired function of circulating dendritic cells in patients with pancreatic cancer. *Clin Immunol* 114:52–60.

Yang, A. S., and Lattime, E. C. (2003). Tumor-induced interleukin 10 suppresses the ability of splenic dendritic cells to stimulate CD4 and CD8 T-cell responses. *Cancer Res* 63:2150–2157.

Yang, L., and Carbone, D. P. (2004). Tumor–host immune interactions and dendritic cell dysfunction. *Adv Cancer Res* 92:13–27.

Yang, L., Yamagata, N., Yadav, R., Brandon, S., Courtney, R. L., Morrow, J. D., Shyr, Y., Boothby, M., Joyce, S., Carbone, D. P., and Breyer, R. M. (2003). Cancer-associated immunodeficiency and dendritic cell abnormalities mediated by the prostaglandin EP2 receptor. *J Clin Invest* 111:727–735.

Yang, T., Witham, T. F., Villa, L., Erff, M., Attanucci, J., Watkins, S., Kondziolka, D., Okada, H., Pollack, I. F., and Chambers, W. H. (2002). Glioma-associated hyaluronan induces apoptosis in dendritic cells via inducible nitric oxide synthase: implications for the use of dendritic cells for therapy of gliomas. *Cancer Res* 62:2583–2591.

Zou, W., Machelon, V., Coulomb-L'Hermin, A., Borvak, J., Nome, F., Isaeva, T., Wei, S., Krzysiek, R., Durand-Gasselin, I., Gordon, A., Pustilnik, T., Curiel, D. T., Galanaud, P., Capron, F., Emilie, D., and Curiel, T. J. (2001). Stromal-derived factor-1 in human tumors recruits and alters the function of plasmacytoid precursor dendritic cells. *Nat Med* 7: 1339–1346.

Macrophages and Tumor Development

Suzanne Ostrand-Rosenberg and Pratima Sinha

1 Introduction

In 1882–1886 the Ukrainian biologist Elie Metchnikoff observed that if he inserted dye particles or small splinters into bipinnaria starfish, cells of the starfish would respond by engulfing the foreign materials. He named these cells "phagocytes", from the Greek "makros" (large) and "phagein" (to eat). Metchnikoff went on to describe similar phagocytic cells in the blood of humans, and in 1892 proposed his "cellular (phagocytic) theory of immunity" which stated that white blood cells were critical elements of the immune system which protected individuals from invading pathogenic organisms (Metchnikoff, 1905; Tauber and Chernyak, 1991). At the time, this theory was quite controversial since the prevailing concept was the "humoral" theory. According to the humoral theory, immunity was provided exclusively by body fluids and soluble substances such as antibodies. Despite the controversy surrounding the cellular theory, Metchnikoff's discovery of phagocytic cells was recognized in 1908 by a Nobel Prize in Physiology or Medicine. The co-recipient of that 1908 Nobel Prize was the German microbiologist, Paul Ehrlich, who was recognized for his discovery of antitoxins. In retrospect, it is ironic that the Nobel committee brought together the cellular and humoral theories of immunity in one Nobel Prize, since it was not until the 1940s that the scientific community as a whole appreciated that both antibodies and cells were essential for protective immunity.

The phagocytic cells identified by Metchnikoff were a heterogenous population of cells that we now know included macrophages as well as other cells with phagocytic activity (e.g., neutrophils, dendritic cells, etc.). Macrophages themselves are a heterogeneous mixture of cells (Gordon and Taylor, 2005; Taylor and Gordon, 2003) which mediate their effects not only through phagocytosis, but also through the production of various soluble factors such as cytokines and chemokines, as well

---

S. Ostrand-Rosenberg
Department of Biological Sciences, University of Maryland Baltimore County, 1000 Hilltop Circle, Baltimore, MD 21250, USA
e-mail: srosenbe@umbc.edu

as by direct cellular contact with other cells. They are present in virtually all tissues and are involved in all aspects of immunity. Macrophages play a critical role in the onset and progression of malignant tumors and in immune surveillance against established tumors and can determine tumor growth vs. tumor regression. Much of their ability to promote transformation and tumor progression is mediated by their ability to cause inflammation, which has long been associated with tumor development (Balkwill and Mantovani, 2001; Balkwill et al., 2005). This chapter will focus on the function of macrophages in tumor immunity. Excellent reviews of the functions of macrophages in non-tumor settings can be found in both text books (Gordon, 2003b, 2006) and the scientific literature (Gordon and Taylor, 2005; Taylor et al., 2005).

2 Macrophages Are a Diverse and Heterogeneous Cell Population with Varied Functions

As blood monocytes migrate through the circulatory system, they are distributed to virtually all tissues of the body. Depending on their ultimate location, blood monocytes can differentiate into tissue macrophages, mature dendritic cells or osteoclasts. The markers F4/80 (mouse) and CD68 (macrosialin; human) distinguish the precursor blood monocyte from the tissue-resident macrophage, although marker expression differs between mouse and human derivatives (Fig. 1). The monocytes that transform into macrophages gain specific and unique properties based on the specific tissue in which they are located and the factors present in their immediate microenvironment. Therefore, macrophages are a diverse and heterogenous population of cells that have a broad variety of functions. They are key players in inflammation, tissue remodeling and repair and phagocytosis. They can function as

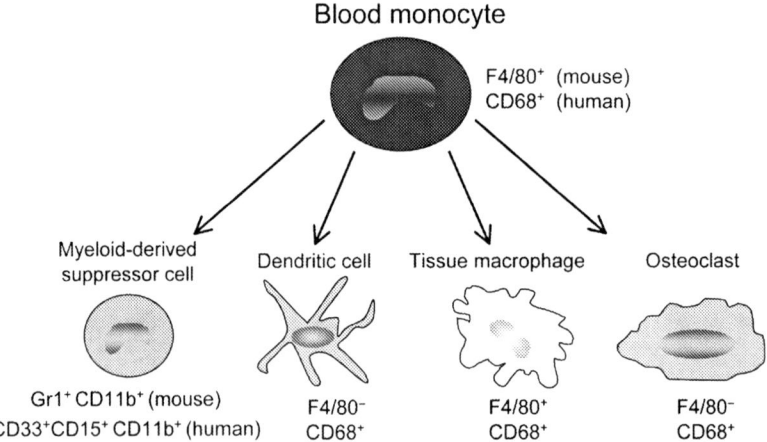

Fig. 1 Tissue-resident macrophages are derived from blood monocytes. When monocytes exit the blood they migrate to tissues and become macrophages

antigen-presenting cells for the activation of adaptive or innate immunity or they can be activated by natural killer (NK) and T cells and become effector cells. They also induce immune suppression and contribute to the induction of autoimmunity and tolerance, or they can enhance acquired humoral immunity by facilitating antibody and complement-mediated phagocytosis and cytotoxicity.

3 Tumor-Associated Macrophages Have a Distinct Phenotype and Promote Tumor Progression

Macrophages have exceptional plasticity and their ultimate function and protein expression profile are dictated by the local environment in which they reside (Lewis and Pollard, 2006). Therefore, macrophages attracted to a site of local infection caused by bacteria will have a very different phenotype and gene expression pattern than macrophages recruited to the site of tumor growth.

It has long been recognized that macrophages accumulate within growing solid tumors, and there is a direct correlation between the quantity of tumor-associated macrophages (TAMS) and a poor prognosis (Mantovani et al., 2006). The hypoxic and inflammatory tumor microenvironment contains chemotactic factors that inhibit macrophage mobility and maintains the TAMS at the tumor site (Grimshaw and Balkwill, 2001), while other factors in the microenvironment induce TAMS to enhance tumor progression. Therefore, TAMS are not merely correlated with a poor prognosis, but actively contribute to tumor progression (Bingle et al., 2002; Pollard, 2004). TAMS facilitate tumor growth through multiple mechanisms that suppress adaptive immunity by blocking the activation of T cells and inducing apoptosis of $CD8^+$ T cells (Saio et al., 2001), by promoting angiogenesis (Lin et al., 2006) and by facilitating the tissue remodeling and repair that tumors require for their continued growth (Mantovani et al., 2002). In contrast, macrophages residing in healthy tissues and macrophages that migrate to sites of inflammation in non-malignant tissues are efficient antigen-presenting cells that activate $CD4^+$ and $CD8^+$ T cells and NK cells and scavengers of cellular debris generated by infectious agents (Fig. 2).

4 Hypoxia Drives the Promotion of Angiogenesis by TAMS

Small deposits of tumor cells can obtain their required nutrients and dispose of their metabolic by-products directly from adjoining vasculature. However, as primary and metastatic tumor cells form larger masses, diffusion limits their access to existing blood vessels. Therefore, progressively growing tumors must induce new vasculature through angiogenesis. TAMS contribute to this process through several mechanisms.

Vascular endothelial growth factor (VEGF) is a key molecule for inducing neo-vascularization, and TAMS produce high levels of VEGF (Leek et al., 2000; Lewis et al., 2000). Several pathways contribute to the production of VEGF by macrophages. TAMS tend to accumulate in regions of tumors that have limiting

Fig. 2 Macrophages from healthy or pathogen-infected tissues and tumor-associated macrophages (TAMS) have distinct functions

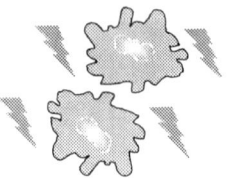

Macrophages residing in healthy or inflamed tissue

Macrophages residing in tumors (TAMS)

• Efficient antigen-presenting cells
• Produce cytokines that promote the activation of CD4$^+$ and CD8$^+$ T cells and NK cells

• Promote tumor angiogenesis
• Produce factors that block activation of CD4$^+$ and CD8$^+$ T cells
• Promote tissue remodeling and repair

amounts of oxygen (Leek et al., 1996; Negus et al., 1997). Hypoxia changes the expression levels of many proteins synthesized by macrophages (Murdoch et al., 2005). Of particular note is the induction of the transcription factors hypoxia-inducible factor (HIF) 1α and 2α (Burke et al., 2002; Talks et al., 2000), since HIFs upregulate VEGF (Jung et al., 2003). Since VEGF itself is a chemoattractant for macrophages, its increased synthesis leads to the accumulation of additional macrophages in hypoxic regions thereby amplifying the production of VEGF (Barleon et al., 1996; Leek et al., 2000) (Fig. 3).

VEGF is also upregulated in macrophages through a hypoxia-independent pathway that is mediated by the pro-inflammatory cytokine IL-1 which itself is produced by macrophages. This pathway involves cyclooxygenase-2 (COX-2), a central mediator of inflammation (Jung et al., 2003). Elegant in vivo studies have demonstrated that VEGF-producing TAMS directly contribute to the vascularization of growing tumors (Bingle et al., 2006).

In addition to VEGF, macrophages promote angiogenesis through other factors that are upregulated by hypoxia (Grimshaw et al., 2002). For example, urokinase-type plasminogen activator and its receptor are upregulated on TAMS and this enzyme is associated with increased vascularization (Foekens et al., 2000; Hildenbrand et al., 1995). Macrophages also produce fibroblast growth factor-2, CXCL8 (IL-8), angiopoietin, inducible nitric oxide synthase (iNOS) and leptin, all of which promote angiogenesis (Lewis and Murdoch, 2005). Gene array studies have demonstrated that macrophages exposed to hypoxia upregulate more than 30 genes associated with angiogenesis (White et al., 2004) (Table 1).

5 TAMS Facilitate Tumor Cell Invasiveness and Metastasis

In addition to their ability to induce angiogenesis, TAMS enhance tumor progression by facilitating tumor cell invasiveness into normal tissue and thereby promoting

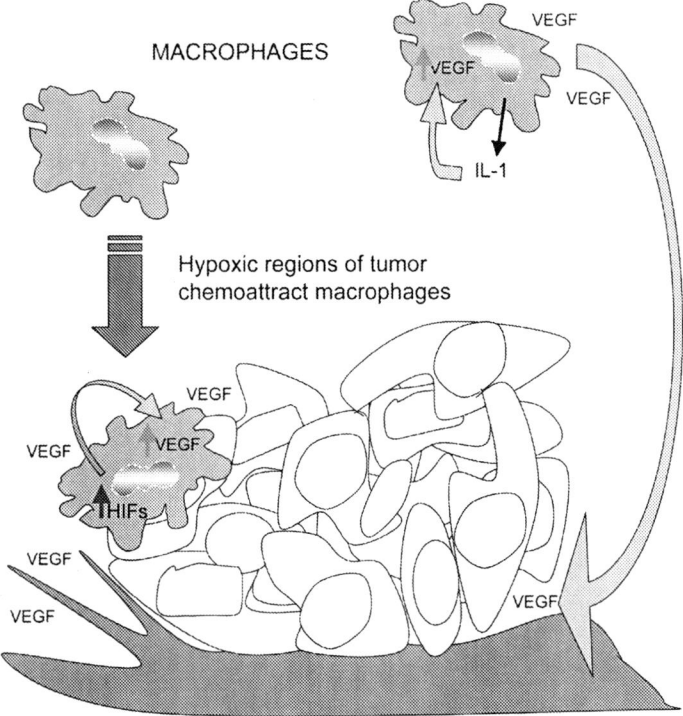

Fig. 3 TAMS stimulate angiogenesis through hypoxia-dependent and hypoxia-independent pathways. Hypoxic regions of tumors attract macrophages and upregulate macrophage expression of HIF-1α and HIF-2α. The HIFs then turn on synthesis of VEGF which induces neo-vascularization. Angiogenesis is also induced via a hypoxia-independent pathway in which IL-1 produced by macrophages in oxygen-sufficient environments upregulates macrophage expression of VEGF

metastasis. These functions are accomplished through diverse mechanisms that are mediated by multiple factors.

Studies with transgenic mice that spontaneously develop mammary tumors and are deficient for or over-express the receptor for a cytokine essential for macrophage development (CSF-1) support the concept that macrophages play a role in regulating metastasis and invasiveness. Such mice deficient for the CSF-1 receptor in their mammary epithelium have a delayed onset of metastatic disease, while mice that over-express CSF-1 in the mammary epithelium have an increased rate of metastasis (Lin et al., 2001). Therefore the presence of activated macrophages in the local environment of a primary tumor facilitate metastasis, while their absence delays metastasis.

Pollard and colleagues have observed that macrophages are frequently present in regions where primary mammary carcinomas are evolving from carcinoma in situ to invasive carcinoma in transgenic mice that spontaneously develop tumors

Table 1 Macrophage genes regulated by hypoxia

Molecule	Activity	Protein Expression
VEGF	Angiogenesis	Increased
MMP1, MMP7	Angiogenesis, metastasis	Increased
TNFα	Inflammation, angiogenesis, cytotoxicity	Increased
IL-1	Inflammation	Increased
PGE2	Inflammation, immunosuppression	Increased
IL-10	Immunosuppression	Increased
CCL3	Chemokine	Increased
CXCR4	Angiogenesis	Increased
IFNγ	T- and NK-cell activation	Increased
CXCL8	Angiogenesis	Increased
MIP-2 (CXCL1)	Chemokine	Increased
CCL2	Chemokine	Increased
CCR5	Chemokine receptor	Decreased
CD80	T-cell activation	Decreased
MIF	Metastasis, anti-migration	Increased
IL-6	Inflammation	Increased
Arginase	Immune suppression, cytotoxicity	Increased

Macrophages were cultured in vitro under hypoxic conditions and analyzed by a variety of protein expression assays. Adapted from Murdoch et al. (2005)

(Pollard, 2004). These observations led to the hypothesis that TAMS contribute to tumor invasion through the breakdown of the basement membrane surrounding the primary tumor, thereby allowing tumor cells to escape the immediate environment and invade neighboring normal tissue. This hypothesis is consistent with the finding that macrophages produce proteases capable of degrading basement membrane (Hagemann et al., 2004). For example, macrophage inhibitory factor (MIF), which is produced by activated macrophages, promotes tumor cell mobility (Sun et al., 2005). MIF also induces expression and secretion of matrix metalloproteinase-9 (MMP9), a protease that degrades basement membrane and extracellular matrix. Additional matrix metalloproteinases (MMP2 and MMP7) are also upregulated in TAMS and/or in macrophages maintained under hypoxic conditions (Hagemann et al., 2004). Epidermal growth factor (EGF), which is produced by macrophages in response to tumor-produced CSF-1, has also been shown in vitro to be a chemoattractant for tumor cells and mediates invasiveness (Goswami et al., 2005). Therefore, TAMS contribute to tumor metastasis by stimulating and/or producing factors that make the basement membrane leaky, so tumor cells can escape.

In addition to indirectly affecting tumor progression, TAMS also secrete factors that directly enhance tumor cell growth and facilitate tumor cell migration that results in metastasis. Aside from its ability to facilitate invasiveness, EGF also induces tumor cell proliferation (O'Sullivan et al., 1993; Wyckoff et al., 2000), as do platelet-derived growth factor, placental growth factor and hepatocyte growth factor, all of which are produced by macrophages (White et al., 2004).

6 TAMS Inhibit Anti-tumor Immunity

Macrophages that localize to sites of infection have different properties from TAMS. This class of macrophages has been called "classically activated" or M1 macrophages because they are activated by bacterial products such as lipopolysaccharides derived from pathogenic bacteria. This activation pathway is described in Sect. 8. Classically activated or M1 macrophages are a major component of the inflammatory milieu that contributes to the elimination of infectious agents. Their ability to destroy bacteria is the result of their pro-inflammatory characteristics including their phagocytic activity for cellular debris, their ability to present antigen and activate $CD4^+$ T helper and $CD8^+$ cytotoxic T lymphocytes and their production of cytokines that promote T-cell activation and the recruitment of additional inflammatory mediators. Because of their upregulated expression of MHC and costimulatory molecules and their potent phagocytic activity, classically activated macrophages can be directly cytotoxic for tumor cells and they can efficiently serve as antigen-presenting cells (APC) to activate tumor-specific cytotoxic T cells (Evans and Alexander, 1972a,b).

In contrast, TAMS are not cytotoxic for tumor cells and they are poor APC, are unable to activate cytotoxic T lymphocytes and actively suppress immune surveillance and anti-tumor immunity (Sinha et al., 2005b). Multiphoton microscopy studies graphically demonstrated that TAMS also promote tumor cell intravasation into surrounding normal tissue and blood vessels (Wyckoff et al., 2007). Whereas in vitro cultured macrophages and macrophages at sites of infection and inflammation produce high levels of reactive oxygen and nitrogen species that promote target cell cytotoxicity, when macrophages localize to tumor sites, particularly to hypoxic regions, their expression of these molecules is drastically reduced (Siegert et al., 1999).

The poor APC activity of macrophages is probably due to their downregulated MHC class II and costimulatory (CD80) molecule levels (Elgert et al., 1998; Lahat et al., 2003). This downregulation is accentuated by hypoxia (Lahat et al., 2003), as is their phagocytic activity. TAMS also impair tumor immunity by their production of cytokines and other factors that polarize immunity away from a tumor-rejecting Type 1 response and toward a tumor-promoting Type 2 response. Notably, TAMS produce high levels of the hallmark Type 2 cytokine IL-10, which induces Type 2 $CD4^+$ and $CD8^+$ T cells, and are impaired for production of the key Type 1 cytokine, IL-12 (Sinha et al., 2007a). The net result is that TAMS are not only unable to directly eliminate tumor cells, but they subvert immune surveillance and suppress immunity mediated by other cells. The pro-tumor effects of TAMS on tumor-bearing individuals has been confirmed by in vivo studies in which mice depleted for macrophages were shown to be resistant to spontaneous metastatic mammary carcinoma as compared to macrophage-bearing mice which were susceptible (Sinha et al., 2005a,b).

The functions associated with tumor-promoting macrophages and TAMS are generally consistent with macrophages that are activated through the so-called "alternative activation" pathway and are frequently called "M2" macrophages (Sica et al., 2006). This activation pathway is described in Sects. 8 and 9.

7 TAMS Facilitate Malignant Transformation by Promoting Inflammation

Pollard (2004) and Mantovani et al. (2005) have proposed that macrophages may not only promote progression of established tumors, but also contribute to the transformation of normal cells through their induction of a pro-inflammatory environment. It was first observed in the late 1800s that cancers are frequently preceded and/or accompanied by inflammation (Balkwill and Mantovani, 2001), and experimental and epidemiological studies support the concept that inflammation promotes tumor progression (Balkwill et al., 2005; Coussens and Werb, 2002; Greten et al., 2004; Pikarsky et al., 2004). Since macrophages produce many pro-inflammatory mediators, they are a central component of the inflammatory milieu. One of the possible mechanisms by which macrophage-induced inflammation promotes tumorigenesis is by the production of reactive oxygen and nitrogen species. These radicals produce peroxynitrite which can mutate the DNA of neighboring cells (Fulton et al., 1984; Maeda and Akaike, 1998).

The major intracellular regulator of the inflammatory program is NF-κB, a transcription factor that is not active unless it is uncoupled from its inhibitors. In addition to its expression in macrophages, NF-κB is also expressed in many tumor cells where its activation produces a more aggressive phenotype and the production of chemokines and cytokines that attract more macrophages to the tumor site, and in turn promote tumor growth (Karin and Lin, 2002). In contrast to its clear role in tumor cells, studies have shown conflicting results for the consequences of NF-κB expression in macrophages. In agreement with its pro-inflammatory role in malignant cells, inhibition of NF-κB in macrophages of mice that spontaneously develop tumors results in delayed tumor onset, consistent with the concept that blocking inflammation is beneficial (Colombo and Mantovani, 2005). However, these findings are in conflict with the observation that defective NF-κB in TAMS of late stage tumors promotes tumor progression, and that restoring NF-κB activity in these TAMS delays tumor growth (Saccani et al., 2006). Sica has suggested that this apparent inconsistency is due to differences in the tumor microenvironments of developing vs. late stage tumors (Sica and Bronte, 2007), a concept that remains to be tested.

Therefore, TAMS use multiple, diverse mechanisms to promote tumor initiation and progression (Fig. 4).

8 "Classically Activated" Vs. "Alternatively Activated" Macrophages

Early studies on macrophage activation demonstrated that mice infected with certain bacteria developed macrophages with heightened indiscriminate antimicrobial activity (Mackaness, 1964). Macrophage activation was due to stimulation with lipopolysaccharide (LPS) derived from the outer membrane of the bacteria

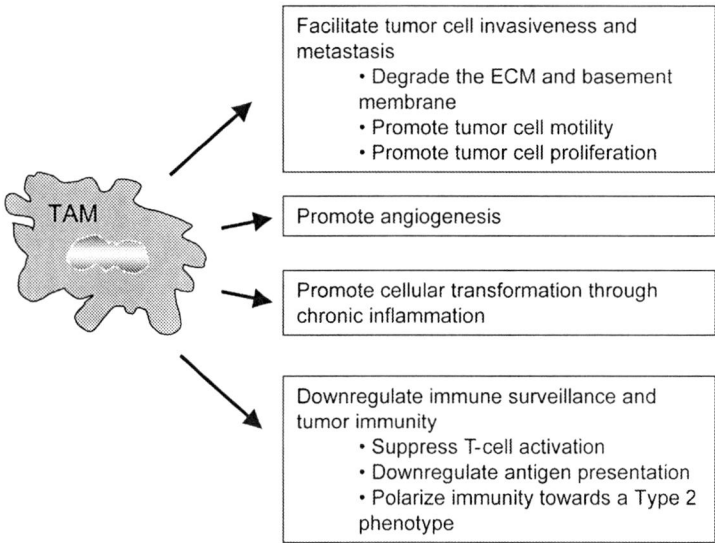

Fig. 4 TAMS promote tumor progression by multiple, diverse mechanisms

in conjunction with the Type 1 cytokines, IFNγ, produced by CD4+ Th1 and NK cells (Dalton et al., 1993) and IL-12 and IL-18 produced by antigen-presenting cells (Dalton et al., 1993). IFNγ in particular synergizes with LPS by upregulating macrophage levels of TLR-4 which are downregulated when LPS binds to its receptor (CD14) (Bosisio et al., 2002). Following LPS and IFNγ binding, 2′ signals are transmitted through an NF-κB-dependent intracellular pathway that results in the transcription of a broad array of inflammatory mediators (Karin and Greten, 2005). Because this was the first well-defined mechanism for macrophage activation, it is known as the "classical" pathway. A schematic of this activation pathway is shown in Fig. 5.

Subsequent studies demonstrated that macrophages are also activated through an "alternative" pathway in response to the Type 2 cytokines IL-4 and IL-13 (Fig. 5). Recognition that macrophages are activated by certain Type 2 cytokines was initially controversial, since the Type 2 cytokine, IL-10, was originally thought to deactivate/inactivate macrophages (Stein et al., 1992).

Classically activated and alternatively activated macrophages share certain surface markers (F4/80 and CD11b or CD68 in the mouse; CD11b or CD68 in humans); however, the two categories of macrophages also have distinct markers. For example, classically activated macrophages induced by IFNγ display increased expression of MHC class II and the costimulatory molecule CD86 (de Waal Malefyt et al., 1993) and reduced expression of the mannose receptor (Stein et al., 1992). In contrast, alternatively activated macrophages induced by IL-4 and IL-13 have heightened expression of the mannose receptor, CD23 (the Fc receptor for IgE) (Becker and Daniel, 1990), CD163 (hemoglobin scavenger receptor) (Kristiansen

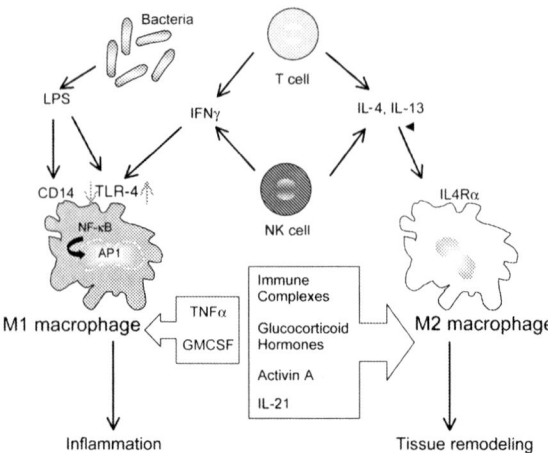

Fig. 5 Cytokines and bacterial products polarize macrophages toward an M1 or an M2 phenotype. M1 or classically activated macrophages are induced by LPS produced by bacteria and IFNγ produced by activated T and NK cells. LPS binds to its receptor (CD14) which interacts with TLR-4 and by signaling through NF-κB, pro-inflammatory mediators are produced. M1 macrophages are also activated by TNFα and/or GM-CSF. M2 macrophages are activated by IL-4 and IL-13 through IL-4Rα or by immune complexes, glucocorticoid hormones, activin A or IL-21

et al., 2001) and Dectin-1 (β-glucan receptor) (Brown et al., 2002) and reduced expression of the lipopolysaccharide receptor CD14.

In addition to different cell surface markers, classically activated and alternatively activated macrophages have distinct functions that are characterized by the production of specific cytokines and chemokines. Classically activated macrophages produce high levels of iNOS and display the characteristic respiratory burst. They also produce pro-inflammatory cytokines and chemokines such as TNFα, IL-12, IL-1, IL-6, IP-10 and MIP-1α. As a result, classically activated macrophages are pro-inflammatory, very effective in killing intracellular pathogens and efficient antigen-presenting cells (Taylor et al., 2005).

Alternatively activated macrophages produce a very different array of cytokines and chemokines and mediate different functions. They produce the Type 2 cytokine IL-10 (Kambayashi et al., 1996), the IL-1 receptor antagonist (Mantovani et al., 2001), TGFβ (Lee et al., 2001), CCL22 (also known as macrophage-derived chemokine MDC) (Bonecchi et al., 1998), CCL17 (thymus- and activation-regulated chemokine) (Imai et al., 1999) and CCL2 (also known as monocyte chemoattractant protein-1 or MCP-1) (Gu et al., 2000). These chemokines and cytokines block the effects of IL-1 and other pro-inflammatory mediators and therefore create a non-pro-inflammatory environment. This phenotype facilitates the production of collagen, fibronectin and other proteins associated with the extracellular matrix, which in turn promote cell growth, tissue repair and tissue remodeling. Alternatively activated macrophages also promote humoral (antibody-mediated) immunity and the elimination of parasites (Gordon, 2003a; Taylor et al., 2005).

Fig. 6 "Classically activated" and "alternatively activated" macrophages have different phenotypes that are characterized by distinct patterns of cell surface markers, cytokines, chemokines and other cytosolic or secreted molecules

	CLASSICALLY ACTIVATED	ALTERNATIVELY ACTIVATED
Cell surface markers		
	MHCII	Mannose receptor
	CD80	CD23 (FcR for IgE)
Chemokines / Cytokines		
	IL-1	AMAC-1 (MIP-1α-like)
	IL-6	CCL22
	TNFα	CCL17
	MIP-1α	CCL2 (MCP1)
	IL-12	IL-1Rα
	IP-10	IL-1 decoy receptor
		IL-10
		TGFβ
Cytosolic / Secreted		
	Collagenase?	Fizz-1
	MMP-9	Vm ½
		L-arginase

Therefore, although classically activated and alternatively activated macrophages are derived from a common precursor, they are molded by their environments to give rise to two distinct cell populations that have specific cell surface markers and specific functions (Fig. 6).

9 Classicially Activated and Alternatively Activated Macrophages and the M1 and M2 Paradigm

During the 1980s, Mosmann and colleagues demonstrated that mature CD4$^+$ T cells fall into one of two categories based on their production of cytokines. The so-called Th1 cells produced IFNγ, while Th2 cells produced IL-4, IL-5 and IL-10 (Mosmann and Coffman, 1989). Additional cytokines characteristic of Th1 vs. Th2 cells have subsequently been identified, and the Type 1 vs. Type 2 terminology has been extended to CD8$^+$ T cells (Tc1 vs. Tc2) (Mosmann and Sad, 1996) and dendritic cells (DC1 vs. DC2) (Rissoan et al., 1999). Type 1 cytokines typically activate macrophages with characteristics of classically activated macrophages, while Type 2 cytokines activate macrophages with characteristics of alternatively activated macrophages. This pattern of activation is most pronounced in the response to *Leishmania major*. C57BL/6 mice are resistant to *Leishmania* and when infected produce IFNγ which induces macrophages that produce nitric oxide (NO) and eliminate the parasite. In contrast, BALB/c mice are susceptible to *Leishmania* and when infected with *Leishmania* produce IL-4 which activates macrophages that produce high levels of arginase (Heinzel et al., 1989; Scott et al., 1988). Since this pattern of cytokine production by macrophages is similar to the polarized production of cytokines by

CD4+ Th1 and Th2 cells, Mills proposed that the two types of macrophages which correspond to classically activated and alternatively activated macrophages be called M1 and M2 macrophages, respectively (Mills et al., 2000).

The M1/M2 paradigm for macrophage activation has been greatly expanded by Mantovani, Sica and colleagues (Mantovani et al., 2006; Mantovani et al., 2002, 2005), and additional factors have been shown to polarize macrophages toward an M1 vs. an M2 phenotype. Generally, induction of M1 macrophages is mediated by IFNγ alone or in combination with LPS, TNFα or GM-CSF, and induction of M2 macrophages is mediated by IL-4, IL-13, immune complexes, IL-10, glucocorticoid hormones, activin A (a TGFβ family member) (Ogawa et al., 2006) and/or IL-21 (Pesce et al., 2006). M1 macrophages are typically pro-inflammatory, mediate tissue destruction, kill intracellular parasites and kill tumor cells. In contrast, M2 macrophages mediate tissue remodeling and angiogenesis, parasite encapsulation and promote tumor progression (Mantovani et al., 2007).

10 Arginine Metabolism in M1 and M2 Macrophages

A major characteristic of M1 vs. M2 macrophages is their metabolism of the amino acid arginine. Arginine is taken up from the surrounding environment by cells such as macrophages through the cationic amino acid transporter (cat) in their plasma membrane. Once in the cytosol, there are two basic pathways by which arginine is metabolized: the nitric oxide (NO) pathway which is generally used by M1 macrophages or the arginase pathway which is used by M2 macrophages. If arginine is metabolized through the NO pathway, then NO and L-citruline are produced; if it is metabolized through the arginase pathway, then L-ornithine and urea are produced. Since this chapter focuses on the role of macrophages in immunity, the following discussion will only deal with those aspects of arginine metabolism relevant to macrophages in tumor immunity; however, both pathways are also involved in non-tumor-related functions of macrophages (Mills, 2001). The intricacies of arginine metabolism were first worked out in macrophages (Mills, 2001). However, it is now clear that the same pathways are involved in myeloid-derived suppressor cells (MDSC) (Bronte and Zanovello, 2005; Bronte et al., 2003), another population of cells that promotes tumor progression and is discussed elsewhere in this monograph.

The NO pathway uses inducible nitric oxide synthase (iNOS) to generate NO. iNOS is upregulated in M1 macrophages by Type 1 cytokines, such as IFNγ and TNFα. Since iNOS expression is persistent, NO can accumulate to more than 1000 fold of the normal level in M1 macrophages (Bogdan, 2001). NO is cytotoxic and destroys target cells because it inhibits cell replication (MacMicking et al., 1997), blocks mitochondrial respiration (Nathan et al., 1991) and may induce apoptosis (Albina et al., 1993; Saio et al., 2001). NO can also interact with superoxide (O_2^-), an oxygen radical also made by macrophages, to generate peroxynitrite ($ONOO^-$) (Albina et al., 1993; Koppenol et al., 1992). Although superoxide may contribute to the toxic effects of M1 macrophages by generating peroxynitrite, O_2^- in some cases

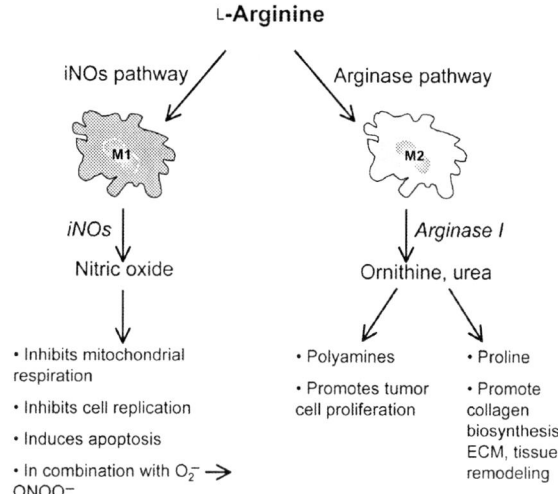

Fig. 7 M1 and M2 macrophages metabolize arginine through different pathways. M1 macrophages metabolize arginine through the iNOS pathway which converts arginine to nitric oxide. M2 macrophages metabolize arginine through the arginase pathway and produce polyamines and proline

can block NO toxicity, and inactivation of O_2^- by superoxide dismutase may reverse this effect (Tohyama et al., 1996).

In contrast to the NO pathway, the arginase pathway used by M2 macrophages promotes tumor progression. Macrophages contain both arginase I and arginase II. Arginase I is present in the cytosol and arginase II is expressed in mitochondria. Arginase I is the relevant enzyme in macrophages and is induced in response to the Type 2 cytokines IL-4, IL-13, TGFβ and/or IL-10 (Morris et al., 1998; Wu and Morris, 1998). It converts arginine to ornithine which in turn is converted to polyamines. Polyamines can favor tumor cell proliferation because they are required for DNA replication (Pegg, 1988). Interestingly, the arginase pathway also favors lymphocyte proliferation (Bowlin et al., 1987); however, its effects on promoting tumor growth appear to over-ride any anti-tumor effects of increased numbers of lymphocytes. In addition to its serving as a precursor for polyamines, ornithine is also a precursor of the amino acid proline, which is a major constituent of collagen. Since collagen is an important component of the extracellular matrix, ornithine may also promote tumor progression by facilitating the tumor infrastructure (Fig. 7).

11 Transcriptional Signatures of M1 and M2 Macrophages

The previous cataloging of molecules and markers expressed by macrophages was done by conventional biochemical and immunological techniques. With the advent of gene arrays and real-time polymerase chain reaction assays, it was possible to analyze the transcriptional program of TAMS. Many of the same molecules that had previously been identified at the protein level were also upregulated at the transcriptional level in TAMS as compared to peritoneal macrophages. These transcripts encoded molecules associated with immune suppression, angiogenesis

Table 2 RNA signatures of tumor-associated macrophages (TAMS, M2) as compared to peritoneal exudate cells (PEC; M1)

Molecule	Activity	No LPS[a]	+LPS[a]	
		TAM[b]	PEC[b]	TAM[b]
Mif	Inhibits macrophage migration	2	NA	NA
IL-12 p40	Inflammation	ND	11,059	150
TNFα	Inflammation	NA	5	3
IL-1β	Inflammation	ND	395	75
IL-6	Inflammation	ND	120	55
CCL3 (MIP-1α)	Inflammation	ND	85	20
TGFβ	Immune suppression	ND	2	18
IL-10	Immune suppression	4	20	100
CCL2 (MCP-1)	Macrophage chemoattractant	6	9	10
CCL5 (RANTES)	T-cell chemoattractant	16	100	180
CXCL9 (MIG)	T-cell chemoattractant	18	20	35
CXCL10	T-cell chemoattractant	8	15	95
CXCL16	Leukocyte chemoattractant	4	15	20

Macrophages were isolated from thioglycolate-induced peritoneal exudates (PEC) from tumor-free mice or from tumor of tumor-bearing mice (TAM). Adapted from Biswas et al. (2006). *ND* not determined

[a] Activated or not activated in vitro with LPS
[b] mRNA fold increase relative to PEC cultured in medium without LPS

and monocyte recruitment. Unexpectedly, transcripts of some IFN-inducible pro-inflammatory genes associated with M1 macrophages were also found (Table 2). Therefore, the transcriptional signature of TAMS is closely related to that of alternatively activated M2 macrophages; however, TAMS also express some IFNγ-inducible genes indicating that they also have distinctive characteristics (Biswas et al., 2006; Ghassabeh et al., 2006).

12 Caveats of the M1/M2 Nomenclature

The M1/M2 nomenclature is a useful shorthand for categorizing the two most prevalent phenotypes of macrophages. However, as seen by their transcriptional signatures and other studies, macrophage populations do not always neatly fit into one of these categories, making the nomenclature convenient, but an over-simplification. Whether there are additional discrete subpopulations of macrophages or whether the M1 and M2 phenotypes represent the polarized extremes with a continuum of phenotypes in between is not known. However, these discrepancies have led Mantovani et al. [2004] to subdivide M2 macrophages into additional categories. Accordingly, they use M1 to refer to all macrophages that are "classically activated" by IFNγ± LPS ± other Type I cytokines, and that as a result are IL-10^{low}, good antigen-presenting cells, IL-12^{high} and IL-23^{high}, so they promote a Type 1 response and produce high levels of NO and other reactive oxygen intermediates. In their terminology, the M2 category is a catch-all for all macrophages that are not M1 and includes macrophages with the common properties of IL-12^{low}, IL-10^{high}, promoting Type 2 responses and enhancing tissue remodeling. They further classify M2 macrophages as M2a for macrophages activated by IL-4 and/or IL-13; M2b for macrophages activated by IC, TLRs or IL-R; and M2c for macrophages activated by IL-10 or glucocorticoid hormones (Mantovani et al., 2004).

Future studies may identify additional categories or subsets of macrophages with distinct phenotypes and functions. Genomic and proteomic analyses may be particularly informative, and it is likely that they will identify new populations of macrophages whose phenotype and function partially overlap with existing populations. To remain useful, the M1/M2 nomenclature will have to incorporate these new populations, and new categories may have to be designated.

13 Cancer Therapies Targeting TAMS

Because TAMS are strongly correlated with poor prognosis, therapeutic strategies designed to eliminate, inactivate or re-polarize them are attractive (Mantovani et al., 2006; Sica and Bronte, 2007). One approach has been to reduce or eliminate TAMS that have already accumulated at the tumor site. Using multiple mouse tumor systems, mice with primary and/or metastatic disease were immunized with a DNA vaccine encoding legumain, an asparaginyl endopeptidase that is over-expressed in TAMS. Treated mice had significantly extended survival times and/or rejected tumors. Enhanced tumor resistance was accompanied by strong $CD8^+$ T-cell responses to TAMS, reduction in the quantity of TAMS and reduced tumor angiogenesis (Luo et al., 2006).

Another therapeutic strategy has been to re-polarize TAMS from an M2 to an M1 phenotype. Mice with large established tumors were inoculated intratumorally with adenovirus encoding the chemokine CCL16 to induce accumulation of macrophages, along with CpG, the ligand for Toll-like receptor 9 and antibodies to the IL-10 receptor to skew the induced macrophages toward an M1 phenotype. Mice developed both innate and adaptive (T-cell) responses against their tumors, reduced or eliminated metastatic disease and rejected primary tumors (Guiducci et al., 2005).

Other studies have identified particular genes that may regulate macrophage polarization and therefore have the potential to be therapeutic targets. One such factor is p50 NF-κB inhibitory homodimer. p50 accumulates to very high levels in the nuclei of TAMS in wild-type mice, and over-expression of p50 blocks IL-12 expression by macrophages from tumor-free mice. In contrast, macrophages from tumor-bearing mice deficient for p50 have an M1 phenotype and reduced tumor progression (Saccani et al., 2006). Therefore, inhibition of p50 accumulation in the nucleus may re-polarize macrophages from an M2 to an M1 phenotype.

Signal transducer and activator of transcription 6 (STAT6) also determines macrophage polarization. IL-4 and IL-13 induce M2 macrophages by binding to the plasma membrane IL-4Rα and signaling through the JAK2/STAT6 pathway. Therefore, mice that are deficient for the STAT6 gene produce M1 and not M2 macrophages and reject established metastatic disease and survive if their primary tumors are surgically removed (Sinha et al., 2005b). STAT6$^{-/-}$ mice are also resistant to recurrence of primary fibrosarcomas (Terabe et al., 2000).

Polarization of macrophages toward an M1 phenotype is also influenced by the MHC class I non-classical CD1d gene. CD1$^{-/-}$ mice are deficient for IL-13 because they lack IL-13-producing NKT cells. In the absence of IL-13, their macrophages are polarized toward an M1 phenotype. Similar to STAT6$^{-/-}$ mice, CD1$^{-/-}$ mice reject established metastatic disease and survive if their primary tumors are surgically removed (Sinha et al., 2005a) and are resistant to recurring fibrosarcomas (Terabe et al., 2003).

Another gene that appears to regulate macrophage phenotype is Src homology 2-containing inositol-5′ phosphatase (SHIP), an enzyme that downregulates the phosphatidylinositol 3-kinase pathway in myeloid cells. Transplanted tumors grow more quickly in SHIP knockout mice, and macrophages from SHIP knockout mice produce elevated levels of arginase I and low levels of NO, indicative of an M2 phenotype. Therefore, SHIP appears to contribute to an M1 phenotype and drugs aimed at maintaining or increasing SHIP expression may impact the development of TAMS (Rauh et al., 2005).

Although not directly affecting TAMS, another therapeutic strategy is based on the observation that macrophages home to hypoxic regions of tumors. Since tumor cells residing in these regions are frequently not affected by standard chemotherapy or radiation therapy, it has been proposed to use macrophages as delivery vehicles for targeting gene expression to resilient areas of tumors (Griffiths et al., 2000; Murdoch et al., 2005)

14 Indolamine 2,3-Dioxygenase, Macrophages and Suppression of Tumor Immunity

Although the enzyme indolamine 2,3-dioxygenase (IDO) was first identified more than 40 years ago, its role in immune suppression has only recently been discovered (Grohmann et al., 2003). IDO is an intracellular enzyme that degrades tryptophan which is taken up by cells through one of two plasma membrane transporters (Seymour et al., 2006). The resulting microenvironment is therefore depleted of tryptophan and surrounding cells are unable to progress through the cell cycle (Munn et al., 1999). Tumor cells themselves secrete IDO which to some extent reduces tumor growth. However, dendritic cells and macrophages also produce IDO, and the IDO produced by these antigen-presenting cells profoundly suppresses T-cell activation and this suppression outweighs the direct effects of IDO on tumor cells. The net result is that lymphocytes in the lymph nodes draining solid tumors are tolerized or non-responsive to tumor antigens thereby eliminating adaptive antitumor immunity (Munn and Mellor, 2004, 2006). Consistent with the concept that M2 macrophages promote tumor growth, one would expect that IDO would be produced by M2, and not by M1, macrophages. However, macrophage (and dendritic cell) synthesis of IDO is induced by the Type 1 cytokine IFNγ (Koide and Yoshida, 1994), which activates M1 macrophages. Studies examining the production of IDO by M1 and M2 macrophages are needed to clarify which macrophages produce IDO and if IDO-mediated T-cell suppression fits into the M1/M2 paradigm.

15 TAMS and Myeloid-Derived Suppressor Cells

In addition to macrophages, another population of cells of myeloid origin, myeloid-derived suppressor cells (MDSC) (Gabrilovich et al., 2007), promote tumor progression. MDSC are the subject of another chapter in this monograph, and hence a detailed description of their function is not included here. However, MDSC and TAMS are histologically related and because they both are potent inhibitors of anti-tumor immunity, a brief comparison of these two myeloid cell populations is warranted.

The phenotypic markers expressed by TAMS were discussed in Sect. 8 and are shown in Fig. 6. MDSC express the macrophage marker CD11b and also Gr1 (mouse) or CD15 and CD33 (human), markers of granulocytes and monocytes that are not expressed by TAMS. Although some studies report that MDSC also express the macrophage marker F4/80, this finding is not consistently reported. It has been suggested that when MDSC enter the tumor microenvironment they differentiate into TAMS and become F4/80$^+$; however, this hypothesis remains to be proven (Kusmartsev et al., 2005; Kusmartsev et al., 2005). If TAM and MDSC are to be distinguished by cell surface markers, then it is necessary to test for multiple markers to avoid mis-identification. Gene array studies assaying for a limited number of genes expressed at the mRNA level confirm that TAM and MDSC not only share many molecules, but also have distinctive expression patterns (Biswas et al., 2006) (Table 3).

TAMS and MDSC both inhibit anti-tumor immunity by interfering with the generation of tumor-specific T lymphocytes. As discussed above, TAMS are poor APC and they produce cytokines that polarize immunity toward a tumor-promoting Type 2 response. MDSC also skew immunity toward a Type 2 response through their release of IL-10. Interestingly, cross-talk between macrophages and MDSC exacerbates this polarization since macrophages induce MDSC to produce elevated levels of the prototype Type 2 cytokine, IL-10, and MDSC downregulate IL-12 production by macrophages (Sinha et al., 2007a) (Fig. 8). MDSC also block T-cell activation by their production of arginase and/or nitric oxide. Although arginase metabolism is similar in MDSC and TAMS, TAMS do not use this pathway for blocking T-cell activation (Bronte and Zanovello, 2005; Kusmartsev et al., 2004; Rodriguez and Ochoa, 2006).

TAMS and MDSC also share the characteristic of facilitating tumor progression by promoting angiogenesis. Similar to TAMS, MDSC mediate this effect through their production of MMP9 (Yang et al., 2004). Whether hypoxia plays a role in MDSC-induced angiogenesis as it does in TAMS is not known.

Table 3 RNA signatures of tumor-associated macrophages (TAMS, M2) vs. myeloid-derived suppressor cells

Molecule	TAM[a]		MDSC[a]
	IL-4[b]	No IL-4	IL-4[b]
TGFβ	2+	2+	–
IL-10	–	1+	1+
IL-1Rα	1+	–	1+
MSr1	–	–	1+
CCL22	2+	1+	1+
Ym1	2+	–	2+
Fizz1	4+	2+	4+
Arg I	5+	2+	3+
Arg II	–	–	1+
	+LPS[c]	No LPS	+LPS[c]
CCL2	1+	1+	1+
CCL5	3+	2+	3+
CXCL9	1+	1+	1+
CXCL10	3+	2+	3+
CXCL16	2+	2+	1+

Macrophages were isolated tumors. MDSC are an established cell line obtained by immortalizing splenic Gr1+ cells from a tumor-bearing mouse. Adapted from Biswas et al. (2006)
[a] Fold increase in mRNA level relative to non-IL-4 or LPS-treated MDSC: + ≤ 10 fold; 2+ ≤ 100 fold; 3+ ≤ 1000 fold; 4+ ≤ 10,000 fold; 5+ ≤ 100,000 fold
[b] Activated by in vitro treatment with IL-4
[c] Activated by in vitro treatment with LPS

As discussed previously, macrophages may promote tumor progression because they are a critical component of the inflammatory microenvironment. Although MDSC do not secrete pro-inflammatory mediators, they are induced by inflammation (Bunt et al., 2006; Rodriguez et al., 2005; Sinha et al., 2007b; Song et al., 2005), and it has been proposed that inflammation therefore contributes to tumor

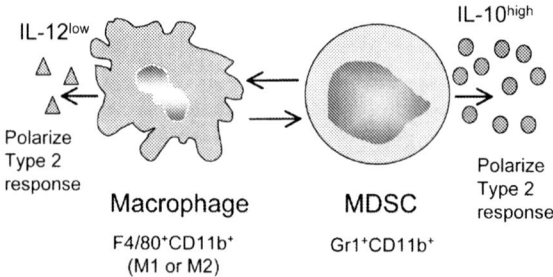

Fig. 8 Cross-talk between macrophages and myeloid-derived suppressor cells (MDSC) polarize immunity toward a tumor-promoting Type 2 response. M1 and M2 macrophages increase the production of IL-10 by MDSC, while MDSC downregulate IL-12 production by macrophages. Both interactions require cell–cell contact and result in a net increase in IL-10 and a decrease in IL-12 which polarizes immunity toward a tumor-promoting Type 2 response

progression by inducing MDSC which suppress adaptive immune surveillance, thereby facilitating the expansion of malignant cells (Sinha et al., 2007b).

Both TAMS and MDSC are heterogeneous populations of cells that consist of a mixture of myeloid cells at different stages of differentiation. Although we currently consider these cells to fall into distinct populations they may represent the extremes of a large family of cells and there may be a spectrum of cells between these extremes that have properties of both populations. Alternatively, TAMS and MDSC may represent alternative differentiation pathways that diverge from a common myeloid progenitor. Regardless of their origin, TAMS and MDSC have many similarities in function and marker expression, and both populations are potent promoters of tumor cell progression.

16 Conclusions

Macrophages are a diffuse and heterogenous population of cells whose ultimate phenotype and function are dictated by the environment in which they exist. Their affinity for hypoxic regions of tumors enables them to home to these locations where they become a major component and take on functions driven by the chemokines and other factors produced by tumor cells and other cells residing in the tumor stroma. Their ability to stimulate angiogenesis, inhibit T-cell activation and promote invasiveness provides them with multiple approaches for promoting the growth of primary tumors and encouraging metastatic disease. In addition to their ability to facilitate progression of established tumors, macrophages also contribute to tumor initiation through their production of pro-inflammatory mediators. Therefore, tumor-associated macrophages are a significant component of malignant diseases and the development of any preventive or therapeutic treatments for cancer must take into account this population of cells.

References

Albina, J. E., Cui, S., Mateo, R. B., and Reichner, J. S. (1993). Nitric oxide-mediated apoptosis in murine peritoneal macrophages. *J Immunol* 150:5080–5085.

Balkwill, F., Charles, K. A., and Mantovani, A. (2005). Smoldering and polarized inflammation in the initiation and promotion of malignant disease. *Cancer Cell* 7:211–217.

Balkwill, F., and Mantovani, A. (2001). Inflammation and cancer: back to Virchow? *Lancet* 357:539–545.

Barleon, B., Sozzani, S., Zhou, D., Weich, H. A., Mantovani, A., and Marme, D. (1996). Migration of human monocytes in response to vascular endothelial growth factor (VEGF) is mediated via the VEGF receptor flt-1. *Blood* 87:3336–3343.

Becker, S., and Daniel, E. G. (1990). Antagonistic and additive effects of IL-4 and interferon-gamma on human monocytes and macrophages: effects on Fc receptors, HLA-D antigens, and superoxide production. *Cell Immunol* 129:351–362.

Bingle, L., Brown, N. J., and Lewis, C. E. (2002). The role of tumour-associated macrophages in tumour progression: implications for new anticancer therapies. *J Pathol* 196:254–265.

Bingle, L., Lewis, C. E., Corke, K. P., Reed, M. W., and Brown, N. J. (2006). Macrophages promote angiogenesis in human breast tumour spheroids in vivo. *Br J Cancer* 94:101–107.
Biswas, S. K., Gangi, L., Paul, S., Schioppa, T., Saccani, A., Sironi, M., Bottazzi, B., Doni, A., Vincenzo, B., Pasqualini, F., Vago, L., Nebuloni, M., Mantovani, A., and Sica, A. (2006). A distinct and unique transcriptional program expressed by tumor-associated macrophages (defective NF-kappaB and enhanced IRF-3/STAT1 activation). *Blood* 107:2112–2122.
Bogdan, C. (2001). Nitric oxide and the immune response. *Nat Immunol* 2:907–916.
Bonecchi, R., Sozzani, S., Stine, J. T., Luini, W., D'Amico, G., Allavena, P., Chantry, D., and Mantovani, A. (1998). Divergent effects of interleukin-4 and interferon-gamma on macrophage-derived chemokine production: an amplification circuit of polarized T helper 2 responses. *Blood* 92:2668–2671.
Bosisio, D., Polentarutti, N., Sironi, M., Bernasconi, S., Miyake, K., Webb, G. R., Martin, M. U., Mantovani, A., and Muzio, M. (2002). Stimulation of toll-like receptor 4 expression in human mononuclear phagocytes by interferon-gamma: a molecular basis for priming and synergism with bacterial lipopolysaccharide. *Blood* 99:3427–3431.
Bowlin, T. L., McKown, B. J., and Sunkara, P. S. (1987). Increased ornithine decarboxylase activity and polyamine biosynthesis are required for optimal cytolytic T lymphocyte induction. *Cell Immunol* 105:110–117.
Bronte, V., Serafini, P., Mazzoni, A., Segal, D. M., and Zanovello, P. (2003). L-Arginine metabolism in myeloid cells controls T-lymphocyte functions. *Trends Immunol* 24: 302–306.
Bronte, V., and Zanovello, P. (2005). Regulation of immune responses by L-arginine metabolism. *Nat Rev Immunol* 5:641–654.
Brown, G. D., Taylor, P. R., Reid, D. M., Willment, J. A., Williams, D. L., Martinez-Pomares, L., Wong, S. Y., and Gordon, S. (2002). Dectin-1 is a major beta-glucan receptor on macrophages. *J Exp Med* 196:407–412.
Bunt, S. K., Sinha, P., Clements, V. K., Leips, J., and Ostrand-Rosenberg, S. (2006). Inflammation induces myeloid-derived suppressor cells that facilitate tumor progression. *J Immunol* 176: 284–290.
Burke, B., Tang, N., Corke, K. P., Tazzyman, D., Ameri, K., Wells, M., and Lewis, C. E. (2002). Expression of HIF-1alpha by human macrophages: implications for the use of macrophages in hypoxia-regulated cancer gene therapy. *J Pathol* 196:204–212.
Colombo, M. P., and Mantovani, A. (2005). Targeting myelomonocytyic cells to revert inflammatin-dependent cancer promotion. *Cancer Res* 65:9113–9116.
Coussens, L. M., and Werb, Z. (2002). Inflammation and cancer. *Nature* 420:860–867.
Dalton, D. K., Pitts-Meek, S., Keshav, S., Figari, I. S., Bradley, A., and Stewart, T. A. (1993). Multiple defects of immune cell function in mice with disrupted interferon-gamma genes. *Science* 259:1739–1742.
de Waal Malefyt, R., Figdor, C. G., Huijbens, R., Mohan-Peterson, S., Bennett, B., Culpepper, J., Dang, W., Zurawski, G., and de Vries, J. E. (1993). Effects of IL-13 on phenotype, cytokine production, and cytotoxic function of human monocytes. Comparison with IL-4 and modulation by IFN-gamma or IL-10. *J Immunol* 151:6370–6381.
Elgert, K. D., Alleva, D. G., and Mullins, D. W. (1998). Tumor-induced immune dysfunction: the macrophage connection. *J Leukoc Biol* 64:275–290.
Evans, R., and Alexander, P. (1972a). Role of macrophages in tumour immunity. I. Co-operation between macrophages and lymphoid cells in syngeneic tumour immunity. *Immunology* 23: 615–626.
Evans, R., and Alexander, P. (1972b). Role of macrophages in tumour immunity. II. Involvement of a macrophage cytophilic factor during syngeneic tumour growth inhibition. *Immunology* 23:627–636.
Foekens, J. A., Peters, H. A., Look, M. P., Portengen, H., Schmitt, M., Kramer, M. D., Brunner, N., Janicke, F., Meijer-van Gelder, M. E., Henzen-Logmans, S. C., van Putten, W. L., and Klijn, J. G. (2000). The urokinase system of plasminogen activation and prognosis in 2780 breast cancer patients. *Cancer Res* 60:636–643.

Fulton, A. M., Loveless, S. E., and Heppner, G. H. (1984). Mutagenic activity of tumor-associated macrophages in *Salmonella typhimurium* strains TA98 and TA 100. *Cancer Res* 44:4308–4311.
Gabrilovich, D. I., Bronte, V., Chen, S. H., Colombo, M. P., Ochoa, A., Ostrand-Rosenberg, S., and Schreiber, H. (2007). The terminology issue for myeloid-derived suppressor cells. *Cancer Res* 67:425; author reply 426.
Ghassabeh, G. H., De Baetselier, P., Brys, L., Noel, W., Van Ginderachter, J. A., Meerschaut, S., Beschin, A., Brombacher, F., and Raes, G. (2006). Identification of a common gene signature for type II cytokine-associated myeloid cells elicited in vivo in different pathologic conditions. *Blood* 108:575–583.
Gordon, S. (2003a). Alternative activation of macrophages. *Nat Rev Immunol* 3:23–35.
Gordon, S. (2003b). Macrophages and the immune response. *Fundamental Immunology*. W. Paul. Philadelphia: Lippincott-Raven Press, pp. 481–495.
Gordon, S. (2006). Mononuclear phagocytes in immune defence. *Immunology*. I. Roitt, J. Brostoff, D. Male and D. Roth. Edinburgh: Mosby, pp. 148–162.
Gordon, S., and Taylor, P. R. (2005). Monocyte and macrophage heterogeneity. *Nat Rev Immunol* 5:953–964.
Goswami, S., Sahai, E., Wyckoff, J. B., Cammer, M., Cox, D., Pixley, F. J., Stanley, E. R., Segall, J. E., and Condeelis, J. S. (2005). Macrophages promote the invasion of breast carcinoma cells via a colony-stimulating factor-1/epidermal growth factor paracrine loop. *Cancer Res* 65:5278–5283.
Greten, F. R., Eckmann, L., Greten, T. F., Park, J. M., Li, Z. W., Egan, L. J., Kagnoff, M. F., and Karin, M. (2004). IKKbeta links inflammation and tumorigenesis in a mouse model of colitis-associated cancer. *Cell* 118:285–296.
Griffiths, L., Binley, K., Iqball, S., Kan, O., Maxwell, P., Ratcliffe, P., Lewis, C., Harris, A., Kingsman, S., and Naylor, S. (2000). The macrophage—a novel system to deliver gene therapy to pathological hypoxia. *Gene Ther* 7:255–262.
Grimshaw, M. J., and Balkwill, F. R. (2001). Inhibition of monocyte and macrophage chemotaxis by hypoxia and inflammation—a potential mechanism. *Eur J Immunol* 31:480–489.
Grimshaw, M. J., Wilson, J. L., and Balkwill, F. R. (2002). Endothelin-2 is a macrophage chemoattractant: implications for macrophage distribution in tumors. *Eur J Immunol* 32:2393–2400.
Grohmann, U., Fallarino, F., and Puccetti, P. (2003). Tolerance, DCs and tryptophan: much ado about IDO. *Trends Immunol* 24:242–248.
Gu, L., Tseng, S., Horner, R. M., Tam, C., Loda, M., and Rollins, B. J. (2000). Control of TH2 polarization by the chemokine monocyte chemoattractant protein-1. *Nature* 404:407–411.
Guiducci, C., Vicari, A. P., Sangaletti, S., Trinchieri, G., and Colombo, M. P. (2005). Redirecting in vivo elicited tumor infiltrating macrophages and dendritic cells towards tumor rejection. *Cancer Res* 65:3437–3446.
Hagemann, T., Robinson, S. C., Schulz, M., Trumper, L., Balkwill, F. R., and Binder, C. (2004). Enhanced invasiveness of breast cancer cell lines upon co-cultivation with macrophages is due to TNF-alpha dependent up-regulation of matrix metalloproteases. *Carcinogenesis* 25:1543–1549.
Heinzel, F. P., Sadick, M. D., Holaday, B. J., Coffman, R. L., and Locksley, R. M. (1989). Reciprocal expression of interferon gamma or interleukin 4 during the resolution or progression of murine leishmaniasis. Evidence for expansion of distinct helper T cell subsets. *J Exp Med* 169:59–72.
Hildenbrand, R., Dilger, I., Horlin, A., and Stutte, H. J. (1995). Urokinase and macrophages in tumour angiogenesis. *Br J Cancer* 72:818–823.
Imai, T., Nagira, M., Takagi, S., Kakizaki, M., Nishimura, M., Wang, J., Gray, P. W., Matsushima, K., and Yoshie, O. (1999). Selective recruitment of CCR4-bearing Th2 cells toward antigen-presenting cells by the CC chemokines thymus and activation-regulated chemokine and macrophage-derived chemokine. *Int Immunol* 11:81–88.
Jung, Y. J., Isaacs, J. S., Lee, S., Trepel, J., and Neckers, L. (2003). IL-1beta-mediated up-regulation of HIF-1alpha via an NFkappaB/COX-2 pathway identifies HIF-1 as a critical link between inflammation and oncogenesis. *FASEB J* 17:2115–2117.

Kambayashi, T., Jacob, C. O., and Strassmann, G. (1996). IL-4 and IL-13 modulate IL-10 release in endotoxin-stimulated murine peritoneal mononuclear phagocytes. *Cell Immunol* 171:153–158.

Karin, M., and Greten, F. R. (2005). NF-kappaB: linking inflammation and immunity to cancer development and progression. *Nat Rev Immunol* 5:749–759.

Karin, M., and Lin, A. (2002). NF-kappaB at the crossroads of life and death. *Nat Immunol* 3: 221–227.

Koide, Y., and Yoshida, A. (1994). The signal transduction mechanism responsible for gamma interferon-induced indoleamine 2,3-dioxygenase gene expression. *Infect Immun* 62:948–955.

Koppenol, W. H., Moreno, J. J., Pryor, W. A., Ischiropoulos, H., and Beckman, J. S. (1992). Peroxynitrite, a cloaked oxidant formed by nitric oxide and superoxide. *Chem Res Toxicol* 5:834–842.

Kristiansen, M., Graversen, J. H., Jacobsen, C., Sonne, O., Hoffman, H. J., Law, S. K., and Moestrup, S. K. (2001). Identification of the haemoglobin scavenger receptor. *Nature* 409: 198–201.

Kusmartsev, S., and Gabrilovich, D. I. (2005). STAT1 signaling regulates tumor-associated macrophage-mediated T cell deletion. *J Immunol* 174:4880–4891.

Kusmartsev, S., Nagaraj, S., and Gabrilovich, D. I. (2005). Tumor-associated CD8+ T cell tolerance induced by bone marrow-derived immature myeloid cells. *J Immunol* 175:4583–4592.

Kusmartsev, S., Nefedova, Y., Yoder, D., and Gabrilovich, D. I. (2004). Antigen-specific inhibition of CD8+ T cell response by immature myeloid cells in cancer is mediated by reactive oxygen species. *J Immunol* 172:989–999.

Lahat, N., Rahat, M. A., Ballan, M., Weiss-Cerem, L., Engelmayer, M., and Bitterman, H. (2003). Hypoxia reduces CD80 expression on monocytes but enhances their LPS-stimulated TNF-alpha secretion. *J Leukoc Biol* 74:197–205.

Lee, C. G., Homer, R. J., Zhu, Z., Lanone, S., Wang, X., Koteliansky, V., Shipley, J. M., Gotwals, P., Noble, P., Chen, Q., Senior, R. M., and Elias, J. A. (2001). Interleukin-13 induces tissue fibrosis by selectively stimulating and activating transforming growth factor beta(1). *J Exp Med* 194:809–821.

Leek, R. D., Hunt, N. C., Landers, R. J., Lewis, C. E., Royds, J. A., and Harris, A. L. (2000). Macrophage infiltration is associated with VEGF and EGFR expression in breast cancer. *J Pathol* 190:430–436.

Leek, R. D., Lewis, C. E., Whitehouse, R., Greenall, M., Clarke, J., and Harris, A. L. (1996). Association of macrophage infiltration with angiogenesis and prognosis in invasive breast carcinoma. *Cancer Res* 56:4625–4629.

Lewis, C., and Murdoch, C. (2005). Macrophage responses to hypoxia: implications for tumor progression and anti-cancer therapies. *Am J Pathol* 167:627–635.

Lewis, C. E., and Pollard, J. W. (2006). Distinct role of macrophages in different tumor microenvironments. *Cancer Res* 66:605–612.

Lewis, J. S., Landers, R. J., Underwood, J. C., Harris, A. L., and Lewis, C. E. (2000). Expression of vascular endothelial growth factor by macrophages is up-regulated in poorly vascularized areas of breast carcinomas. *J Pathol* 192:150–158.

Lin, E. Y., Li, J. F., Gnatovskiy, L., Deng, Y., Zhu, L., Grzesik, D. A., Qian, H., Xue, X. N., and Pollard, J. W. (2006). Macrophages regulate the angiogenic switch in a mouse model of breast cancer. *Cancer Res* 66:11238–11246.

Lin, E. Y., Nguyen, A. V., Russell, R. G., and Pollard, J. W. (2001). Colony-stimulating factor 1 promotes progression of mammary tumors to malignancy. *J Exp Med* 193:727–740.

Luo, Y., Zhou, H., Krueger, J., Kaplan, C., Lee, S. H., Dolman, C., Markowitz, D., Wu, W., Liu, C., Reisfeld, R. A., and Xiang, R. (2006). Targeting tumor-associated macrophages as a novel strategy against breast cancer. *J Clin Invest* 116:2132–2141.

Mackaness, G. B. (1964). The immunological basis of acquired cellular resistance. *J Exp Med* 120:105–120.

MacMicking, J., Xie, Q. W., and Nathan, C. (1997). Nitric oxide and macrophage function. *Annu Rev Immunol* 15:323–350.

Maeda, H., and Akaike, T. (1998). Nitric oxide and oxygen radicals in infection, inflammation, and cancer. *Biochemistry (Mosc)* 63:854–865.

Mantovani, A. (2005). Cancer: inflammation by remote control. *Nature* 435:752–753.
Mantovani, A. (2006). Macrophage diversity and polarization: in vivo veritas. *Blood* 108:408–409.
Mantovani, A., Locati, M., Vecchi, A., Sozzani, S., and Allavena, P. (2001). Decoy receptors: a strategy to regulate inflammatory cytokines and chemokines. *Trends Immunol* 22:328–336.
Mantovani, A., Schioppa, T., Porta, C., Allavena, P., and Sica, A. (2006). Role of tumor-associated macrophages in tumor progression and invasion. *Cancer Metastasis Rev* 25:315–322.
Mantovani, A., Sica, A., and Locati, M. (2005). Macrophage polarization comes of age. *Immunity* 23:344–346.
Mantovani, A., Sica, A., and Locati, M. (2007). New vistas on macrophage differentiation and activation. *Eur J Immunol* 37:14–16.
Mantovani, A., Sica, A., Sozzani, S., Allavena, P., Vecchi, A., and Locati, M. (2004). The chemokine system in diverse forms of macrophage activation and polarization. *Trends Immunol* 25:677–686.
Mantovani, A., Sozzani, S., Locati, M., Allavena, P., and Sica, A. (2002). Macrophage polarization: tumor-associated macrophages as a paradigm for polarized M2 mononuclear phagocytes. *Trends Immunol* 23:549–555.
Metchnikoff, E. (1905). *Immunity in infective disease.* Cambridge: Cambridge University Press.
Mills, C. D. (2001). Macrophage arginine metabolism to ornithine/urea or nitric oxide/citrulline: a life or death issue. *Crit Rev Immunol* 21:399–425.
Mills, C. D., Kincaid, K., Alt, J. M., Heilman, M. J., and Hill, A. M. (2000). M-1/M-2 macrophages and the Th1/Th2 paradigm. *J Immunol* 164:6166–6173.
Morris, S. M., Jr, Kepka-Lenhart, D., and Chen, L. C. (1998). Differential regulation of arginases and inducible nitric oxide synthase in murine macrophage cells. *Am J Physiol* 275:E740–E747.
Mosmann, T. R., and Coffman, R. L. (1989). TH1 and TH2 cells: different patterns of lymphokine secretion lead to different functional properties. *Annu Rev Immunol* 7:145–173.
Mosmann, T. R., and Sad, S. (1996). The expanding universe of T-cell subsets: Th1, Th2 and more. *Immunol Today* 17:138–146.
Munn, D. H., and Mellor, A. L. (2004). IDO and tolerance to tumors. *Trends Mol Med* 10:15–18.
Munn, D. H., and Mellor, A. L. (2006). The tumor-draining lymph node as an immune-privileged site. *Immunol Rev* 213:146–158.
Munn, D. H., Shafizadeh, E., Attwood, J. T., Bondarev, I., Pashine, A., and Mellor, A. L. (1999). Inhibition of T cell proliferation by macrophage tryptophan catabolism. *J Exp Med* 189:1363–1372.
Murdoch, C., and Lewis, C. E. (2005). Macrophage migration and gene expression in response to tumor hypoxia. *Int J Cancer* 117:701–708.
Murdoch, C., Muthana, M., and Lewis, C. E. (2005). Hypoxia regulates macrophage functions in inflammation. *J Immunol* 175:6257–6263.
Nathan, C. F., and Hibbs, J. B., Jr (1991). Role of nitric oxide synthesis in macrophage antimicrobial activity. *Curr Opin Immunol* 3:65–70.
Negus, R. P., Stamp, G. W., Hadley, J., and Balkwill, F. R. (1997). Quantitative assessment of the leukocyte infiltrate in ovarian cancer and its relationship to the expression of C–C chemokines. *Am J Pathol* 150:1723–1734.
Ogawa, K., Funaba, M., Chen, Y., and Tsujimoto, M. (2006). Activin A functions as a Th2 cytokine in the promotion of the alternative activation of macrophages. *J Immunol* 177:6787–6794.
O'Sullivan, C., Lewis, C. E., Harris, A. L., and McGee, J. O. (1993). Secretion of epidermal growth factor by macrophages associated with breast carcinoma. *Lancet* 342:148–149.
Pegg, A. E. (1988). Polyamine metabolism and its importance in neoplastic growth and a target for chemotherapy. *Cancer Res* 48:759–774.
Pesce, J., Kaviratne, M., Ramalingam, T. R., Thompson, R. W., Urban, J. F., Jr, Cheever, A. W., Young, D. A., Collins, M., Grusby, M. J., and Wynn, T. A. (2006). The IL-21 receptor augments Th2 effector function and alternative macrophage activation. *J Clin Invest* 116:2044–2055.
Pikarsky, E., Porat, R. M., Stein, I., Abramovitch, R., Amit, S., Kasem, S., Gutkovich-Pyest, E., Urieli-Shoval, S., Galun, E., and Ben-Neriah, Y. (2004). NF-kappaB functions as a tumour promoter in inflammation-associated cancer. *Nature* 431:461–466.

Pollard, J. W. (2004). Tumour-educated macrophages promote tumour progression and metastasis. *Nat Rev Cancer* 4:71–78.
Rauh, M. J., Ho, V., Pereira, C., Sham, A., Sly, L. M., Lam, V., Huxham, L., Minchinton, A. I., Mui, A., and Krystal, G. (2005). SHIP represses the generation of alternatively activated macrophages. *Immunity* 23:361–374.
Rissoan, M. C., Soumelis, V., Kadowaki, N., Grouard, G., Briere, F., de Waal Malefyt, R., and Liu, Y. J. (1999). Reciprocal control of T helper cell and dendritic cell differentiation. *Science* 283:1183–1186.
Rodriguez, P. C., Hernandez, C. P., Quiceno, D., Dubinett, S. M., Zabaleta, J., Ochoa, J. B., Gilbert, J., and Ochoa, A. C. (2005). Arginase I in myeloid suppressor cells is induced by COX-2 in lung carcinoma. *J Exp Med* 202:931–939.
Rodriguez, P. C., and Ochoa, A. C. (2006). T cell dysfunction in cancer: role of myeloid cells and tumor cells regulating amino acid availability and oxidative stress. *Semin Cancer Biol* 16:66–72.
Saccani, A., Schioppa, T., Porta, C., Biswas, S. K., Nebuloni, M., Vago, L., Bottazzi, B., Colombo, M. P., Mantovani, A., and Sica, A. (2006). p50 nuclear factor-kappaB overexpression in tumor-associated macrophages inhibits M1 inflammatory responses and antitumor resistance. *Cancer Res* 66:11432–11440.
Saio, M., Radoja, S., Marino, M., and Frey, A. B. (2001). Tumor-infiltrating macrophages induce apoptosis in activated CD8 (+) T cells by a mechanism requiring cell contact and mediated by both the cell-associated form of TNF and nitric oxide. *J Immunol* 167:5583–5593.
Scott, P., Natovitz, P., Coffman, R. L., Pearce, E., and Sher, A. (1988). Immunoregulation of cutaneous leishmaniasis. T cell lines that transfer protective immunity or exacerbation belong to different T helper subsets and respond to distinct parasite antigens. *J Exp Med* 168:1675–1684.
Seymour, R. L., Ganapathy, V., Mellor, A. L., and Munn, D. H. (2006). A high-affinity, tryptophan-selective amino acid transport system in human macrophages. *J Leukoc Biol* 80:1320–1327.
Sica, A., and Bronte, V. (2007). Altered macrophage differentiation and immune dysfunction in tumor development. *J Clin Invest* 117:1155–1166.
Sica, A., Schioppa, T., Mantovani, A., and Allavena, P. (2006). Tumour-associated macrophages are a distinct M2 polarised population promoting tumour progression: potential targets of anti-cancer therapy. *Eur J Cancer* 42:717–727.
Siegert, A., Denkert, C., Leclere, A., and Hauptmann, S. (1999). Suppression of the reactive oxygen intermediates production of human macrophages by colorectal adenocarcinoma cell lines. *Immunology* 98:551–556.
Sinha, P., Bunt, S., Clements, V. K., Albelda, S., and Ostrand-Rosenberg, S. (2007a). Cross-talk between myeloid-derived suppressor cells and macrophages subverts tumor immunity towards a type 2 response. *J Immunol* 179(2):977–83.
Sinha, P., Clements, V. K., Fulton, A. M., and Ostrand-Rosenberg, S. (2007b). Prostaglandin E2 promotes tumor progression by inducing myeloid-derived suppressor cells. *Cancer Res* 67:4507–4513.
Sinha, P., Clements, V. K., and Ostrand-Rosenberg, S. (2005a). Interleukin-13-regulated M2 macrophages in combination with myeloid suppressor cells block immune surveillance against metastasis. *Cancer Res* 65:11743–11751.
Sinha, P., Clements, V. K., and Ostrand-Rosenberg, S. (2005b). Reduction of myeloid-derived suppressor cells and induction of M1 macrophages facilitate the rejection of established metastatic disease. *J Immunol* 174:636–645.
Song, X., Krelin, Y., Dvorkin, T., Bjorkdahl, O., Segal, S., Dinarello, C. A., Voronov, E., and Apte, R. N. (2005). CD11b+/Gr-1+ immature myeloid cells mediate suppression of T cells in mice bearing tumors of IL-1beta-secreting cells. *J Immunol* 175:8200–8208.
Stein, M., Keshav, S., Harris, N., and Gordon, S. (1992). Interleukin 4 potently enhances murine macrophage mannose receptor activity: a marker of alternative immunologic macrophage activation. *J Exp Med* 176:287–292.

Sun, B., Nishihira, J., Yoshiki, T., Kondo, M., Sato, Y., Sasaki, F., and Todo, S. (2005). Macrophage migration inhibitory factor promotes tumor invasion and metastasis via the Rho-dependent pathway. *Clin Cancer Res* 11:1050–1058.

Talks, K. L., Turley, H., Gatter, K. C., Maxwell, P. H., Pugh, C. W., Ratcliffe, P. J., and Harris, A. L. (2000). The expression and distribution of the hypoxia-inducible factors HIF-1alpha and HIF-2alpha in normal human tissues, cancers, and tumor-associated macrophages. *Am J Pathol* 157:411–421.

Tauber, A. I., and Chernyak, L. (1991). *Metchnikoff and the Origins of Immunology: From Metaphor to Theory*. New York: Oxford University Press.

Taylor, P. R., and Gordon, S. (2003). Monocyte heterogeneity and innate immunity. *Immunity* 19:2–4.

Taylor, P. R., Martinez-Pomares, L., Stacey, M., Lin, H. H., Brown, G. D., and Gordon, S. (2005). Macrophage receptors and immune recognition. *Annu Rev Immunol* 23:901–944.

Terabe, M., Matsui, S., Noben-Trauth, N., Chen, H., Watson, C., Donaldson, D. D., Carbone, D. P., Paul, W. E., and Berzofsky, J. A. (2000). NKT cell-mediated repression of tumor immunosurveillance by IL-13 and the IL-4R-STAT6 pathway. *Nat Immunol* 1:515–520.

Terabe, M., Matsui, S., Park, J. M., Mamura, M., Noben-Trauth, N., Donaldson, D. D., Chen, W., Wahl, S. M., Ledbetter, S., Pratt, B., Letterio, J. J., Paul, W. E., and Berzofsky, J. A. (2003). Transforming growth factor-beta production and myeloid cells are an effector mechanism through which CD1d-restricted T cells block cytotoxic T lymphocyte-mediated tumor immunosurveillance: abrogation prevents tumor recurrence. *J Exp Med* 198:1741–1752.

Tohyama, M., Kawakami, K., Futenma, M., and Saito, A. (1996). Enhancing effect of oxygen radical scavengers on murine macrophage anticryptococcal activity through production of nitric oxide. *Clin Exp Immunol* 103:436–441.

White, J. R., Harris, R. A., Lee, S. R., Craigon, M. H., Binley, K., Price, T., Beard, G. L., Mundy, C. R., and Naylor, S. (2004). Genetic amplification of the transcriptional response to hypoxia as a novel means of identifying regulators of angiogenesis. *Genomics* 83:1–8.

Wu, G., and Morris, S. M., Jr (1998). Arginine metabolism: nitric oxide and beyond. *Biochem J* 336(Pt 1):1–17.

Wyckoff, J. B., Segall, J. E., and Condeelis, J. S. (2000). The collection of the motile population of cells from a living tumor. *Cancer Res* 60:5401–5404.

Wyckoff, J. B., Wang, Y., Lin, E. Y., Li, J. F., Goswami, S., Stanley, E. R., Segall, J. E., Pollard, J. W., and Condeelis, J. (2007). Direct visualization of macrophage-assisted tumor cell intravasation in mammary tumors. *Cancer Res* 67:2649–2656.

Yang, L., DeBusk, L. M., Fukuda, K., Fingleton, B., Green-Jarvis, B., Shyr, Y., Matrisian, L. M., Carbone, D. P., and Lin, P. C. (2004). Expansion of myeloid immune suppressor Gr+CD11b+ cells in tumor-bearing host directly promotes tumor angiogenesis. *Cancer Cell* 6:409–421.

Myeloid-Derived Suppressor Cells in Cancer

Paolo Serafini and Vincenzo Bronte

1 History and Nomenclature of Tumor-Conditioned Myeloid Cells with Suppressive Activity on the Immune Response

In the late 1970s, many researchers described the presence of a cellular population that could inhibit different activities of the immune system, both in vivo and in vitro. These cells, named natural suppressor (NS) cells, inhibited the proliferative responses of T-helper lymphocytes to mitogens or alloantigens, antibody production by B lymphocytes and the generation of cytotoxic T lymphocytes (CTLs) independently of antigen and MHC restriction (Strober, 1984). NS cells were also suspected to be involved in pathways of tolerance induction. NS cells were shown to appear only briefly in fetal newborn tissues and the placenta during pregnancy as well as during the neonatal maturation of the lymphoid tissues; however, they could be induced in adults by manipulation of the lymphoid tissues with certain treatments such as total lymphoid irradiation, cyclophosphamide administration and during graft-versus-host disease (Strober, 1984). The presence of these cells in several body environments, all characterized by either enhanced hematopoiesis or an intense immune response, suggested their possible involvement in regulating myeloid cell differentiation and controlling lymphocyte and myeloid expansion.

Assigning a characteristic phenotype to NS cells was an unresolved problem for many years and several discrepancies were found in the marker distribution in cells with suppressive activity on T lymphocytes activated in vitro, even though some evidence pointed to the monocytic/macrophage lineage (Maier et al., 1985, 1989; Strober et al., 1989; Sykes et al., 1990). The phenotype of NS cells was originally defined "null" because they appeared to lack the usual markers of mature macrophages, T, B and natural killer (NK) cells. Purified NS cells did not lose their inhibitory activity during in vitro culture nor kill classic NK targets, nor differentiate into macrophages or mature lymphocytes. Unfortunately, despite the importance of these early findings, many experimental limitations (such as a restricted antibody

---

P. Serafini
Department of Microbiology & Immunology, Dodson Interdisciplinary Immunotherapy Institute, University of Miami, School of Medicine, Miami, Florida, USA

panel to identify their phenotype, the widespread use of culture supernatants with unknown cytokines and growth factors composition and the absence of high-purity techniques to isolate cell subsets) postponed for many year the progress in understanding their biology. These technical restrictions, combined with experimental difficulties in validating some results and the absence of a clear phenotype, made the very existence of NS doubtful. For these reasons until the 1990s the immunosuppressive role of NS/suppressive myeloid cells in tumor-bearing host was still poorly known.

Already Subiza et al. (1989) showed the expansion of NS in Ehrlich tumor-bearing mice, but the first clear involvement of myeloid cells in lowering immune surveillance and in promoting tumor growth was provided in 1995. The administration of an antibody directed against the antigen Gr1 (recognizing the cross-reacting molecules of lymphocyte antigen 6 complex locus C and G) to immunocompetent mice reduced the growth of an ultraviolet light-induced tumor (Seung et al., 1995). The effect of the in vivo anti-Gr1 administration was originally attributed to the elimination of granulocytes, but successive reports from our and other groups suggested that the Gr1$^+$ cells were mostly CD11b$^+$ and comprised both polymorphonuclear and mononuclear cells, including elements at different maturation stages along the myelomonocytic differentiation lineage (Kusmartsev and Gabrilovich, 2006b; Serafini, 2006a).

The clear heterogeneity of the CD11b$^+$/Gr1$^+$ cells has generated some confusion, somehow amplified by the use of different terms to define the same cells (i.e., natural suppressor cells, immature myeloid cells or myeloid suppressor cells). Recently a panel of investigators agreed to use the common term of *myeloid-derived suppressor cells* (MDSCs; Gabrilovich et al., 2007). Even though numerous findings suggest that the main population responsible for the immune dysfunctions induced in CD8$^+$ T cells has a monocytic rather than granulocytic phenotype among the mouse CD11b$^+$/Gr1$^+$ cells (Gallina et al., 2006; Kusmartsev and Gabrilovich, 2005; Van Ginderachter et al., 2006), the use of the myeloid term highlights the common finding of the enhanced myelopoiesis in tumor-bearing hosts and our incomplete understanding of the relationship between the two main components generated by this altered myeloid differentiation (i.e., polymorphonuclear and mononuclear cells).

2 Mouse MDSCs: Biology and Function

MDSCs represent a heterogeneous population of myeloid cells comprising immature macrophages, granulocytes, dendritic cells (DCs) and other myeloid cells at earlier stages of differentiation that can be identified in mice by expression of CD11b and Gr1. Co-expression of these markers, together with the immature marker CD31, and the ability to form colonies in agar is consistent with the phenotype of myeloid progenitors (Bronte et al., 2000; Fu et al., 1990; Melani et al., 2003). In healthy mice CD11b$^+$/Gr1$^+$ cells can be detected in sizeable numbers only in the bone marrow (BM, about 30–40%); however, small numbers of these cells

(<4%) can also be found in the blood and spleen. $CD11b^+/Gr1^+$ cells comprise myeloid precursors that can generate mature granulocytes, macrophages and DCs when cultured in vitro with the appropriate cytokines cocktail (Apolloni et al., 2000; Bronte et al., 2000; Kusmartsev et al., 2003). Disturbances in cytokine and chemokine balance, induced by tumor growth, infection, immune stress and even vaccination, can alter the homeostasis of this population leading to its accumulation in the secondary lymphoid organs and, ultimately, influencing their maturation toward a suppressive phenotype. It must be pointed out that $CD11b^+/Gr1^+$ cells in the BM of naïve mice do not show a relevant suppressive activity ex vivo and suppression of T-cell function can be observed only when supra-physiologic numbers of cells are used in in vitro assays (Bronte, unpublished data); however, they can acquire full suppressive function when cultured for few days in the presence of granulocyte macrophage-colony stimulating factor (GM-CSF) (Rossner et al., 2005) or with activated $CD4^+$ T cells (Serafini, unpublished data). Also, in mice bearing solid or hematologic tumors, ex vivo BM-derived $CD11b^+$ cells show little suppressive activity and low expression of suppressive markers (Serafini, unpublished data). Taken together these findings seem to indicate that the majority of BM $CD11b^+/Gr1^+$ are still pluripotent cells that can differentiate, depending on the kind and/or duration of cytokine/chemokine stimulation, into cells able to either enhance (e.g., myeloid DCs) or restrain (MDSCs) the immune response (Kusmartsev and Gabrilovich, 2006a; Rossner et al., 2005).

Differently from BM, MDSCs in peripheral organs are fully suppressive. In various mouse models, indeed, the dysfunctional immune responses of T lymphocytes in tumor-bearing mice depended almost entirely on the accumulation of MDSCs in the blood and secondary lymphoid organs. Primary tumor resection, $Gr1^+$ depletion, pharmacological inhibition or genetic MDSC inactivation often results, in fact, in a complete correction of T-cell dysfunctions (Bronte et al., 1998, 1999; Danna et al., 2004; De Santo et al., 2005; Gallina et al., 2006; Rodriguez et al., 2005; Salvadori et al., 2000; Serafini et al., 2006b; Seung et al., 1995; 2005a, 2005a; Terabe et al., 2003).

These studies have also unveiled a functional plasticity of the $CD11b^+/Gr1^+$ cells in tumor-bearing hosts that was confirmed by in vitro experiments in which MDSCs extracted from tumor-bearing mice were cultured with Th1- or Th2-derived cytokines. The MDSCs' suppressive phenotype was enhanced, in fact, by the addition of Th2 cytokines (such as interleukin (IL)-4) or IL-10 to these cultures. Conversely, MDSCs co-cultured with Th1 cytokines *enhanced* antigen-specific T-cell cytotoxicity, thereby underscoring the ability of MDSCs to differentiate into functional antigen-presenting cells (APCs) when placed in the appropriate cytokine environment (Bronte et al., 2000). Moreover, MDSC subsets can appear transiently in cultures of BM cells stimulated with GM-CSF to generate myeloid DCs (Rossner et al., 2005): These cells were $CD11c^-$ myeloid precursor cells with ring-shaped nuclei and were $Gr1^{low}$, $CD11b^+$, $CD31^+$, $ER-MP58^+$, $asialoGM1^+$ and $F4/80^+$ (Rossner et al., 2005), a phenotype very similar to MDSCs described in tumor-bearing mice (Kusmartsev et al., 2005). Despite these in vitro observations and the fact that strong signal such as Flt3L, a combination of GM-CSF and IL-4 (Bronte et al., 2000) or all-*trans* retinoic acid (ATRA) (Kusmartsev, et al., 2003) can force MDSCs to differentiate in fully mature DCs, other in vivo data suggest that MDSCs do not simply represent a transitional population along DC maturation but, instead, are fully differentiated cells with a suppressive phenotype (Kusmartsev et al., 2003). In fact, while $CD11b^+/Gr1^+$ from naïve mice adoptive transferred into naïve congenic mice differentiated into mature $CD11c^+MHC$ class II^+ DCs and $Gr1^-$ $F4/80^+$ macrophages within 5 days, MDSCs from tumor-bearing mice retained the phenotype of immature cells for longer time ($Gr1^+CD11b^+$) and the differentiation in macrophages was significantly impaired (Kusmartsev et al., 2003).

$CD11b^+$ $Gr1^+$ MDSCs are strictly related to macrophages and their simple in vitro culture can allow their maturation in macrophage-like cells ($CD11b^+$, $Gr1^-$, $F4/80^+$, class II $MHC^{-/low}$, $CD80^+$) with enhanced immunosuppressive activity (Gallina et al., 2006; Kusmartsev and Gabrilovich 2005; Van Ginderachter et al., 2006). Macrophages can be activated by either the classical or alternative pathways. Classically activated macrophages (M1 type) release pro-inflammatory cytokines and are central elements of the delayed-type hypersensitivity response, which leads to microbicidal activity through the release of nitric oxide (NO) by nitric oxide synthase (NOS) and cellular immunity. By contrast, alternatively activated macrophages (M2 type) are essential for humoral immunity, tissue repair and fibrosis (all of which are mediated through the production of polyamines and L-proline by the enzyme arginase (ARG)), as well as allergic and anti-parasitic responses (Gordon, 2003). The prevalent view is that classical activation by IFN-γ induces the activity of NOS2, whereas alternative activation by T-helper 2 cytokines such as IL-4 and IL-13 induces ARG1 activity (Munder et al., 1998), and that these two pathways are thought to be antithetic. MDSCs challenge this classical dichotomous view and support the idea that M1 and M2 macrophages are the two extremes of a continuum of diverse functional states (Mantovani et al., 2004). In many experimental situations, in fact, MDSCs can co-express ARG1 and NOS2 (Bronte and Zanovello, 2005; Bronte et al., 2003b; Serafini et al., 2004b). We recently demonstrated that, upon activation by IFN-γ, MDSCs produce both IL-13 and IFN-γ, which are utilized in an autocrine manner to enhance the production and activity of both ARG1 and NOS2 enzymes (Gallina et al., 2006). Thus MDSC respond "paradoxically" to a Th1 cytokine in that they produce and secrete both Th2 (IL-13) and Th1 cytokines (IFN-γ). In other models, however, MDSCs were shown to restrain both $CD4^+$ and $CD8^+$ T-cell functions using different mechanisms (see below), whose choice can be determined by the relative contribution of Th1 and Th2 cytokines (Serafini et al., 2004b).

3 MDSCs and Tumor Progression

MDSC appearance has been prevalently reported in transplantable tumor mouse models (Serafini et al., 2004b). These models have often been generated following multiple in vivo passages of the transplantable cells, which ultimately can select for clones able to avoid immune recognition. For these aspects, tumors, induced by chemical carcinogen or by the activation of tissue-restricted, transgenic oncogenes, are often considered more reliable as models for tumor initiation and progression. Indeed, MDSC accumulation has been reported not only in methylcholanthrene- or 1,2-dimethylhydrazine-diHCl-induced tumors (Horiguchi et al., 1999; Talmadge et al., 2007), but also in mice in which the expression of the transforming rat oncogene c-erbB-2, under the control of the mouse mammary tumor virus promoter, drives the spontaneous development of mammary carcinomas with a progression resembling that of human breast cancers (Melani et al., 2003). Interestingly, in these latter mice, tumor multiplicity directly correlated with the accumulation of MDSCs in the peripheral blood and in the spleens. Analogously, in the BW-Sp3 lymphoma model, most of the BW-Sp3-bearing mice mount a $CD8^+$ T-cell-mediated response resulting in tumor regression. Nonetheless, tumor progression occurs in some of the recipients and is associated with MDSC accumulation. Again, in vivo MDSC depletion is sufficient to restore CTL activity (Liu et al., 2003).

Interestingly, in some mouse models, pro-tumoral activity of MDSCs does not require their expansion in the secondary lymphoid organs. In a transformed fibrosarcoma model, in which tumors grow, spontaneously regress and then recur, Terabe et al. [2003] found that IL-13 producing $CD4^+$ NKT cells suppressed $CD8^+$ CTLs to prevent complete tumor elimination. The suppressive mechanism of NKT cells, however, was not direct but involved MDSCs. In fact, IL-13 secreted by NKT cells was sufficient to activate MDSCs to secrete transforming growth factor (TGF)-β that acted as final suppressive molecule (Fig. 1). Blocking TGF-β or depleting $Gr1^+$ cells in vivo prevented tumor recurrence. This negative regulatory circuit was also found to be active in a lung metastasis model of the mouse colon carcinoma CT26 (Park et al., 2005).

As discussed later, recent evidences indicate that MDSCs can promote tumor progression not only by suppressing the antitumor immune response, but also (and possibly more importantly) by promoting tumor angiogenesis through their incorporation in tumor vessel and regulation of vascular endothelial growth factor (VEGF) bioavailability (Yang et al., 2004). These findings were recently confirmed by Young and Cigal (2006) who demonstrated how $CD34^+$ cells cultured in the presence of Lewis lung carcinoma conditioned medium were skewed in their differentiation toward endothelial cells expressing CD31 and CD144. Moreover, a small subset of tumor-infiltrating $CD11b^+$ myeloid cells characterized by the expression of the Tie2, a receptor tyrosine kinase known to be restricted to endothelial cells, was recently characterized (De Palma et al., 2005). This population, called Tie2-expressing monocytes (TEMs), was advanced to represent a new hematopoietic lineage of pro-angiogenic cells, selectively recruited to spontaneous and orthotopic tumor sites and required for their neovascularization (De Palma et al., 2005).

4 Tumor-Derived Factors Regulate the Expansion, Recruitment and Activation of MDSCs

Numerous findings indicate that tumor-derived factors (TDFs) promote not only MDSC recruitment but also maturation toward an immunosuppressive phenotype.

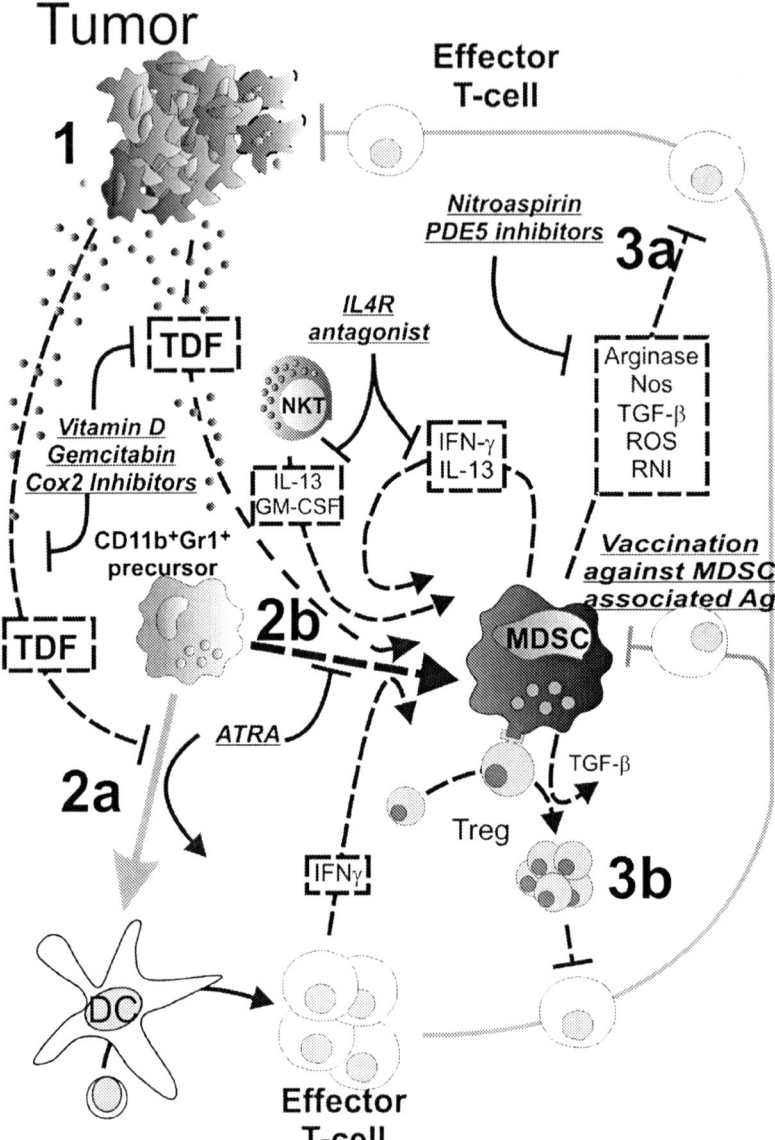

Fig. 1 Pathways of MDSC recruitment and activation. Growing tumors (1) release soluble factors (TDF) that alter the normal myelopoiesis. These factors inhibit DC maturation (2*a*) from $CD11b^+Gr1^+$ precursors and enhance their maturation toward a suppressive phenotype (2*b*) generating MDSCs. Other factors such as IFN-γ, IL-13, GM-CSF released by host cells (NKT, effector T cells or MDSCs themselves) help MDSC maturation/activation. Once fully activated, MDSCs can suppress the antitumor immune response directly (3*a*) or indirectly by expanding the $CD4^+CD25^+$ Treg cells (3*b*). Tumor suppressive pathways are indicated by filled cells and broken black lines; Anti-tumor immunity pathways by unfilled cells and by solid gray lines. The different strategies to restrain MDSC suppressive activity (discussed in the text) are indicate in black solid lines and underlined text

Indeed, conditioned media from tumor cell lines can inhibit the in vitro differentiation of DCs from their precursors (Gabrilovich et al., 1996), and normal BM cells could give rise to immunosuppressive elements simply by culturing them for a few days with supernatants from a highly metastatic Lewis lung carcinoma variant (Young et al., 1990a). For more than 10 years efforts have been made to identify these TDFs (Cirillo et al., 1988; Heldin, 2004; Kunicka et al., 1991; Lim et al., 1991; Moore et al., 1992; Pegoraro et al., 1985; Rodriguez et al., 2005; Serafini, 2006a; Young et al., 1992). Tumors secrete a large panel of cytokines, chemokines or other diffusible molecules that, alone or in combination, can induce MDSCs recruitment and increase their maturation into fully suppressive cells. To date, a number of candidate proteins have been identified and are discussed below.

4.1 Colony Stimulating Factor 1 (CSF-1)

The secretion of CSF-1, also called macrophage-colony stimulating factor (M-CSF), has been described in various cancers including acute myeloblastic leukemia (Haran-Ghera et al., 1997; Rambaldi et al., 1988), renal cell carcinoma (Gerharz et al., 2001), bladder carcinoma (Champelovier et al., 2002) and about 70% of human breast cancers (Lin et al., 2002). Its expression in breast cancer is associated with poor prognosis and is likely involved in tumor progression (Lin et al., 2002). CSF-1 can recruit immunosuppressive macrophages and can alter the normal DC maturation (Menetrier-Caux et al., 1998). Conditioned media from renal carcinoma cell lines could alter the differentiation of DCs into mature APCs and this effect could be abrogated by the use of neutralizing antibody against IL-6 and CSF-1 (Menetrier-Cauz et al., 1998). Interestingly, both IL-4 and IL-13 reversed the inhibitory effects exerted by either renal cell carcinoma-conditioned medium or IL-6 and CSF-1 combination on the phenotypic and functional differentiation of $CD34^+$ cells into DCs. IL-4 downregulated M-CSF and IL-6R-transducing chain expression, decreased the secondary production of CSF-1 and prevented the loss of GM-CSF receptor α-chain expression, which normally occurs during the differentiation of $CD34^+$ cells (Menetrier-Caux et al., 1998). Moreover, human and mouse monocytes exposed to CSF-1 acquire the ability to suppress antigen- and mitogen-driven T-cell proliferation through a mechanism of T-cell "starvation". CSF-1-treated macrophages, in fact, can deplete the microenvironment of the essential amino acid tryptophan through the enzymatic activity of indoleamin-2,3 dioxygenase (IDO; Wing et al., 1986). Even though IDO-mediated immunosuppression has been described in cancer (Mellor and Munn, 2004; Mellor et al., 2003), the current evidence indicates that L-arginine rather than L-tryptophan metabolism is used by MDSCs to alter T-lymphocyte reactivity to the antigen, as further discussed below.

4.2 IL-6

During the late stages of tumor growth, IL-1, IL-6 and acute-phase proteins are increased. High levels of IL-6 have been detected in leukemia, lymphoma, multiple myeloma, melanoma, as well as in breast, lung, ovarian, renal cell and pancreatic cancers (Trikha et al., 2003) and are associated with poor prognosis. The physiological activity of IL-6 is complex, producing both pro-inflammatory and anti-inflammatory effects. In addition, IL-6 affects the differentiation of myeloid lineages, including macrophages and DCs, both in vitro and in vivo (Park et al., 2004) through the activation of the transcription factor STAT3, which exerts a negative regulatory function on the adaptive and innate immune system during tumor development, as described below. The important role of IL-6 in inhibition of DC differentiation has been shown in multiple myeloma (Ratta et al., 2002). Moreover, soluble factors derived from the BM of patients with multiple myeloma inhibited the generation of DCs, and VEGF- and/or IL-6-specific antibodies neutralized this inhibitory effect (Hayashi et al., 2003). The same neutralizing effect can be accomplished by inhibiting the mitogen-activated protein kinase (MAPK) p38, which is activated in the cultured BM cells by co-culture with myeloma cells or exposure to tumor culture conditioning medium. Inhibiting p38 MAPK activity in BM cells cultured in the presence of tumor culture conditioning medium restored the generation of functional, BM-derived DCs (Wang et al., 2006).

4.3 Vascular Endothelial Growth Factor

VEGF plays an important role in the formation of blood vessels during embryogenesis, hematopoiesis and tumor neovascularization (Jain, 2005). It is secreted by most tumors and high levels correlate with a poor prognosis (Toi et al., 1996). Neutralizing antibodies against VEGF restored DCs differentiation from hemopoietic precursors blocked by tumor-conditioned media (Gabrilovich et al., 1998). VEGF has been directly linked with the systemic MDSC expansion. The administration of recombinant VEGF to tumor-free mice, in fact, resulted in inhibition of DC development and was associated with an increase in the number of MDSCs in the spleen (Gabrilovich et al., 1998). Moreover, tumor progression and multiplicity in transgenic female BALB-neuT mice, which spontaneously develop mammary carcinomas as mentioned earlier, correlated with the increased serum levels of VEGF and the progressive accumulation of MDSCs in the blood and spleen (Melani et al., 2003). Furthermore, recent findings showed that the link between MDSCs, VEGF and tumor progression could be complex and suggest that VEGF can be one of the molecules regulating the crosstalk between tumor and tumor-associated MDSCs (Gabrilovich, 2004). Yang et al. [2004], using the MC26 colorectal carcinoma and the Lewis lung carcinoma models, showed that MDSCs could stimulate tumor progression by promoting tumor angiogenesis. Tumor-associated MDSCs, in fact, express high levels of the matrix metalloprotease 9 (MMP-9). Deletion of MMP-9 in these cells completely abolished their tumor-promoting ability. MDSCs were also found to be incorporated directly into tumor endothelium and regulate the bioavailability of VEGF by releasing it from the extracellular matrix (Yang et al., 2004). These findings underlie the importance of MDSCs in tumor growth and progression, not only as effectors of tumor escape, but also by providing molecular and cellular components necessary for tumor neo-angiogenesis.

4.4 GM-CSF

Although GM-CSF has long been considered an immune adjuvant, recent evidence uncovered its dual role in stimulating as well as suppressing the immune system. Almost 31% of tested human tumor cells lines (including breast, cervical ovarian, prostate, colon, renal cancer as well as melanoma) secreted this cytokine (Bronte et al., 1999). GM-CSF is also secreted by many mouse cell lines such as squamous cells carcinoma (Smith et al., 1998), colon and mammary adenocarcinoma (Bronte et al., 1999) and plasmacytoma (Merchav et al., 1987). Moreover, Takeda et al. (1991) showed that the GM-CSF secretion correlated with the capacity to metastasize when various transplantable mouse tumors were injected subcutaneously. We showed that either tumor-transduced GM-CSF or the administration of recombinant GM-CSF protein in mice was sufficient to recruit MDSCs into the secondary lymphoid organs and suppress antigen-specific $CD8^+$ T cells (Bronte et al., 1999). MDSCs induced in vitro by culturing BM cells with Lewis lung carcinoma variant (LN7) supernatants could facilitate tumor engraftment once adoptively transferred into an immunocompetent mouse. The Lewis lung carcinoma-LN7 supernatants that contained the factor inducing the tumor-promoting cells also had colony stimulating factor activity. The ability of Lewis lung carcinoma cells to mediate both effects was completely abrogated by a combination of neutralizing antibodies against GM-CSF and IL-3 (Young et al., 1991). Moreover, mouse GM-CSF and IL-3 can synergize in vitro to induce an immunosuppressive phenotype of cultured BM cells (Young et al., 1990b).

On the other hand, GM-CSF has been shown to elicit powerful immune responses when combined with γ-irradiated tumor cell vaccines, in various mouse models and in the clinical setting (Dranoff, 2002, 2003), which has led to its widespread use as an immune adjuvant to augment antitumor immunity. Utilizing a bystander vaccine strategy in which the antigen dose and steric hindrance could be maintained constant while altering the GM-CSF dose, we assessed the impact of high versus low concentrations of GM-CSF administered in a vaccine formulation on priming of antitumor immunity. We confirmed the efficacy of low doses of GM-CSF secreting vaccine and defined a threshold above which the vaccine not only lost its efficacy but also resulted in significant in vivo immunosuppression mediated by MDSC recruitment (Serafini et al., 2004a). A recent systematic analysis of different clinical trials performed with this cytokine suggests that the same phenomenon can take place in humans. Although in some of these studies GM-CSF appeared to help the generation of an immune response, in others no effect or even a suppressive effect was reported. GM-CSF may increase the vaccine-induced immune response when administered repeatedly at relatively low doses (range 40–80 μg for 1–5 days) whereas an opposite effect was often reported at dosages between 100 and 500 μg (Parmiani et al., 2007). These findings support the dual role of GM-CSF on the immune response and highlight several critical parameters such as dose, systemic concentration, duration of exposure as key factors for GM-CSF effect on the immune system, which need to be considered when utilizing GM-CSF as a vaccine adjuvant.

4.5 IL-10

Elevated IL-10 concentrations have been found in patients with solid tumors and hematological malignancies (Pawelec, 2004) and are used as a marker of tumor progression (De Vita et al., 2000a,2000b). IL-10 production has often been correlated with the induction of T-cell anergy and, together with TGF-β, is considered one of the key immunosuppressive factors released by tumors (Chen et al., 2001). DCs cultured with IL-10 induce T-cell anergy and differentiation of suppressive T cells (Steinbrink et al., 2002). Myeloid DCs propagated from BALB/c (H2^d) mouse BM progenitors in IL-10 and TGF-β expressed lower toll-like receptor (TLR)4 transcripts than lypopolysaccharide (LPS)-stimulated control DCs and were resistant to further maturation (Lan et al., 2006). These DCs also expressed comparatively low levels of surface MHC class II, CD40, CD80, CD86 and programmed death-ligand 2 (B7-DC; CD273) and secreted high levels of IL-10, but low levels of IL-12p70 compared with activated control DCs (Lan et al., 2006). These "alternatively activated DC" induced alloantigen-specific hyporesponsive T-cell proliferation, enhanced IL-10 production by alloactivated T cells, expanded $CD4^+CD25^+Foxp3^+$ T-regulatory (Treg) cells in vitro and prolonged heart allograft survival when administered in vivo (Lan et al., 2006).

To investigate the activity of various cytokines on MDSC phenotypes, we immortalized splenic $CD11b^+/Gr1^+$ cells with a retrovirus encoding the v-*myc* and v-*raf* oncogenes (Apolloni et al., 2000). Two MDSC lines, MSC-1 and MSC-2, were selected based on their ability to inhibit antigen-specific proliferative and functional CTL responses. The MSC-1 line was constitutively inhibitory, whereas the MSC-2 line possessed a more immature phenotype and required further signals to generate a fully suppressive population. Treatment with IL-10 was able to provide such a signal. In fact, IL-10-pretreated MDSC-2 suppressed CTL generation in a mixed leukocyte reaction (MLR) when added as a third party through a mechanism that required the expression of ARG1 and NOS2 (Serafini, unpublished data), enzymes crucial for MDSC suppressive pathways (Bronte and Zanovello, 2005; further discussed below). Despite the clear relationship between IL-10 and the suppressive MDSC behavior, the role of tumor-derived IL-10 on MDSCs is still unclear and is a subject of ongoing investigation.

4.6 IL-13

IL-13 shares its receptor components with the IL-4 receptor, which explains why these cytokines share several (but not all) biological features. The promiscuous receptor for IL-4 and IL-13 (alias IL4R type II) is composed of the IL4Rα chain and IL13Rα1 chain (Terabe et al., 2004), while IL4Rα and the gamma chain (γc), common to the receptors for different members of the cytokine family comprising IL-2, IL-4, IL-7, IL-9, IL-15 and IL-21, associate to compose the IL-4 receptor (alias IL4R type I). Since the IL4Rα chain is the only component that possesses kinase-sensitive tyrosine residues in the cytoplasmic domain, signals from both type I and type II IL4R are transduced by the IL4Rα chain (Keegan et al., 1994). IL4Rα phosphorylation, upon engagement and dimerization, recruits and phosphorylates STAT6 that dimerizes and migrates to the nucleus to activate the transcription of several proteins including ARG1 (Gray et al., 2005). As previously described, the first evidence of IL-13 involvement in MDSC suppressive function derived from the work of Terabe et al. [2003], who showed that tumor recurrence in a fibrosarcoma murine model was dependent on MDSC activation by IL-13-secreting NKT cell. The authors demonstrated that IL-13 activated $CD11b^+Gr1^+$ cells, which in turn directly suppressed $CD8^+$ CTLs (Terabe et al., 2003). Tumor recurrence could be prevented either by MDSC depletion (Terabe et al., 2003) or by IL-13 neutralization (Terabe et al., 2000).

We recently showed that IL4Rα expression on MDSC is required for their suppressive phenotype, and that genetic ablation of this receptor on monocytes and granulocytes is sufficient to revert MDSC-mediated immune-suppression in vivo and, in conjunction with an adoptive cell transfer, eradicate an established colon carcinoma (Gallina et al., 2006). $CD11b^+IL4R\alpha^+$ cells produced IL-13 and IFN-γ and integrated the downstream signals of these cytokines to trigger the molecular pathways suppressing antigen-activated $CD8^+$ T lymphocytes (Gallina et al., 2006).

4.7 IFN-γ

IFN-γ plays a central role in coordinating tumor immunity, being the most relevant cytokine for immunosurveillance and immunoediting (Dunn et al., 2004). IFN-γ exerts its biological effects through interaction with an IFN-γ receptor that is ubiquitously expressed on nearly all cells (Bach et al., 1997). IFN-γ upregulates MHC class I expression as well as the expression of genes needed for antigen processing, including the transporters associated with antigen processing (TAP)-1 and TAP-2, and the proteasomal components named Low Molecular weight Proteins (LMP)-2 and LMP-7 (Seliger et al., 1996). For these reasons, IFN-γ is thought to augment the immunogenicity of many tumors. Moreover IFN-γ can inhibit tumor angiogenesis through direct or indirect mechanisms. In combination with TNF-α, in fact, IFN-γ directly reduces endothelial cell adhesion and survival by down-modulating the activation of $\alpha v \beta 3$ integrin, an adhesion receptor critical for tumor angiogenesis

(Ruegg et al., 1998). On the other hand, IFN-γ can indirectly repress angiogenesis by inducing the production of anti-angiogenic secondary molecules such as IP-10 and MIG (Dias et al., 1998; Sgadari et al., 1996). Although IFN-γ signaling in the tumor cell has been predominantly viewed as an important molecular pathway for effective antitumor immunity, significant evidence now indicates that, in some cases, it may negatively impact on the effectiveness of an antitumor immune response. IFN-γ signaling can, in fact, down-modulate the expression of tumor antigens (Beatty and Paterson, 2000) as well as induce the loss of efficient processing of some tumor antigens by DCs (Morel et al., 2000). Morel et al. [2000] reported that IFN-γ induction of LMP-2 and LMP-7 immunoproteasome results in less efficient processing of melanoma tumor antigens (e.g., MART-1/Melan-A) allowing for evasion of recognition by CTLs and decreased tumor immunogenicity. Moreover, gene expression profiling of tumor-associated macrophages as well as MDSCs indicates the presence of a distinct IFN-γ signature coupled with a M2 phenotype (Biswas et al., 2006). The importance of IFN-γ in activating MDSC suppressive activities has been reported by us and other groups. Synthesis of NO in macrophages is catalyzed by NOS2, whose expression is upregulated by a number of cytokines, including IFN-γ, TNF-α and IL-2 (MacMicking et al., 1997). We showed that IFN-γ together with a cell-mediated signal from activated splenocytes is necessary for generating the full suppressive activity and high levels of NO secretion on both fresh MDSCs and immortalized cell lines (Gallina et al., 2006; Mazzoni et al., 2002). These signals are produced by activated T cells, and in the absence of an activation signal, T cells do not stimulate NO production in MDSCs (Gallina et al., 2006; Mazzoni et al., 2002). More recently Huang et al. [2006] showed that the secretion of IL-10 and TGF-β by Gr1$^+$CD115$^+$ MDSCs was induced and enhanced upon IFN-γ stimulation. These IFN-γ-activated MDSCs, in addition to being able to suppress T-cell proliferation in vitro, were able to induce the development/expansion of forkhead box P3 (Foxp3)$^+$ Treg in vivo, when transferred in tumor-bearing mice (Huang et al., 2006). However, since the adoptive MDSC transfer experiments were conducted in irradiated tumor-bearing recipient, these findings do not exclude that additional tumor-derived factors or cytokines released by the irradiated host might be necessary for Treg expansion (Huang et al., 2006). Using the 4T1 murine mammary carcinoma in recipient mice in which T-cell response is skewed toward a Th1 response and Th2 response as well as IL-4/IL-13 pathways are impaired by genetic ablation of STAT6 (Kaplan and Grusby, 1998), Sinha et al. [2005b] suggested that IFN-γ is not sufficient per se to activate MDSC immunosuppression. In this model more than 60% of STAT6$^{-/-}$ mice immunologically rejected spontaneous metastatic mammary carcinoma and survived indefinitely when their primary tumors were removed, whereas 95% of STAT6-competent BALB/c mice succumbed to metastatic disease. Immunity in post-surgery STAT6-deficient mice was associated with a rapid decrease in the MDSC population and with the IFN-γ-dependent activation of type 1 tumoricidal macrophages. Under peculiar experimental conditions, such as the deletion of STAT6 in all the cells of the host, IFN-γ might thus favor an antitumor response even in the presence of MDSCs. Functional genomic analysis and experiments in cell-type-selective gene knock out mice have unveiled a complex interaction between Th1 and Th2 cytokines to activate MDSC suppressive program. To effectively exert their suppressive function on antigen-activated $CD8^+$ T cells, MDSCs must (1) be activated by IFN-γ production from antigen-stimulated T cells, (2) release their own IFN-γ and (3) be responsive to IL-13 (Gallina et al., 2006). Cooperation between these two cytokines leads to the activation of ARG1 and NOS2 that mediate MDSC suppressive activity (Bronte and Zanovello, 2005).

4.8 Prostaglandins (PGEs)

PGEs are strong immune modulators that are normally secreted in the immune responses by many cells including macrophages and DCs. Cyclooxygenase-2 (COX2) over-expression is a widely recognized feature of human lung, colon, breast cancer and prostate cancers (Gasparini et al., 2003). The products of COX2 enzyme activity, prostaglandins and mainly PGE2 have been implicated in tumor-associated subversion of immune functions, since inhibitors of prostaglandin synthesis typically enhanced antitumor immunity. Freshly excised solid human tumor cells produce substantially more PGE than established tumor cell lines (Sombroek et al., 2002): Interestingly, while primary tumor cell-conditioned media profoundly hampered the in vitro DC differentiation from $CD14^+$ monocytes or $CD34^+$ myeloid precursors, the effects of supernatants derived from established tumor cell lines were minor (Sombroek et al., 2002). In these experiments, COX1- and COX2-regulated prostanoids were found to be the exclusive responsible for the reduced differentiation of monocyte to DCs. In contrast, both PGE and IL-6 contributed to the tumor-induced inhibition of DC differentiation from $CD34^+$ myeloid precursor cells. A recent study showed that tumor-induced DC abnormalities were, at least in part, mediated by the prostaglandin EP2 receptor (Yang et al., 2003). Using the 3LL lung carcinoma model, Rodriguez et al. [2005] showed that ARG1 expression was independent on T–cell-produced cytokines. These tumor cells, in fact, constitutively express COX1 and COX2 and produced high levels of PGE_2 (Rodriguez et al., 2005). Genetic or pharmacological inhibition of COX2, but not COX1, blocked ARG1 induction in MDSCs both in vitro and in vivo. The authors showed that signaling through the PGE_2 receptor E-prostanoid 4 was required to induce ARG1 in MDSCs. Inhibition of this pathway was sufficient to block ARG1 expression, reverse MDSC-mediated immunosuppression and elicit a T-cell-mediated antitumor response (Rodriguez et al., 2005). More recently, celecoxib, a specific inhibitor of COX2, was found to normalize the number of MDSCs in Swiss mice in which intestinal tumor was chemically induced by 1,2-dimethylhydrazine-diHCl (Talmadge et al., 2007). Moreover COX2 inhibition decreased ARG1 and NOS2 expression in the secondary lymphoid organs, promoted T-cell infiltration in the tumor and, overall, reduced tumor multiplicity (Talmadge et al., 2007).

5 Transcription Factors Regulating MDSC Function

A central role in the polarization of myeloid cell functions, as well as in tumor progression and alteration of immune response to cancer, is emerging for selected members of the signal transducer and activator of transcription (STAT) family. In particular, the STAT1, 3 and 6 have been shown to play a major role in transmitting polarizing signals to the nucleus (Yoshimura, 2006) and each of them can play distinct roles in macrophage polarization and MDSC functions. A fundamental component of several signal-transduction pathways associated with STAT is the Janus activated kinase (JAK) family. These molecules are actively involved in cellular survival, proliferation, differentiation and apoptosis. In mammals, four members of the JAK family are known (JAK1, JAK2, JAK3 and TYK2; Rane and Reddy, 2000). Receptor oligomerization induced by cytokine binding triggers JAK activation by either auto- or *trans*-phosphorylation. Subsequent to ligand binding, activated JAKs phosphorylate receptors on target tyrosine residues, generating docking sites for STATs through the STAT Src homology 2 (SH2) domain. Activated JAKs recruit and phosphorylate STATs, which lead to their dimerization and nuclear translocation, where they modulate the expression of target genes.

5.1 STAT1

It is known that STAT1 negatively regulates angiogenesis, tumorigenicity and metastasis of tumor cells (Bromberg, 2002). Since STAT1 mediates IFN-dependent signaling, this transcription factor is an important mediator for antitumor immunity (Shankaran et al., 2001). On the other hand, using analysis of STAT activity in combination with STAT knockout mice, STAT1 emerged as an important player in tumor-associated MDSC suppressive activity (Kusmartsev and Gabrilovich 2005;). Tumor microenvironment and the inflammation, caused by both tumor growth and tissue invasion, in fact, promoted the differentiation and activation of $Gr1^+$ precursor recruited at the tumor site into ARG1 and NOS2 expressing MDSCs through a STAT1-dependent mechanism (Kusmartsev and Gabrilovich 2005;). MDSCs, isolated from the tumor of STAT1-deficient mice, in fact, failed to upregulate ARG1 and NOS2 and were unable to suppress the proliferation of anti-CD3/anti-CD28 stimulated splenocytes (Kusmartsev and Gabrilovich 2005;). Moreover, in a mouse squamous cell carcinoma, STAT1 deficiency enhanced IL-12-mediated tumor regression, by a T-cell-dependent mechanism (Das et al., 2001). In agreement with the role of STAT1 as central mediator of IFN-γ biological activities, administration of neutralizing antibodies against IFN-γ inhibited tumor growth in IL-12-treated, $STAT1^{+/+}$ mice (Das et al., 2001). These data might have an experimental confirmation by the fact that IFN-γ produced by activated T cell and by MDSCs themselves is required to trigger NOS2 activation and synergize with IL4Rα-ARG1 pathways in MDSCs isolated from tumor mass (Gallina et al., 2006). In line with this picture, mice deficient for the suppressor of cytokine signaling-1 factor (SOCS-1), which are characterized by hyperactivation of STAT1, displayed spontaneous development of colorectal carcinomas (Hanada et al., 2006). The negative role of STAT1 activation in cancer seems to be confirmed in some human cancers since, by analyzing tumor associated macrophages (TAMs) derived from 211 patients affected with follicular lymphoma, the presence of STAT1 in TAMs was an important independent prognostic factor that correlated with an adverse outcome (Alvaro et al., 2006).

5.2 STAT3

STAT3 is activated in many human cancers, including 82% of prostate cancers (Mora et al., 2002), 70% of breast cancers (Dolled-Filhart et al., 2003), more than 82% of squamous cell carcinoma of the head and neck (Nagpal et al., 2002) and 71% of nasopharyngeal carcinoma (Hsiao et al., 2003). STAT3 participates in oncogenesis through the upregulation of genes encoding apoptosis inhibitors (Bcl-x_L, Mcl-1 and survivin), cell cycle regulators (cyclin D_1 and c-Myc) and inducers of angiogenesis such as VEGF (Buettner et al., 2002). Recent studies suggest that STAT3 activation in tumors might play an important role not only in maintaining the transformed phenotype in tumor cells, but also in inhibiting immune surveillance (Wang et al., 2004). The STAT3 signaling pathway in tumor cells can, in fact, inhibit production of pro-inflammatory danger signals and induce expression of factors that inhibit DC functional maturation (Wang et al., 2004). Moreover, STAT3 expression in macrophages has been associated with their ability to induce T-cell tolerance, whereas targeted disruption of *Stat3* gene in these cells stimulated production of pro-inflammatory cytokines and abrogated their tolerogenic features (Cheng et al., 2003). It is noteworthy that ablating STAT3 in hematopoietic cells triggers an intrinsic immune-surveillance system that inhibits tumor growth and metastasis (Kortylewski et al., 2005). Incubation of hematopoietic progenitor cells with tumor cell-conditioned medium resulted in the activation of JAK2 and STAT3 and was associated with an accumulation of MDSCs. Importantly, MDSCs derived from tumor-bearing mice demonstrated constitutive activation of JAK2/STAT3 pathway (Nefedova et al., 2004). Inhibition of STAT3 activation in hematopoietic progenitor cells via dominant negative STAT3D retroviral vector or with the use of JAK2/STAT3-specific small molecule inhibitors abrogated the effect of tumor-derived factors on the generation/activation of MDSCs (Nefedova et al., 2004, 2005).

STAT3 is also essential for signaling through the IL-10 receptor since mice lacking STAT3 in macrophages and neutrophils have a strikingly similar phenotype to IL-10-deficient mice (Takeda et al., 1999). It is interesting to note that these mice develop chronic enterocolitis with age, likely through a complex contributory network including polarized immune response toward the Th1 phenotype, overexpression of pro-inflammatory cytokines and deficiency of the immunosuppressive action of macrophages and neutrophils (Takeda et al., 1999). Activation of STAT3, via IL-10, upregulates the α-chain of the IL-4 receptor that leads to an increased

IL-4-dependent expression of ARG1 (Lang et al., 2002). Furthermore, IL-10 synergizes with LPS in inducing NOS2. However, NOS2 regulation by activation of STAT3 needs further study since STAT3 has been reported either to activate (Finder et al., 2001; Yu et al., 2003) or repress (Yu et al., 2002) NOS2 expression.

5.3 STAT6

STAT6 is another member of the STAT family which has attracted attention since mice deficient for the STAT6 gene have enhanced immunosurveillance against primary and metastatic tumors. Moreover, STAT6 is a downstream transcription factor for IL4R and IL13R whose role in MDSCs activation has been established by different studies. More than 60 % of STAT6$^{-/-}$ mice immunologically reject spontaneous metastatic mammary carcinoma and survive indefinitely if their primary tumors are removed, whereas 95 % of STAT6-competent BALB/c mice succumb to metastatic disease (Ostrand-Rosenberg et al., 2002; Sinha et al., 2005b). The authors suggested that STAT6 deficiency prevents signaling through the type 2 IL4R, thereby blocking the production of ARG1 and promoting the synthesis of NO by myeloid cells. The importance of this pathway in MDSC-mediated suppression has been further demonstrated in the fibrosarcoma model described before. In this model, in fact, tumor recurrence was completely prevented in STAT6$^{-/-}$ mice (Terabe et al., 2000).

6 Mechanisms of MDSC-Dependent Immune Suppression

Although it is clear that MDSCs can inhibit the immune response against cancer, it must be pointed out that MDSCs can be present in various functional differentiation grades that might explain the prevalence of the different immunosuppressive mechanisms described in different tumor models (Serafini, 2006a). These functional and phenotypic differences can be related to the status of the disease, MDSC localization in different anatomical compartments or the different microenvironments that each tumor can establish. MDSC suppressive/tolerizing activity in vivo can be dependent on the expression of MHC class I on their surface as elegantly shown by Kusmartsev et al. [2005]. Using an experimental system based on the adoptive transfer of transgenic T cells into naïve recipients, the authors showed that Gr1$^+$ MDSCs as well as DCs from tumor-bearing mice were able to uptake and process tumor-associated antigens. However, while DCs did not reduce the generation of tumor-specific T-cell CD8$^+$ producing IFN-γ, MDSCs were able to induce anergy of CD8$^+$ T cell that no longer responded to peptide stimulation (Kusmartsev et al., 2005). This tolerogenic state could be rescued in vivo through immunization with mature DCs. Taken together, these data suggest that the tolerogenic state is reversible and that the balance between mature DCs and MDSCs in secondary lymphoid organs can determine the final outcome of the immune response. Although these cells can produce high level of ARG1 and ROS (Kusmartsev et al., 2005) their tolerogenic activity seemed to be dependent mostly on NO production since L-NMMA, an inhibitor of various NOS, completely reverted the ability of these MDSCs to tolerize $CD8^+$ T cells (Gabrilovich et al., 2001).

In other situations, however, MDSCs were shown to be powerful and unselective inhibitors since they not only inhibited peptide, mitogen or anti-CD3/CD28 activated $CD4^+$ and $CD8^+$ T cells (Serafini et al., 2004b) but also activated NK and NK T cells (Liu et al., 2007; Suzuki et al., 2005). Altogether these data indicate that MDSCs can also suppress T and NK cells in an antigen- and MHC-independent fashion. The concept of antigen independence might be misleading, since MDSCs inhibit only activated T lymphocytes, either naïve or memory, whereas resting lymphocytes are spared. The necessity of activation of effector T cells, combined with the fact that MDSCs need to be in strict contact with T cells to deliver the inhibitory signals (Serafini, 2006a), suggest that MDSC suppressive activities are endowed with some degree of selectivity, even in the absence of an MHC-restricted suppression. MDSCs can restrain the immune response through different mechanisms that operate singularly or in combination. Such mechanisms can be direct or indirect, in this latter case involving the generation or the expansion of other regulatory population such as $CD4^+CD25^+$ Treg cells (Fig. 1).

6.1 Direct Mechanisms of Immune Suppression

6.1.1 TGF-β

The link between TGF-β and MDSCs was shown first by Young et al. [1996] that demonstrated that myeloid progenitor cells derived from tumor-bearing mice produced increased amounts of TGF-β, NO, IL-10 and PGE_2. NO and TGF-β, however, were the mediators by which MDSCs inhibited in vitro the anti-CD3 antibody-induced T-cell proliferation. $Gr1^+$ cells were proposed to bind the IgG-TGF-β complex on their Fc receptors and the binding could trigger these immature myeloid cells to suppress CTL response (Beck et al., 2003). Moreover, as described above, MDSCs can be activated by IL-13 to secrete TGF-β (Terabe et al., 2003). TGF-β, per se, possesses anti-proliferative effect on T cells (Kehrl et al., 1986), arresting their cell cycle typically in the G1 phase (Morris et al., 1989; Stoeck et al., 1989) by inducing the expression of the cell cycle inhibitors p27KIP1 and p21CIP1 (Wolfraim et al., 2004) or by inhibiting IL-2 secretion (Brabletz et al., 1993). Importantly TGF-β has been shown to inhibit the differentiation of $CD4^+$ T cells into Th1 or Th2 cells by suppressing the expression of T-bet and GATA-3 master regulators of Th1 and Th2 conversion, respectively (Becker et al., 2006). Despite the evidence that TGF-β is essential for the maintenance of peripheral tolerance, the mechanism by which TGF-β acts remains unclear. TGF-β may directly be important for the induction of peripheral tolerance by downregulating the differentiation and function of auto-reactive effector T cells as described above. Alternatively, TGF-β may play a role in the induction of Treg cells, which then inhibit T-cell effector function, as further discussed below (Becker et al., 2006).

6.1.2 L-Arginine Metabolism

Many of the inhibitory pathways involved in MDSC-mediated immune suppression are related to the metabolism of the amino acid L-arginine (L-Arg) (reviewed in Bronte and Zanovello, 2005, and discussed by A. Ochoa in another chapter of this book). L-Arg is metabolized mainly by two enzymes: NOS, which oxidizes L-Arg in two steps that generate NO and citrulline; and ARG, which converts L-Arg into urea and L-ornithine (Bogdan, 2001; Wu and Morris, 1998).

ARG-Dependent Suppression

MDSCs infiltrating a mouse lung carcinoma expressed high levels of ARG1 and the L-Arg transporter CAT2B (Rodriguez et al., 2004). These myeloid cells readily consumed L-Arg and inhibited re-expression of the ζ-chain of CD3 complex in T lymphocytes thereby impairing their function. The CD3 ζ chain is the main signal-transduction component of the TCR complex and is required for the correct assembly of the receptor, and altered expression has been detected in peripheral blood T cells in patients with cancer, chronic infections and autoimmune diseases (Baniyash, 2004). This mechanism of T-cell inactivation by ARG-induced deregulation of CD3 ζ chain was shown to be relevant for tumor escape in vivo, because injection of the ARG inhibitor *N*-hydroxy-nor-L-arginine (nor-NOHA) delayed the growth of transplantable lung carcinoma in a dose-dependent manner (Rodriguez et al., 2004).

NOS-Dependent Suppression

The ability of NOS inhibitors to reverse MDSC-induced immunosuppression, both in vivo and in vitro, confirms the immuno-regulatory role of NO (Bogdan, 2001; Bronte and Zanovello, 2005). NO-mediated suppression of T-cell activation is not associated with the early events triggered by TCR recognition but, instead, with the signaling cascade downstream of the IL-2 receptor (Mazzoni et al., 2002). NO is known to negatively regulate intracellular-signaling proteins either directly by S-nitrosylation of crucial cysteine residues or indirectly by activation of soluble guanylate cyclase and cyclic-GMP-dependent protein kinases (Bingisser et al., 1998; Duhe et al., 1998; Fischer et al., 2001). In T cells the phosphorylation, and thus the activation of important signaling proteins in the IL-2-receptor pathway including JAK1, JAK3, STAT5, extracellular-signal-regulated kinase (ERK) and AKT, is blocked by NO (Bingisser et al., 1998; Mazzoni et al., 2002). Persistent release of NO by MDSCs might also be associated with the transcriptional loss of STAT5A and STAT5B in T and B cells, observed in mice bearing large mammary carcinomas and in individuals infected with HIV (Pericle et al., 1997, 1998) and might, therefore, be responsible for the impaired T-cell function observed under these conditions. A direct pro-apoptotic effect has also been observed in T cells exposed to high concentrations of NO likely mediated by numerous factors such as the accumulation of the tumor-suppressor protein p53, signaling through CD95 (also known as Fas) or TNF-receptor family members, or signaling through caspase-independent pathways (Macphail et al., 2003; Mannick et al., 1999).

ARG and NOS Cooperation in Suppression

Synergism between these enzymes in MDSCs was difficult to envision considering reports indicating that ARG activation limits the availability of L-Arg as a substrate for NOS and thus negatively regulates its enzymatic activity (Munder et al., 1999). However, many reports recently showed that these two enzymes can be co-expressed within the same population or microenvironment (Bronte et al., 2003a, 2005; Bruch-Gerharz et al., 2003; Brys et al., 2005; Currie et al., 1979; De Santo et al., 2005; Gallina et al., 2006; Kusmartsev and Gabrilovich 2005;; Serafini et al., 2006b). When these two enzymes are co-expressed, ARG1, by lowering the L-Arg concentration in the local environment, operates to switch NOS2 activity, shifting its function from the production of NO to O_2^- (Bronte et al., 2003a; Xia and Zweier, 1997; Xia et al., 1998). When L-Arg concentrations are suboptimal, the reductase and oxygenase domains of NOS2 transfer electrons to the co-substrate O_2 and produce O_2^-, which reacts with other molecules, thereby generating several reactive nitrogen intermediates (RNI), such as peroxynitrite, and reactive oxygen species (ROS), such as hydrogen peroxide (H_2O_2). These species have multiple inhibitory effects on T cells. The combined activity of ARG and NOS was recently shown to be important for the suppressive activity of tumor-infiltrating CD11b$^+$ myeloid cells (De Santo et al., 2005; Kusmartsev et al., 2005) and splenic CD11b$^+$Gr1$^+$ cells from mice bearing subcutaneous tumors or in models of chronic helminthic infections (Bronte et al., 2003a; Brys et al., 2005).

6.1.3 Reactive Oxygen Species

In addition to amino acid starvation, MDSCs can block T-cell function through the production of highly oxidative ROS, as previously mentioned. H_2O_2 production by macrophages infiltrating metastatic melanoma induced the loss of CD3 ζ chain in naive T cells (Kono et al., 1996a,1996b; Otsuji et al., 1996). Moreover an increase in CD11b$^+$ CD15$^+$ granulocytes was observed in patients with pancreatic cancer (Schmielau and Finn, 2001). These cells reduced CD3 ζ expression and decreased cytokine production in T cells through a H_2O_2-mediated mechanism (Schmielau and Finn, 2001). Moreover, MDSCs freshly isolated from tumor-bearing mice but not their control counterparts were able to inhibit antigen-specific response of CD8$^+$ T cells (Kusmartsev et al., 2004). These MDSCs obtained from tumor-bearing mice had significantly higher levels of ROS than Gr1$^+$ cell isolated from tumor-free animals. Since ROS production could be blocked by ARG inhibitors, these data suggest that ARG could be involved in the mechanisms of T-cell inhibition through generation of ROS and may link ARG1 to T-cell dysfunction observed at the tumor site (Kusmartsev et al., 2004).

The mechanism underlying the preferential H_2O_2 generation following ARG activation is currently not known but it might be linked to the contemporaneous activation of different NOS isoforms. Under conditions of limited availability of L-Arg not only NOS2 but also NOS1 (also called nNOS) and NOS3 (also called eNOS) produce O_2^- (Andrew and Mayer, 1999). The only major difference between the NOS isoforms in terms of the reactions performed lies in the rate of this NADPH-dependent oxidation, termed the "uncoupled reaction". Under these conditions, NOS1 continues to transfer electrons to the heme and hence oxidize NADPH at a high rate, whereas in NOS3 and NOS2, this reaction occurs at a much slower rate (Andrew and Mayer, 1999). While NOS2 at low concentration of L-Arg produces both NO and O_2^- (Andrew and Mayer, 1999), NOS1 in the same condition produces O_2^- and H_2O_2 (Tsai et al., 2005), but not NO (Que et al., 2002). These data, which await to be confirmed in MDSCs, suggest a scenario where NOS isoform expression determines the final molecular mediator of ARG-dependent suppression.

6.2 Indirect Mechanism of Immune Suppression: Regulation of $CD4^+$ $CD25^+$ Treg Homeostasis

Considerable interest was recently raised by the hypothesis about a link between MDSCs and $CD4^+CD25^+$ Treg cells. MDSCs, in fact, share many features with immature DCs (e.g., low expression of MHC class II, CD80 expression, antigen uptake capacity, etc.) that have often been proposed to be associated with either T-cell tolerization or Treg cell expansion. Mahnke et al. [2003] demonstrated that specific in vivo targeting of immature DCs with the mAb anti-DEC-205 coupled to various antigens resulted in the presentation of the antigens in a tolerizing context. Using ovalbumin (OVA) as a model antigen, the initial expansion of OVA-specific T cells was followed by anergy and appearance of T cells expressing CD25 and CTLA-4. Functional analysis of this T-cell population revealed that $CD25^+$ T cells from the anti-DEC-OVA complex-injected animals suppressed proliferation and IL-2 production of conventional $CD4^+$ T cells in a cell-contact-dependent way. Depletion of $CD25^+$ T cells from bulk T-cell cultures restored T-cell proliferation (Mahnke et al., 2003). The first evidence of a connection between MDSCs and Treg was provided in a model of allogeneic BM transplantation (MacDonald et al., 2005). $CD11b^+/Gr1^+$ MDSCs, expanded in vivo by Progenipoietin-1 (a synthetic G-CSF/Flt-3 ligand molecule) administration, were found to suppress the initiation of graft-versus-host disease (GVHD) after allogeneic BM transplantation by inducing a population of MHC class-II-restricted Treg producing IL-10 (MacDonald et al., 2005). Moreover since either plasmocytoid or myeloid DCs, expanded with the same molecules, were unable to affect GVHD, these experiments suggested a prominent role of MDSCs in Treg induction and unveiled a new role of MDSCs in regulating peripheral tolerance. The importance of tumor-conditioned infiltrating cells in controlling Treg homeostasis was recently shown in a melanoma mouse model and a colon carcinoma rat model. Ghiringhelli et al. [2005] reported that, during tumor progression, Treg cells accumulate in tumors and secondary lymphoid organs through a mechanism that mainly required the proliferation of pre-existing natural Treg in the draining lymph nodes and in the tumor bed. In both rodent models this proliferation was dependent on the accumulation of TGF-β-secreting CD11b$^+$CD11c$^+$MHC-Cl2^{low} cells in the tumor draining lymph nodes (Ghiringhelli et al., 2005). This proliferation was significantly reduced in TGF-β RII$^{-/-}$ animals underscoring the importance of the TGF-β pathway in Treg proliferation and tumor-tolerance induction (Ghiringhelli et al., 2005). Importantly, the tolerogenic TGF-β secreting APCs could be recreated in vitro by incubating CD11c from tumor-free mice with the supernatant of tumor-conditioned media, suggesting that these tolerogenic APCs might be related to MDSCs (Ghiringhelli et al., 2005).

By using the colon carcinoma mouse model MCA stably transformed with the influenza hemoagglutinin (HA) antigen, Huang et al. [2006] showed that MDSCs from tumor-bearing mice could suppress the expansion of effector HA-specific CD25$^-$CD4$^+$ T cell through a NOS2-mediated mechanism and also generate or expand the pool of CD4$^+$CD25$^+$ Foxp3$^+$ Treg cells (Huang et al., 2006). In vitro experiments performed by this group showed, in fact, that while HA-specific CD4$^+$CD25$^-$ T cells cultured with MDSCs failed to proliferate to the cognate antigen, the percentage of HA-specific CD4$^+$CD25$^+$Foxp3$^+$ T cells in culture was significantly augmented. However, since the number of cells recovered per well was not reported, it is difficult to determine whether this is a real conversion of effector cell into Treg cells or whether the Treg percentage increase was an indirect consequence of effector T cell death (Huang et al., 2006). The in vivo experiments, however, clearly showed that adoptive co-transfer of MDSCs and HA-specific T cells into irradiated tumor-bearing mice resulted in an increase in the number of CD4$^+$, Foxp3$^+$, HA-specific, T cells with regulatory capacity (Huang et al., 2006). This increase was mediated by TGF-β and IL-10 production by MDSCs as well as MDSC activation by IFN-γ (Huang et al., 2006). However, since adoptively transferred HA-specific T cells contain usually 5–10 % of natural Treg, it is still not clear whether MDSCs can mediate conversion of Treg from effector T cells or they play a role only in the expansion of pre-existing natural Treg population.

Recent findings, however, suggest that the relationship between MDSCs and Treg is not limited to the homeostatic control of the CD4$^+$ regulatory population. Yang et al. (2006) showed that a mouse ovarian carcinoma (MOSEC line 1D8) triggered the accumulation of MDSCs and CD4$^+$CD25$^+$ Treg cells in spleen, ascites and tumor tissue. Since genetic ablation and antibody blockade of either CD80 or its ligand CD152 significantly retarded tumor growth (Yang et al., 2006), the authors suggested that tumor-mediated CD80 upregulation on MDSC was important for immune evasion and tumor progression. Interestingly, in vitro experiments examining the suppressive activity of Treg cells and MDSCs revealed that both populations were simultaneously necessary to inhibit IFN-γ production from antigen-specific T cells stimulated with the cognate peptide. Moreover CD80 neutralization experiments showed that the engagement of CD80 on MDSC with CD152 expressed by Treg cells was required for MDSC-Treg cell cooperation in inducing IFN-γ suppression. Since binding of CD152-Ig to DCs was shown to induce T-cell anergy by upregulating the expression of IDO (Mellor and Munn, 2004), binding of CD80 and CD152 may also activate MDSC suppressive program, suggesting that in some cases MDSCs and not Treg cells are the final effectors of immune suppression.

7 Human MDSCs

MDSCs have been described in patients affected by different tumors. As in the case of mouse MDSCs, however, the phenotype of these cells is not fully defined and both immature and mature myeloid cells have been described. In head and neck cancer patients, for example, the release GM-CSF and the tumor infiltration with $CD34^+$ were determined to be negative prognostic factors because they were associated with an increased rate of tumor and metastasis recurrence (Young et al., 1997). Moreover the increased number of $CD34^+$ cells in the PBMCs of these patients was associated with the suppression of the anamnestic responses to recall antigens, a frequent characteristic in head and neck cancer patients (Pak et al., 1995b). Interestingly, exposure of $CD34^+$ suppressors to the cytokine combination GM-CSF + IL-4 induced the maturation of the immature suppressor cells into DCs, with a parallel reversal of their immunosuppressive properties (Garrity et al., 1997). A more extensive study identified human MDSCs in the peripheral blood of patients with squamous cell carcinoma, head and neck cancer, breast cancer and non-small-cell lung cancer. In these cases, MDSCs were described as immature cells positive for the marker CD34, CD33 and CD13, but negative for the myelomonocytic marker CD15. The variable expression of HLA-DR and CD11c molecules allowed the identification of two main populations: one third of the cells were immature monocyte/DCs, and the remaining cells encompassed earlier myeloid differentiation stages. Like mouse MDSCs, human immature cells caused suppression of antigen- and mitogen-induced T-lymphocyte proliferation, and the combination of GM-CSF and IL-4 drove their differentiation to mature DCs (Almand, 2001).

This phenotypic characterization, however, has not been confirmed in other malignancies or in different disease stages. Analysis of PBMCs, from patients affected by metastatic adenocarcinomas of the pancreas, colon and breast cancer, revealed an increase of the oxidative activity of $CD15^+$ granulocytes that resulted in an elevated ROS production. Granulocyte activation correlated with the inhibition of TCR ζ chain expression and cytokine production (Schmielau and Finn, 2001). PBMCs from 123 patients with metastatic renal cell carcinoma had an increase in ARG activity that was associated with the downregulation of the CD3 ζ chain expression and reduction of IL-2 and IFN-γ production by anti-CD3/anti-CD28 stimulated PBMCs (Zea et al., 2005). Cell fractionation studies revealed that ARG activity was limited to $CD11b^+CD15^+CD14^-$ polymorphonuclear granulocytes and depletion of $CD11b^+$ cell from PBMCs was sufficient to restore ζ-chain expression, cytokine production and proliferation of otherwise anergic T cells present among PBMCs (Zea et al., 2005). These data suggest that granulocytes can, in some situations, act as the mouse MDSCs but other data indicate that $CD14^+$ cells might also contribute to tumor-induced suppression. By analyzing multiple myeloma and head and neck cancer patients with a similar strategy, we recently identified a subset of $CD14^+$ monocytes characterized by an elevated ARG activity (Serafini et al., 2006b). Depletion of $CD14^+$ cells or pharmacologic inhibition of ARG1 and NOS2 activity was sufficient to restore the proliferation capacity of PBMCs upon stimulation with anti-CD3/anti-CD28 coated beads (Serafini et al., 2006b). Similar findings were recently shown in a clinical trial in which stage IV melanoma patients were vaccinated with the heat shock protein gp96, with or without GM-CSF as adjuvant to better prime the immune response (Parmiani et al., 2006). Similar to what we reported in mice, where high doses of GM-CSF secreting vaccine restrained the immune response through the recruitment of MDSCs (Serafini et al., 2004a), GM-CSF was shown to lower instead of increase the frequency of melanoma antigen-specific T cells, as well as their capacity to secrete IFN-γ (Parmiani et al., 2006). Increased frequency of immature $CD14^+$ $HLA-DR^-$ TGF-β producing myeloid cell was found in the PBMCs of GM-CSF-treated patients and was correlated with the lack of anti-melanoma T-cell response (Parmiani et al., 2006). Taken together, the existing data on human MDSCs indicate that these cells share many of the functional properties found in mice. However, it is still very problematic to associate a unique panel of markers to human MDSCs. This difficulty can depend on the great plasticity and accepted heterogeneity that characterize MDSCs. Phenotypic differences in MDSCs can, in fact, reflect intrinsic differences in human cancers such as tumor stage, patients' age and therapeutic history or simply the genetic variation, which is much higher in humans than in laboratory mouse strains.

8 Therapeutic Approaches

In the last few years, it has become widely accepted that it is necessary not only to stimulate the antitumor T-cell response but also to subvert the suppressive network and tolerogenic microenvironment associated with cancer progression, in order to accomplish an immune-mediated eradication of cancer. Since MDSCs play an important role in tumor-induced immunosuppression, many groups exploited different strategies to achieve their depletion, differentiation into mature APCs or their pharmacological inhibition (Fig. 1).

8.1 In Vivo Depletion of MDSCs

The treatment of mice with the monoclonal anti-Gr1 antibody (clone RB6-8C5) was shown to slow tumor progression of an aggressive variant of an UV-induced tumor (Seung et al., 1995). This variant cell line differed from the parental tumor by its ability to secrete leukocyte chemotactic factors. Moreover, while the parental line induced a CTL response resulting in tumor eradication, the variant line continued to grow, even though it was still recognized by tumor-specific CTLs in vitro. $Gr1^+$ cell depletion enhanced $CD8^+$ T-cell-mediated immune responsiveness and resulted in the rejection of the variant tumor cell line (Seung et al., 1995). The same anti-Gr1 antibody has been used for mechanistic studies to underscore the importance of MDSCs in the suppressive pathways of mice acutely infected with vaccinia virus or in mice bearing various tumors (Bronte et al., 2000; Seung et al., 1995; Terabe et al., 2003). However, its use in a therapeutic setting seems untenable. First, Gr1 is not solely expressed on MDSCs but also found on neutrophils, which, if depleted, could increase the incidence of opportunistic infections. Second, the discontinuation of the antibody treatment results in a rapid increase of MDSC numbers and, thereby, restores immune suppression (unpublished observations). As such, antibody treatment requires a careful dose titration and constant monitoring of the immune function in the treated hosts.

A more effective strategy to deplete MDSCs was recently proposed (Luo et al., 2006). DNA-based vaccination against legumain (a member of the asparaginyl endopeptidase family over-expressed by TAMs and MDSCs) effectively depleted tumor-infiltrating myelomonocytic cells and, more importantly, induced tumor regression by reducing both tumor angiogenesis and the suppressive properties of tumor microenvironment (Luo et al., 2006). This study suggests a new, promising strategy by which immune-mediated depletion of TAMs and MDSCs in the tumor stroma might decrease the release of factors promoting tumor growth, angiogenesis and metastatic capacity (Colombo and Mantovani, 2005). Moreover since vaccination is direct against accessory cells and not malignant cells, this strategy should circumvent tumor-escape mechanisms based on antigenic loss that, until now, represents one of the most important barriers for an affective immunotherapy.

Other therapeutic strategies to deplete MDSC in vivo include the use of Gemcitabine or host irradiation. The chemotherapeutic drug Gemcitabine, in fact, given at dose similar or equivalent to that used in patients, reduced substantially the number of splenic MDSCs in tumor-bearing mice, whereas the number of $CD4^+$ T cells, $CD8^+$ T cells, NK cells, macrophages or B cells were not apparently affected (Suzuki et al., 2005). MDSC depletion increased the antitumor activity of $CD8^+$ T cells and activated NK cells (Suzuki et al., 2005). However, a note of caution about the interpretation of these data is necessary since T cells undergo homeostatic proliferation after Gemcitabine treatment (Serafini, unpublished data).

8.2 MDSC Differentiation in Fully Mature APCs

MDSC differentiation into functional APCs by cytokines or small molecules can be an intriguing strategy. In fact, this differentiation would not only remove MDSC suppressive mechanisms, but also provide tumor-antigen-loaded APCs that should potentiate the immune response against the malignancies, since tumor-associated MDSCs can uptake tumor antigens (Kusmartsev et al., 2005; Zhang et al., 2007). In vivo administration of all-*trans*-retinoic acid (ATRA) reduced the presence of MDSCs in different tumor models and this effect was not a consequence of a direct ATRA-antitumor effect (Kusmartsev, 2003). ATRA, indeed, differentiated MDSCs in vivo into mature DCs, macrophages and granulocytes and significantly increased the efficacy of antitumor vaccine. Additional combinations have been used to force MDSC maturation. Low doses of IFN-γ plus TNF, for example, reduced the number of MDSCs in a metastatic Lewis lung carcinoma model by forcing MDSC differentiation into mature macrophages (Pak et al., 1995a). When these cytokines were administered in vivo with a regimen of high-dose IL-2, the numbers of $CD4^+$ and $CD8^+$ T cells within the tumor mass increased and the combined treatment reduced the size of the primary tumor and the number of pulmonary metastases more efficiently than the individual treatments (Pak et al., 1995a). Whereas ATRA and IFN-γ plus TNF directly affected MDSC maturation, 1α25-dihydroxyvitamin D_3 had a more indirect action since it reduced MDSC number and function in tumor-bearing mice by interfering with GM-CSF production by malignant cells (Young et al., 1996). This finding was confirmed in a phase IB clinical trial conducted with head and neck cancer patients. Treatment with 25-hydroxyvitamin D_3 reduced the number of immune-suppressive $CD34^+$ cells, increased HLA-DR expression, augmented the serological concentration of IL-12 and IFN-γ and improved T-cell blastogenesis (Lathers et al., 2004).

Cyclooxygenase2 (COX2) is another pharmacologic target that is expressed by tumor-associated fibroblast as well as many cancerous cells including colon, cervical, lung, skin, bladder and pancreas carcinomas as well as Burkitt-type B-cell lymphoma (Baglole et al., 2006). Rodriguez et al. [2005] recently showed that PGE_2 produced by COX2, expressed by the 3LL murine lung carcinoma, induced ARG1 in MDSCs. Genetic and pharmacological inhibition of COX2 in vivo blocked ARG1 induction in MDSCs and it was sufficient to stimulate an antitumor immune response able to eradicate the neoplastic lesions (Rodriguez et al., 2005).

8.3 STAT3 Inhibition

STAT3 activation plays an important role both in tumor progression for its anti-apoptotic effect on neoplastic cells and its ability to alter tumor immune surveillance by blocking DC maturation and promoting the accumulation of MDSCs. Inhibition of STAT3 phosphorylation could, thus, be a useful therapeutic approach for both its direct antitumor effect (Buettner et al., 2002) and its indirect immune-mediated role (Nefedova et al., 2004). To date, most of the studies have focused on the direct pro-apoptotic role of STAT3 inhibition on tumors (Nikitakis et al., 2004). In contrast, its role on immune function is less understood. The effect of STAT3 inhibition on DC maturation has been explored in vitro using retroviral vectors encoding the dominant negative form of STAT3 (STAT3D; Nefedova et al., 2004). Since STAT3D protein bears a mutation in the DNA binding domain, it is still capable to dimerize with wild-type STAT3 protein but this heterodimer no longer promotes DNA transcription. BM hematopoietic cells infected with the STAT3D-encoding virus and cultured with GM-CSF + IL-4 and TDFs failed to generate MDSCs but evolved into fully mature DCs (Nefedova et al., 2004). These results support the rationale for targeting STAT3 as a tool to overcome tumor-associated suppression. While prevention of STAT3-mediated induction of MDSCs has shown interesting preclinical results, it is less clear whether this approach can be utilized to overcome the immune suppression of terminally differentiated MDSCs and further experiments are required to understand the effect of these strategies at later points during tumor progression. Preliminary data in our laboratory indicate that the phenotype and the function of MDSCs may vary with the time of exposure and/or the concentration of TDFs (Serafini, unpublished data). Moreover, tumor-infiltrating MDSCs present a more suppressive phenotype (Kusmartsev and Gabrilovich 2005;) that could be more difficult to revert.

8.4 Pharmacological Inhibition of MDSC Suppressive Pathways

As discussed above and in greater detail in another chapter of this book by Augusto Ochoa and collaborators, MDSC suppressive activity is mainly dependent on the metabolism of the semi-essential amino acid L-Arg. In particular, two enzymes, NOS2 and ARG1, are responsible for the mechanisms by which MDSCs can restrain immune function (Bronte and Zanovello, 2005). L-Arg metabolism by ARG and NOS can thus represent a promising target to overcome MDSC-induced immune suppression. MDSCs recovered from melanoma-bearing NOS2 knock out (C57BL/6-NOS2$^{-/-}$) mice do not suppress the alloantigen- or peptide-stimulated T cells, whereas CTL generation was impaired when the same cells were recovered from melanoma-bearing C57BL/6-NOS2$^{+/+}$ mice (Serafini et al., 2006b). As described above, however, the prevalence of either ARG or NOS pathway in MDSCs depends on multiple factors that are not completely understood. In CT26 tumor-bearing BALB/c mice, for example, MDSC suppressive activities are mediated by the co-expression of ARG1 and NOS2 (Bronte et al., 2003a). In this model, functional CTLs can be recovered only through the inhibition of both enzymes by the combination of NOHA and l-NMMA (Bronte et al., 2003a) or by peroxynitrite scavengers (De Santo et al., 2005). In some tumors, thus, contemporaneous targeting of ARG and NOS might be required to achieve a therapeutic effect. This is also true for human tumors, such as prostate cancer, that over-express both ARG2 and NOS2 which cause a functional paralysis of tumor-infiltrating CD8$^+$ T lymphocytes (Bronte, 2005). We found that, by culturing small tumor samples of prostate cancer in medium containing a combination of NOHA and L-NMMA, tumor-infiltrating lymphocytes recovered their functions, which paralleled the reduction in the nitrotyrosine-containing proteins (Bronte, 2005). Unfortunately, the pharmacological in vivo inhibition of ARG1 can present serious side effects considering its critical role in the urea cycle, and L-NMMA was found to be toxic in humans since it could induce myocardial depression and significantly increase mortality (Freeman et al., 2001). We recently attempted to overcome these barriers, with two classes of drugs that effectively affect L-Arg metabolism in MDSCs: nitro-aspirin (De Santo et al., 2005) and PDE-5 inhibitors (Serafini et al., 2006b). For both classes of compounds, safety has been confirmed in humans: nitro-aspirins have demonstrated less toxicity than normal aspirin (Fiorucci et al., 2003), and PDE-5 inhibitors are currently used by millions of people for the treatment of erectile dysfunction, pulmonary hypertension and cardiac hypertrophy (Briganti et al., 2005). When administered to tumor-bearing mice, nitro-aspirins increased the number and function of tumor-antigen-specific T cells, thus enhancing cancer vaccine efficacy (De Santo et al., 2005). In vivo PDE-5 inhibitors administration, by down-modulating ARG1 and NOS2 expression, reduced MDSCs suppressive machinery, enhanced intratumoral T-cell infiltration and activation, reduced tumor outgrowth and improved the antitumor efficacy of adoptive T-cell therapy (Serafini et al., 2006b). PDE-5 inhibition also restored in vitro T-cell proliferation of anti-CD3/anti-CD28 stimulated PBMCs from multiple myeloma and head and neck cancer patients suggesting its efficacy also in human malignancies (Serafini et al., 2006b). Pharmacological inhibition of MDSC suppressive pathways is thus a promising strategy to overcome tumor-induced immune defects, which will likely play a critical role in enhancing antitumor efficacy of immune-based strategies.

8.5 IL-13/IL4Rα/STAT6 Pathway as a New Target to Restrain MDSC Suppressive Function

In recent years, the importance of the IL-13/IL4Rα/STAT6 pathway in MDSC-mediated immune suppression has been confirmed by different groups (Gallina et al., 2006; Sinha et al., 2005c; Terabe et al., 2004). As discussed before, IL4Rα expression correlated with MDSC suppressive capacity (Gallina et al., 2006), and the genetic ablation of IL4Rα limited only to macrophages and neutrophils, was sufficient, not only to revert MDSC-mediated immune suppression but also, when coupled with the adoptive tumor-specific T-cell transfer, to eradicate a pre-existing mouse colon carcinoma (Gallina et al., 2006). Moreover the development of spontaneous lung metastasis in a murine mammary carcinoma was reduced by genetic ablation of STAT6 (Sinha et al., 2005b). Finally the engagement of IL4Rα by IL-13 induced STAT6 phosphorylation and triggered ARG activation and/or TGF-β secretion in MDSCs. IL4Rα$^{-/-}$ mice did not present the MDSC-assisted recurrence of a mouse fibrosarcoma observed in wild-type mice (Terabe et al., 2000), and similar results could be obtained by in vivo administration of soluble IL13Rα 2-Fc, a molecule able to neutralize IL-13 (Terabe et al., 2000). This molecule belongs to a class of drugs developed to treat asthma and allergy by targeting the IL-4 and IL-13 pathways. This class of drugs can be divided into three groups: soluble receptors (such IL13Rα2-Fc and sIL4R-fc) able to bind and neutralize these Th2 cytokines in vivo; mutated forms of IL-4 or IL-13 that bind the receptor without triggering the downstream signal machinery; and fusion proteins in which IL-13 or IL-4 is coupled with toxin able to induce apoptosis of target cells. All these compounds were shown to be active in vivo and many of them are now in clinical trials for the treatment of different pathologies including asthma, cancer (glioma and lung cancer) and HIV

(Cutler and Brombacher, 2005). IL4Rα can thus be a relevant target to inhibit or deplete MDSC suppressive activity and the availability of compounds already in clinic for other disease could allow a rapid translation of preclinical positive findings into novel therapeutic approaches for human malignancies.

9 Conclusions

Despite the bulk of evidence describing MDSC's importance in cancer, several issues concerning MDSC biology are not completely defined and will require further studies. The in vivo MDSC antigen-specific, tolerogenic ability is in sharp contrast with the antigen-independent suppression described in many (if not all) in vitro assays; this dichotomy, common also for other regulatory elements of the immune response such as Treg cells, need to be further investigated. MDSCs employ different mechanisms to suppress T lymphocytes in different tumor models suggesting a tumor-type-related influence on the biology and/or activity of these cells; however, the molecular factors responsible for these functional differences are not completely known. Cell-specific knock out genes essential for either MDSC development/activation or effector function (such as ARG1) need to be generated to better understand MDSC physiological functions not only in tumor but also under other conditions. The interplay between the granulocytic and the monocytic components of the $CD11b^+/Gr1^+$ cells needs to be addressed; moreover, the in vitro culture conditions to differentiate mouse and human MDSC need to be optimized to achieve phenotype and functional activity consistent with tumor-derived MDSCs. Since the phenotype and the prevalence of MDSCs in human cancer remain uncertain, efforts are required to identify functional phenotypic markers that, most likely, are those better conserved among species and cancer types. It is important to stress that incongruence in this field likely reflects the multifaceted alteration of myelopoiesis underlying the MDSC appearance. MDSCs are not terminally differentiated myelomonocytic cells and preserve a degree of plasticity that makes them more susceptible than other cell types to the influence of the experimental settings. Furthermore, it must be pointed out that the definition of subsets among MDSCs has just begun, differently from what happened for other cell types such as DCs, extensively investigated in recent years.

In contrast to these uncertainties, preclinical evidence strongly supports the idea that new approaches targeting MDSCs can have beneficial effects on anti-tumor immune responses, either spontaneous or elicited by active or passive immunotherapy.

References

Almand, B., Clark, J. I., Nikitina, E., van Beynen, J., English, N. R., Knight, S. C., Carbone, D. P., and Gabrilovich, D. I. (2001). Increased production of immature myeloid cells in cancer patients: a mechanism of immunosuppression in cancer. *J Immunol* 166(1):678–689.

Alvaro, T., Lejeune, M., Camacho, F. I., Salvado, M. T., Sanchez, L., Garcia, J. F., Lopez, C., Jaen, J., Bosch, R., Pons, L. E., Bellas, C., and Piris, M. A. (2006). The presence of STAT1-positive tumor-associated macrophages and their relation to outcome in patients with follicular lymphoma. *Haematologica* 91(12):1605–1612.

Andrew, P. J., and Mayer, B. (1999). Enzymatic function of nitric oxide synthases. *Cardiovasc Res* 43(3):521–531.

Apolloni, E., Bronte, V., Mazzoni, A., Serafini, P., Cabrelle, A., Segal, D. M., Young, H. A., and Zanovello, P. (2000). Immortalized myeloid suppressor cells trigger apoptosis in antigen-activated T lymphocytes. *J Immunol* 165(12):6723–6730.

Bach, E. A., Aguet, M., and Schreiber, R. D. (1997). The IFN gamma receptor: a paradigm for cytokine receptor signaling. *Annu Rev Immunol* 15:563–591.

Baglole, C. J., Ray, D. M., Bernstein, S. H., Feldon, S. E., Smith, T. J., Sime, P. J., and Phipps, R. P. (2006). More than structural cells, fibroblasts create and orchestrate the tumor microenvironment. *Immunol Invest* 35(3–4):297–325.

Baniyash, M. (2004). TCR zeta-chain downregulation: curtailing an excessive inflammatory immune response. *Nat Rev Immunol* 4(9):675–687.

Beatty, G. L., and Paterson, Y. (2000). IFN-gamma can promote tumor evasion of the immune system in vivo by down-regulating cellular levels of an endogenous tumor antigen. *J Immunol* 165(10):5502–5508.

Beck, C., Schreiber, K., Schreiber, H., and Rowley, D. A. (2003). C-kit+ FcR+ myelocytes are increased in cancer and prevent the proliferation of fully cytolytic T cells in the presence of immune serum. *Eur J Immunol* 33(1):19–28.

Becker, C., Fantini, M. C., and Neurath, M. F. (2006). TGF-beta as a T cell regulator in colitis and colon cancer. *Cytokine Growth Factor Rev* 17(1–2):97–106.

Bingisser, R. M., Tilbrook, P. A., Holt, P. G., and Kees, U. R. (1998). Macrophage-derived nitric oxide regulates T cell activation via reversible disruption of the Jak3/STAT5 signaling pathway. *J Immunol* 160(12):5729–5734.

Biswas, S. K., Gangi, L., Paul, S., Schioppa, T., Saccani, A., Sironi, M., Bottazzi, B., Doni, A., Vincenzo, B., Pasqualini, F., Vago, L., Nebuloni, M., Mantovani, A., and Sica, A. (2006). A distinct and unique transcriptional program expressed by tumor-associated macrophages (defective NF-kappaB and enhanced IRF-3/STAT1 activation). *Blood* 107(5):2112–2122.

Bogdan, C. (2001). Nitric oxide and the immune response. *Nat Immunol* 2(10):907–916.

Brabletz, T., Pfeuffer, I., Schorr, E., Siebelt, F., Wirth, T., and Serfling, E. (1993). Transforming growth factor beta and cyclosporin A inhibit the inducible activity of the interleukin-2 gene in T cells through a noncanonical octamer-binding site. *Mol Cell Biol* 13(2):1155–1162.

Briganti, A., Salonia, A., Gallina, A., Sacca, A., Montorsi, P., Rigatti, P., and Montorsi, F. (2005). Drug insight: oral phosphodiesterase type 5 inhibitors for erectile dysfunction. *Nat Clin Pract Urol* 2(5):239–247.

Bromberg, J. (2002). Stat proteins and oncogenesis. *J Clin Invest* 109(9):1139–1142.

Bronte, V., Apolloni, E., Cabrelle, A., Ronca, R., Serafini, P., Zamboni, P., Restifo, N. P., and Zanovello, P. (2000). Identification of a CD11b(+)/Gr-1(+)/CD31(+) myeloid progenitor capable of activating or suppressing CD8(+) T cells. *Blood* 96(12):3838–3846.

Bronte, V., Chappell, D. B., Apolloni, E., Cabrelle, A., Wang, M., Hwu, P., and Restifo, N. P. (1999). Unopposed production of granulocyte-macrophage colony-stimulating factor by tumors inhibits CD8+ T cell responses by dysregulating antigen-presenting cell maturation. *J Immunol* 162(10):5728–5737.

Bronte, V., Kasic, T., Gri, G., Gallana, K., Borsellino, G., Marigo, I., Battistini, L., Iafrate, M., Prayer-Galetti, T., Pagano, F., and Viola, A. (2005). Boosting antitumor responses of T lymphocytes infiltrating human prostate cancers. *J Exp Med* 201(8):1257–1268.

Bronte, V., Serafini, P., De Santo, C., Marigo, I., Tosello, V., Mazzoni, A., Segal, D. M., Staib, C., Lowel, M., Sutter, G., Colombo, M. P., and Zanovello, P. (2003a). IL-4-induced arginase 1 suppresses alloreactive T cells in tumor-bearing mice. *J Immunol* 170(1):270–278.

Bronte, V., Serafini, P., Mazzoni, A., Segal, D. M., and Zanovello, P. (2003b). L-arginine metabolism in myeloid cells controls T-lymphocyte functions. *Trends Immunol* 24(6):302–306.

Bronte, V., Wang, M., Overwijk, W. W., Surman, D. R., Pericle, F., Rosenberg, S. A., and Restifo, N. P. (1998). Apoptotic death of CD8+ T lymphocytes after immunization: induction of a suppressive population of Mac-1+/Gr-1+ cells. *J Immunol* 161(10):5313–5320.

Bronte, V., and Zanovello, P. (2005). Regulation of immune responses by L-arginine metabolism. *Nat Rev Immunol* 5(8):641–654.

Bruch-Gerharz, D., Schnorr, O., Suschek, C., Beck, K. F., Pfeilschifter, J., Ruzicka, T., and Kolb-Bachofen, V. (2003). Arginase 1 overexpression in psoriasis: limitation of inducible nitric oxide synthase activity as a molecular mechanism for keratinocyte hyperproliferation. *Am J Pathol* 162(1):203–211.

Brys, L., Beschin, A., Raes, G., Ghassabeh, G. H., Noel, W., Brandt, J., Brombacher, F., and De Baetselier, P. (2005). Reactive oxygen species and 12/15-lipoxygenase contribute to the antiproliferative capacity of alternatively activated myeloid cells elicited during helminth infection. *J Immunol* 174(10):6095–6104.

Buettner, R., Mora, L. B., and Jove, R. (2002). Activated STAT signaling in human tumors provides novel molecular targets for therapeutic intervention. *Clin Cancer Res* 8(4):945–954.

Champelovier, P., Boucard, N., Levacher, G., Simon, A., Seigneurin, D., and Praloran, V. (2002). Plasminogen- and colony-stimulating factor-1-associated markers in bladder carcinoma: diagnostic value of urokinase plasminogen activator receptor and plasminogen activator inhibitor type-2 using immunocytochemical analysis. *Urol Res* 30(5):301–309.

Chen, M. L., Wang, F. H., Lee, P. K., and Lin, C. M. (2001). Interleukin-10-induced T cell unresponsiveness can be reversed by dendritic cell stimulation. *Immunol Lett* 75(2):91–96.

Cheng, F., Wang, H. W., Cuenca, A., Huang, M., Ghansah, T., Brayer, J., Kerr, W. G., Takeda, K., Akira, S., Schoenberger, S. P., Yu, H., Jove, R., and Sotomayor, E. M. (2003). A critical role for Stat3 signaling in immune tolerance. *Immunity* 19(3):425–436.

Cirillo, C., Montaldo, P., Lanciotti, M., Parodi, M. T., Castagnola, E., and Ponzoni, M. (1988). [Immunosuppressive factors produced by a T cell line derived from acute lymphoblastic leukemia]. *Boll Ist Sieroter Milan* 67(4):295–308.

Colombo, M. P., and Mantovani, A. (2005). Targeting myelomonocytic cells to revert inflammation-dependent cancer promotion. *Cancer Res* 65(20):9113–9116.

Currie, G. A., Gyure, L., and Cifuentes, L. (1979). Microenvironmental arginine depletion by macrophages in vivo. *Br J Cancer* 39(6):613–620.

Cutler, A., and Brombacher, F. (2005). Cytokine therapy. *Ann N Y Acad Sci* 1056:16–29.

Danna, E. A., Sinha, P., Gilbert, M., Clements, V. K., Pulaski, B. A., and Ostrand-Rosenberg, S. (2004). Surgical removal of primary tumor reverses tumor-induced immunosuppression despite the presence of metastatic disease. *Cancer Res* 64(6):2205–2211.

Das, J., Chen, C. H., Yang, L., Cohn, L., Ray, P., and Ray, A. (2001). A critical role for NF-kappa B in GATA3 expression and TH2 differentiation in allergic airway inflammation. *Nat Immunol* 2(1):45–50.

De Palma, M., Venneri, M. A., Galli, R., Sergi Sergi, L., Politi, L. S., Sampaolesi, M., and Naldini, L. (2005). Tie2 identifies a hematopoietic lineage of proangiogenic monocytes required for tumor vessel formation and a mesenchymal population of pericyte progenitors. *Cancer Cell* 8(3):211–226.

De Santo, C., Serafini, P., Marigo, I., Dolcetti, L., Bolla, M., Del Soldato, P., Melani, C., Guiducci, C., Colombo, M. P., Iezzi, M., Musiani, P., Zanovello, P., and Bronte, V. (2005). Nitroaspirin corrects immune dysfunction in tumor-bearing hosts and promotes tumor eradication by cancer vaccination. *Proc Natl Acad Sci USA* 102(11):4185–4190.

De Vita, F., Orditura, M., Galizia, G., Romano, C., Lieto, E., Iodice, P., Tuccillo, C., and Catalano, G. (2000a). Serum interleukin-10 is an independent prognostic factor in advanced solid tumors. *Oncol Rep* 7(2):357–361.

De Vita, F., Orditura, M., Galizia, G., Romano, C., Roscigno, A., Lieto, E., and Catalano, G. (2000b). Serum interleukin-10 levels as a prognostic factor in advanced non-small cell lung cancer patients. *Chest* 117(2):365–373.

Dias, S., Boyd, R., and Balkwill, F. (1998). IL-12 regulates VEGF and MMPs in a murine breast cancer model. *Int J Cancer* 78(3):361–365.

Dolled-Filhart, M., Camp, R. L., Kowalski, D. P., Smith, B. L., and Rimm, D. L. (2003). Tissue microarray analysis of signal transducers and activators of transcription 3 (Stat3) and phospho-Stat3 (Tyr705) in node-negative breast cancer shows nuclear localization is associated with a better prognosis. *Clin Cancer Res* 9(2):594–600.

Dranoff, G. (2002). GM-CSF-based cancer vaccines. *Immunol Rev* 188:147–154.

Dranoff, G. (2003). GM-CSF-secreting melanoma vaccines. *Oncogene* 22(20):3188–3192.

Duhe, R. J., Evans, G. A., Erwin, R. A., Kirken, R. A., Cox, G. W., and Farrar, W. L. (1998). Nitric oxide and thiol redox regulation of Janus kinase activity. *Proc Natl Acad Sci USA* 95(1):126–131.

Dunn, G. P., Old, L. J., and Schreiber, R. D. (2004). The immunobiology of cancer immunosurveillance and immunoediting. *Immunity* 21(2):137–148.

Finder, J. D., Petrus, J. L., Hamilton, A., Villavicencio, R. T., Pitt, B. R., and Sebti, S. M. (2001). Signal transduction pathways of IL-1beta-mediated iNOS in pulmonary vascular smooth muscle cells. *Am J Physiol Lung Cell Mol Physiol* 281(4):L816–L823.

Fiorucci, S., Santucci, L., Gresele, P., Faccino, R. M., Del Soldato, P., and Morelli, A. (2003). Gastrointestinal safety of NO-aspirin (NCX-4016) in healthy human volunteers: a proof of concept endoscopic study. *Gastroenterology* 124(3):600–607.

Fischer, T. A., Palmetshofer, A., Gambaryan, S., Butt, E., Jassoy, C., Walter, U., Sopper, S., and Lohmann, S. M. (2001). Activation of cGMP-dependent protein kinase Ibeta inhibits interleukin 2 release and proliferation of T cell receptor-stimulated human peripheral T cells. *J Biol Chem* 276(8):5967–5974.

Freeman, B. D., Danner, R. L., Banks, S. M., and Natanson, C. (2001). Safeguarding patients in clinical trials with high mortality rates. *Am J Respir Crit Care Med* 164(2):190–192.

Fu, Y. X., Watson, G., Jimenez, J. J., Wang, Y., and Lopez, D. M. (1990). Expansion of immunoregulatory macrophages by granulocyte-macrophage colony-stimulating factor derived from a murine mammary tumor. *Cancer Res* 50:227–234.

Gabrilovich, D. I. (2004). Mechanisms and functional significance of tumour-induced dendritic-cell defects. *Nat Rev Immunol* 4(12):941–952.

Gabrilovich, D. I., Bronte, V., Chen, S. H., Colombo, M. P., Ochoa, A., Ostrand-Rosenberg, S., and Schreiber, H. (2007). The terminology issue for myeloid-derived suppressor cells. *Cancer Res* 67(1):425; author reply 426.

Gabrilovich, D. I., Chen, H. L., Girgis, K. R., Cunningham, H. T., Meny, G. M., Nadaf, S., Kavanaugh, D., and Carbone, D. P. (1996). Production of vascular endothelial growth factor by human tumors inhibits the functional maturation of dendritic cells. *Nat Med* 2:1096–1103.

Gabrilovich, D. I., Ishida, T., Oyama, T., Ran, S., Kravtsov, V., Nadaf, S., and Carbone, D. P. (1998). Vascular endothelial growth factor inhibits the development of dendritic cells and dramatically affects the differentiation of multiple hematopoietic lineages in vivo. *Blood* 92(11):4150–4166.

Gabrilovich, D. I., Velders, M. P., Sotomayor, E. M., and Kast, W. M. (2001). Mechanism of immune dysfunction in cancer mediated by immature gr-1(+) myeloid cells. *J Immunol* 166(9):5398–5406.

Gallina, G., Dolcetti, L., Serafini, P., Santo, C. D., Marigo, I., Colombo, M. P., Basso, G., Brombacher, F., Borrello, I., Zanovello, P., Bicciato, S., and Bronte, V. (2006). Tumors induce a subset of inflammatory monocytes with immunosuppressive activity on CD8 T cells. *J Clin Invest* 116(10):2777–2790.

Garrity, T., Pandit, R., Wright, M. A., Benefield, J., Keni, S., and Young, M. R. (1997). Increased presence of CD34+ cells in the peripheral blood of head and neck cancer patients and their differentiation into dendritic cells. *Int J Cancer* 73(5):663–669.

Gasparini, G., Longo, R., Sarmiento, R., and Morabito, A. (2003). Inhibitors of cyclo-oxygenase 2: a new class of anticancer agents? *Lancet Oncol* 4(10):605–615.

Gerharz, C. D., Reinecke, P., Schneider, E. M., Schmitz, M., and Gabbert, H. E. (2001). Secretion of GM-CSF and M-CSF by human renal cell carcinomas of different histologic types. *Urology* 58(5):821–827.

Ghiringhelli, F., Puig, P. E., Roux, S., Parcellier, A., Schmitt, E., Solary, E., Kroemer, G., Martin, F., Chauffert, B., and Zitvogel, L. (2005). Tumor cells convert immature myeloid dendritic cells into TGF-beta-secreting cells inducing CD4+CD25+ regulatory T cell proliferation. *J Exp Med* 202(7):919–929.

Gordon, S. (2003). Alternative activation of macrophages. *Nat Rev Immunol* 3(1):23–35.

Gray, M. J., Poljakovic, M., Kepka-Lenhart, D., and Morris, S. M., Jr (2005). Induction of arginase I transcription by IL-4 requires a composite DNA response element for STAT6 and C/EBPbeta. *Gene* 353(1):98–106.

Hanada, T., Kobayashi, T., Chinen, T., Saeki, K., Takaki, H., Koga, K., Minoda, Y., Sanada, T., Yoshioka, T., Mimata, H., Kato, S., and Yoshimura, A. (2006). IFNgamma-dependent, spontaneous development of colorectal carcinomas in SOCS1-deficient mice. *J Exp Med* 203(6):1391–1397.

Haran-Ghera, N., Krautghamer, R., Lapidot, T., Peled, A., Dominguez, M. G., and Stanley, E. R. (1997). Increased circulating colony-stimulating factor-1 (CSF-1) in SJL/J mice with radiation-induced acute myeloid leukemia (AML) is associated with autocrine regulation of AML cells by CSF-1. *Blood* 89(7):2537–2545.

Hayashi, T., Hideshima, T., Akiyama, M., Raje, N., Richardson, P., Chauhan, D., and Anderson, K. C. (2003). Ex vivo induction of multiple myeloma-specific cytotoxic T lymphocytes. *Blood* 102(4):1435–1442.

Heldin, C. H. (2004). Development and possible clinical use of antagonists for PDGF and TGF-beta. *Ups J Med Sci* 109(3):165–178.

Horiguchi, S., Petersson, M., Nakazawa, T., Kanda, M., Zea, A. H., Ochoa, A. C., and Kiessling, R. (1999). Primary chemically induced tumors induce profound immunosuppression concomitant with apoptosis and alterations in signal transduction in T cells and NK cells. *Cancer Res* 59(12):2950–2956.

Hsiao, J. R., Jin, Y. T., Tsai, S. T., Shiau, A. L., Wu, C. L., and Su, W. C. (2003). Constitutive activation of STAT3 and STAT5 is present in the majority of nasopharyngeal carcinoma and correlates with better prognosis. *Br J Cancer* 89(2):344–349.

Huang, B., Pan, P. Y., Li, Q., Sato, A. I., Levy, D. E., Bromberg, J., Divino, C. M., and Chen, S. H. (2006). Gr-1+CD115+ immature myeloid suppressor cells mediate the development of tumor-induced T regulatory cells and T-cell anergy in tumor-bearing host. *Cancer Res* 66(2): 1123–1131.

Jain, R. K. (2005). Normalization of tumor vasculature: an emerging concept in antiangiogenic therapy. *Science* 307(5706):58–62.

Kaplan, M. H., and Grusby, M. J. (1998). Regulation of T helper cell differentiation by STAT molecules. *J Leukoc Biol* 64(1):2–5.

Keegan, A. D., Nelms, K., White, M., Wang, L. M., Pierce, J. H., and Paul, W. E. (1994). An IL-4 receptor region containing an insulin receptor motif is important for IL-4-mediated IRS-1 phosphorylation and cell growth. *Cell* 76(5):811–820.

Kehrl, J. H., Roberts, A. B., Wakefield, L. M., Jakowlew, S., Sporn, M. B., and Fauci, A. S. (1986). Transforming growth factor beta is an important immunomodulatory protein for human B lymphocytes. *J Immunol* 137(12):3855–3860.

Kono, K., Ressing, M. E., Brandt, R. M., Melief, C. J., Potkul, R. K., Andersson, B., Petersson, M., Kast, W. M., and Kiessling, R. (1996a). Decreased expression of signal-transducing zeta chain in peripheral T cells and natural killer cells in patients with cervical cancer. *Clin Cancer Res* 2(11):1825–1828.

Kono, K., Salazar-Onfray, F., Petersson, M., Hansson, J., Masucci, G., Wasserman, K., Nakazawa, T., Anderson, B., and Kiessling, R. (1996b). Hydrogen peroxide secreted by tumor-derived macrophages down-modulates signal-transducing zeta molecules and inhibits tumor-specific T cell- and natural killer cell-mediated cytotoxicity. *Eur J Immunol* 26(6):1308–1313.

Kortylewski, M., Kujawski, M., Wang, T., Wei, S., Zhang, S., Pilon-Thomas, S., Niu, G., Kay, H., Mule, J., Kerr, W. G., Jove, R., Pardoll, D., and Yu, H. (2005). Inhibiting Stat3 signaling in the hematopoietic system elicits multicomponent antitumor immunity. *Nat Med* 11(12): 1314–1321.

Kunicka, J. E., Fox, F. E., Seki, H., Oleszak, E. L., and Platsoucas, C. D. (1991). Hybridoma-derived human suppressor factors: inhibition of growth of tumor cell lines and effect on cytotoxic cells. *Hum Antibodies Hybridomas* 2(3):160–169.

Kusmartsev, S., Cheng, F., Yu, B., Nefedova, Y., Sotomayor, E., Lush, R., and Gabrilovich, D. (2003). All-trans-retinoic acid eliminates immature myeloid cells from tumor-bearing mice and improves the effect of vaccination. *Cancer Res* 63(15):4441–4449.

Kusmartsev, S., and Gabrilovich, D. I. (2003). Inhibition of myeloid cell differentiation in cancer: the role of reactive oxygen species. *J Leukoc Biol* 74(2):186–196.

Kusmartsev, S., and Gabrilovich, D. I. (2005). STAT1 signaling regulates tumor-associated macrophage-mediated T cell deletion. *J Immunol* 174(8):4880–4891.

Kusmartsev, S., and Gabrilovich, D. I. (2006a). Effect of tumor-derived cytokines and growth factors on differentiation and immune suppressive features of myeloid cells in cancer. *Cancer Metastasis Rev* 25(3):323–331.

Kusmartsev, S., and Gabrilovich, D. I. (2006b). Role of immature myeloid cells in mechanisms of immune evasion in cancer. *Cancer Immunol Immunother* 55(3):237–245.

Kusmartsev, S., Nagaraj, S., and Gabrilovich, D. I. (2005). Tumor-associated CD8+ T cell tolerance induced by bone marrow-derived immature myeloid cells. *J Immunol* 175(7):4583–4592.

Kusmartsev, S., Nefedova, Y., Yoder, D., and Gabrilovich, D. I. (2004). Antigen-specific inhibition of CD8+ T cell response by immature myeloid cells in cancer is mediated by reactive oxygen species. *J Immunol* 172(2):989–999.

Lan, Y. Y., Wang, Z., Raimondi, G., Wu, W., Colvin, B. L., de Creus, A., and Thomson, A. W. (2006). "Alternatively activated" dendritic cells preferentially secrete IL-10, expand Foxp3+CD4+ T cells, and induce long-term organ allograft survival in combination with CTLA4-Ig. *J Immunol* 177(9):5868–5877.

Lang, R., Patel, D., Morris, J. J., Rutschman, R. L., and Murray, P. J. (2002). Shaping gene expression in activated and resting primary macrophages by IL-10. *J Immunol* 169(5):2253–2263.

Lathers, D. M., Clark, J. I., Achille, N. J., and Young, M. R. (2004). Phase 1B study to improve immune responses in head and neck cancer patients using escalating doses of 25-hydroxyvitamin D3. *Cancer Immunol Immunother* 53(5):422–430.

Lim, S. H., Worman, C. P., Jewell, A., and Goldstone, A. H. (1991). Production of tumour-derived suppressor factor in patients with acute myeloid leukaemia. *Leuk Res* 15(4):263–268.

Lin, E. Y., Gouon-Evans, V., Nguyen, A. V., and Pollard, J. W. (2002). The macrophage growth factor CSF-1 in mammary gland development and tumor progression. *J Mammary Gland Biol Neoplasia* 7(2):147–162.

Liu, Y., Van Ginderachter, J. A., Brys, L., De Baetselier, P., Raes, G., and Geldhof, A. B. (2003). Nitric oxide-independent CTL suppression during tumor progression: association with arginase-producing (M2) myeloid cells. *J Immunol* 170(10):5064–5074.

Liu, C., Yu, S., Kappes, J., Wang, J., Grizzle, W. E., Zinn, K. R., and Zhang, H. G. (2007). Expansion of spleen myeloid suppressor cells represses NK cell cytotoxicity in tumor bearing host. *Blood* 109:4336–4342.

Luo, Y., Zhou, H., Krueger, J., Kaplan, C., Lee, S. H., Dolman, C., Markowitz, D., Wu, W., Liu, C., Reisfeld, R. A., and Xiang, R. (2006). Targeting tumor-associated macrophages as a novel strategy against breast cancer. *J Clin Invest* 116(8):2132–2141.

MacDonald, K. P., Rowe, V., Clouston, A. D., Welply, J. K., Kuns, R. D., Ferrara, J. L., Thomas, R., and Hill, G. R. (2005). Cytokine expanded myeloid precursors function as regulatory antigen-presenting cells and promote tolerance through IL-10-producing regulatory T cells. *J Immunol* 174(4):1841–1850.

MacMicking, J., Xie, Q. W., and Nathan, C. (1997). Nitric oxide and macrophage function. *Annu Rev Immunol* 15:323–350.

Macphail, S. E., Gibney, C. A., Brooks, B. M., Booth, C. G., Flanagan, B. F., and Coleman, J. W. (2003). Nitric oxide regulation of human peripheral blood mononuclear cells: critical time dependence and selectivity for cytokine versus chemokine expression. *J Immunol* 171(9):4809–4815.

Mahnke, K., Qian, Y., Knop, J., and Enk, A. H. (2003). Induction of CD4+/CD25+ regulatory T cells by targeting of antigens to immature dendritic cells. *Blood* 101(12):4862–4869.

Maier, T., Holda, J. H., and Claman, H. N. (1985). Graft-vs-host reactions (GVHR) across minor murine histocompatibility barriers. II. Development of natural suppressor cell activity. *J Immunol* 135:1644–1651.

Maier, T., Holda, J. H., and Claman, H. N. (1989). Natural suppressor cells. *Prog Clin Biol Res* 288:235–244.

Mannick, J. B., Hausladen, A., Liu, L., Hess, D. T., Zeng, M., Miao, Q. X., Kane, L. S., Gow, A. J., and Stamler, J. S. (1999). Fas-induced caspase denitrosylation. *Science* 284(5414):651–654.

Mantovani, A., Sica, A., Sozzani, S., Allavena, P., Vecchi, A., and Locati, M. (2004). The chemokine system in diverse forms of macrophage activation and polarization. *Trends Immunol* 25(12):677–686.

Mazzoni, A., Bronte, V., Visintin, A., Spitzer, J. H., Apolloni, E., Serafini, P., Zanovello, P., and Segal, D. M. (2002). Myeloid suppressor lines inhibit T cell responses by an NO-dependent mechanism. *J Immunol* 168(2):689–695.

Melani, C., Chiodoni, C., Forni, G., and Colombo, M. P. (2003). Myeloid cell expansion elicited by the progression of spontaneous mammary carcinomas in c-erbB-2 transgenic BALB/c mice suppresses immune reactivity. *Blood* 102(6):2138–2145.

Mellor, A. L., and Munn, D. H. (2004). IDO expression by dendritic cells: tolerance and tryptophan catabolism. *Nat Rev Immunol* 4(10):762–774.

Mellor, A. L., Munn, D. H., Chandler, P., Keskin, D., Johnson, T., Marshall, B., Jhaver, K., and Baban, B. (2003). Tryptophan catabolism and T cell responses. *Adv Exp Med Biol* 527:27–35.

Menetrier-Caux, C., Montmain, G., Dieu, M. C., Bain, C., Favrot, M. C., Caux, C., and Blay, J. Y. (1998). Inhibition of the differentiation of dendritic cells from CD34(+) progenitors by tumor cells: role of interleukin-6 and macrophage colony-stimulating factor. *Blood* 92(12):4778–4791.

Merchav, S., Apte, R. N., Tatarsky, I., and Ber, R. (1987). Effect of plasmacytoma cells on the production of granulocyte-macrophage colony-stimulating activity (GM-CSA) in the spleen of tumor-bearing mice. *Exp Hematol* 15(9):995–1000.

Moore, S. C., Shaw, M. A., and Soderberg, L. S. (1992). Transforming growth factor-beta is the major mediator of natural suppressor cells derived from normal bone marrow. *J Leukoc Biol* 52(6):596–601.

Mora, L. B., Buettner, R., Seigne, J., Diaz, J., Ahmad, N., Garcia, R., Bowman, T., Falcone, R., Fairclough, R., Cantor, A., Muro-Cacho, C., Livingston, S., Karras, J., Pow-Sang, J., and Jove, R. (2002). Constitutive activation of Stat3 in human prostate tumors and cell lines: direct inhibition of Stat3 signaling induces apoptosis of prostate cancer cells. *Cancer Res* 62(22):6659–6666.

Morel, S., Levy, F., Burlet-Schiltz, O., Brasseur, F., Probst-Kepper, M., Peitrequin, A. L., Monsarrat, B., Van Velthoven, R., Cerottini, J. C., Boon, T., Gairin, J. E., and Van den Eynde, B. J. (2000). Processing of some antigens by the standard proteasome but not by the immunoproteasome results in poor presentation by dendritic cells. *Immunity* 12(1):107–117.

Morris, D. R., Kuepfer, C. A., Ellingsworth, L. R., Ogawa, Y., and Rabinovitch, P. S. (1989). Transforming growth factor-beta blocks proliferation but not early mitogenic signaling events in T-lymphocytes. *Exp Cell Res* 185(2):529–534.

Munder, M., Eichmann, K., and Modolell, M. (1998). Alternative metabolic states in murine macrophages reflected by the nitric oxide synthase/arginase balance: competitive regulation by CD4+ T cells correlates with Th1/Th2 phenotype. *J Immunol* 160(11):5347–5354.

Munder, M., Eichmann, K., Moran, J. M., Centeno, F., Soler, G., and Modolell, M. (1999). Th1/Th2-regulated expression of arginase isoforms in murine macrophages and dendritic cells. *J Immunol* 163(7):3771–3777.

Nagpal, J. K., Mishra, R., and Das, B. R. (2002). Activation of Stat-3 as one of the early events in tobacco chewing-mediated oral carcinogenesis. *Cancer* 94(9):2393–2400.

Nefedova, Y., Cheng, P., Gilkes, D., Blaskovich, M., Beg, A. A., Sebti, S. M., and Gabrilovich, D. I. (2005). Activation of dendritic cells via inhibition of Jak2/STAT3 signaling. *J Immunol* 175(7):4338–4346.

Nefedova, Y., Huang, M., Kusmartsev, S., Bhattacharya, R., Cheng, P., Salup, R., Jove, R., and Gabrilovich, D. (2004). Hyperactivation of STAT3 is involved in abnormal differentiation of dendritic cells in cancer. *J Immunol* 172(1):464–474.

Nikitakis, N. G., Siavash, H., and Sauk, J. J. (2004). Targeting the STAT pathway in head and neck cancer: recent advances and future prospects. *Curr Cancer Drug Targets* 4(8):637–651.

Ostrand-Rosenberg, S., Clements, V. K., Terabe, M., Park, J. M., Berzofsky, J. A., and Dissanayake, S. K. (2002). Resistance to metastatic disease in STAT6-deficient mice requires hemopoietic and nonhemopoietic cells and is IFN-gamma dependent. *J Immunol* 169(10):5796–5804.

Otsuji, M., Kimura, Y., Aoe, T., Okamoto, Y., and Saito, T. (1996). Oxidative stress by tumor-derived macrophages suppresses the expression of CD3 zeta chain of T-cell receptor complex and antigen-specific T-cell responses. *Proc Natl Acad Sci USA* 93(23):13119–13124.

Pak, A. S., Ip, G., Wright, M. A., and Young, M. R. (1995a). Treating tumor-bearing mice with low-dose gamma-interferon plus tumor necrosis factor alpha to diminish immune suppressive granulocyte-macrophage progenitor cells increases responsiveness to interleukin 2 immunotherapy. *Cancer Res* 55(4):885–890.

Pak, A. S., Wright, M. A., Matthews, J. P., Collins, S. L., Petruzzelli, G. J., and Young, M. R. I. (1995b). Mechanisms of immune suppression in patients with head and neck cancer: presence of CD34(+) cells which suppress immune functions within cancers that secrete granulocyte-macrophage colony-stimulating factor. *Clin Cancer Res* 1(1):95–103.

Park, J. M., Terabe, M., van den Broeke, L. T., Donaldson, D. D., and Berzofsky, J. A. (2005). Unmasking immunosurveillance against a syngeneic colon cancer by elimination of CD4+ NKT regulatory cells and IL-13. *Int J Cancer* 114(1):80–87.

Park, S. J., Nakagawa, T., Kitamura, H., Atsumi, T., Kamon, H., Sawa, S., Kamimura, D., Ueda, N., Iwakura, Y., Ishihara, K., Murakami, M., and Hirano, T. (2004). IL-6 regulates in vivo dendritic cell differentiation through STAT3 activation. *J Immunol* 173(6):3844–3854.

Parmiani, G., Castelli, C., Pilla, L., Santinami, M., Colombo, M. P., and Rivoltini, L. (2007). Opposite immune functions of GM-CSF administered as vaccine adjuvant in cancer patients. *Ann Oncol* 18(2):226–32.

Pawelec, G. (2004). Tumour escape: antitumour effectors too much of a good thing? *Cancer Immunol Immunother* 53(3):262–274.

Pegoraro, L., Fierro, M. T., Lusso, P., Giovinazzo, B., Lanino, E., Giovarelli, M., Matera, L., and Foa, R. (1985). A novel leukemia T-cell line (PF-382) with phenotypic and functional features of suppressor lymphocytes. *J Natl Cancer Inst* 75(2):285–290.

Pericle, F., Kirken, R. A., Bronte, V., Sconocchia, G., DaSilva, L., and Segal, D. M. (1997). Immunocompromised tumor-bearing mice show a selective loss of STAT5a/b expression in T and B lymphocytes. *J Immunol* 159:2580–2585.

Pericle, F., Pinto, L., Hicks, S., Kirken, R. A., Sconocchia, G., Rusnak, J., Dolan, M. J., Sherear, G. M., and Segal, D. (1998). HIV-1 infection induces a selective reduction in STAT5 protein expression. *J Immunol* 160:28–31.

Que, L. G., George, S. E., Gotoh, T., Mori, M., and Huang, Y. C. (2002). Effects of arginase isoforms on NO Production by nNOS. *Nitric Oxide* 6(1):1–8.

Rambaldi, A., Wakamiya, N., Vellenga, E., Horiguchi, J., Warren, M. K., Kufe, D., and Griffin, J. D. (1988). Expression of the macrophage colony-stimulating factor and c-fms genes in human acute myeloblastic leukemia cells. *J Clin Invest* 81(4):1030–1035.

Rane, S. G., and Reddy, E. P. (2000). Janus kinases: components of multiple signaling pathways. *Oncogene* 19(49):5662–5679.

Ratta, M., Fagnoni, F., Curti, A., Vescovini, R., Sansoni, P., Oliviero, B., Fogli, M., Ferri, E., Della Cuna, G. R., Tura, S., Baccarani, M., and Lemoli, R. M. (2002). Dendritic cells are functionally defective in multiple myeloma: the role of interleukin-6. *Blood* 100(1):230–237.

Rodriguez, P. C., Hernandez, C. P., Quiceno, D., Dubinett, S. M., Zabaleta, J., Ochoa, J. B., Gilbert, J., and Ochoa, A. C. (2005). Arginase I in myeloid suppressor cells is induced by COX-2 in lung carcinoma. *J Exp Med* 202(7):931–939.

Rodriguez, P. C., Quiceno, D. G., Zabaleta, J., Ortiz, B., Zea, A. H., Piazuelo, M. B., Delgado, A., Correa, P., Brayer, J., Sotomayor, E. M., Antonia, S., Ochoa, J. B., and Ochoa, A. C. (2004). Arginase I production in the tumor microenvironment by mature myeloid cells inhibits T-cell receptor expression and antigen-specific T-cell responses. *Cancer Res* 64(16):5839–5849.

Rossner, S., Voigtlander, C., Wiethe, C., Hanig, J., Seifarth, C., and Lutz, M. B. (2005). Myeloid dendritic cell precursors generated from bone marrow suppress T cell responses via cell contact and nitric oxide production in vitro. *Eur J Immunol* 35(12):3533–3544.

Ruegg, C., Yilmaz, A., Bieler, G., Bamat, J., Chaubert, P., and Lejeune, F. J. (1998). Evidence for the involvement of endothelial cell integrin alphaVbeta3 in the disruption of the tumor vasculature induced by TNF and IFN-gamma. *Nat Med* 4(4):408–414.

Salvadori, S., Martinelli, G., and Zier, K. (2000). Resection of solid tumors reverses T cell defects and restores protective immunity. *J Immunol* 164(4):2214–2220.

Schmielau, J., and Finn, O. J. (2001). Activated granulocytes and granulocyte-derived hydrogen peroxide are the underlying mechanism of suppression of T-cell function in advanced cancer patients. *Cancer Res* 61(12):4756–4760.

Seliger, B., Hohne, A., Knuth, A., Bernhard, H., Meyer, T., Tampe, R., Momburg, F., and Huber, C. (1996). Analysis of the major histocompatibility complex class I antigen presentation machinery in normal and malignant renal cells: evidence for deficiencies associated with transformation and progression. *Cancer Res* 56(8):1756–1760.

Serafini, P., Borrello, I., and Bronte, V. (2006a). Myeloid suppressor cells in cancer: recruitment, phenotype, properties, and mechanisms of immune suppression. *Semin Cancer Biol* 16(1): 53–65.

Serafini, P., Carbley, R., Noonan, K. A., Tan, G., Bronte, V., and Borrello, I. (2004a). High-dose granulocyte-macrophage colony-stimulating factor-producing vaccines impair the immune response through the recruitment of myeloid suppressor cells. *Cancer Res* 64(17):6337–6343.

Serafini, P., De Santo, C., Marigo, I., Cingarlini, S., Dolcetti, L., Gallina, G., Zanovello, P., and Bronte, V. (2004b). Derangement of immune responses by myeloid suppressor cells. *Cancer Immunol Immunother* 53(2):64–72.

Serafini, P., Meckel, K., Kelso, M., Noonan, K., Califano, J., Koch, W., Dolcetti, L., Bronte, V., and Borrello, I. (2006b). Phosphodiesterase-5 inhibition augments endogenous antitumor immunity by reducing myeloid-derived suppressor cell function. *J Exp Med* 203(12):2691–2702.

Seung, L. P., Rowley, D. A., Dubey, P., and Schreiber, H. (1995). Synergy between T-cell immunity and inhibition of paracrine stimulation causes tumor rejection. *Proc Natl Acad Sci USA* 92:6254–6258.

Sgadari, C., Angiolillo, A. L., and Tosato, G. (1996). Inhibition of angiogenesis by interleukin-12 is mediated by the interferon-inducible protein 10. *Blood* 87(9):3877–3882.

Shankaran, V., Ikeda, H., Bruce, A. T., White, J. M., Swanson, P. E., Old, L. J., and Schreiber, R. D. (2001). IFNgamma and lymphocytes prevent primary tumour development and shape tumour immunogenicity. *Nature* 410(6832):1107–1111.

Sinha, P., Clements, V. K., Miller, S., and Ostrand-Rosenberg, S. (2005a). Tumor immunity: a balancing act between T cell activation, macrophage activation and tumor-induced immune suppression. *Cancer Immunol Immunother* 54(11):1137–1142.

Sinha, P., Clements, V. K., and Ostrand-Rosenberg, S. (2005b). Reduction of myeloid-derived suppressor cells and induction of M1 macrophages facilitate the rejection of established metastatic disease. *J Immunol* 174(2):636–645.

Sinha, P., Clements, V. K., and Ostrand-Rosenberg, S. (2005c). Interleukin-13-regulated M2 macrophages in combination with myeloid suppressor cells block immune surveillance against metastasis. *Cancer Res* 65(24):11743–11751.

Smith, C. W., Chen, Z., Dong, G., Loukinova, E., Pegram, M. Y., Nicholas-Figueroa, L., and Van Waes, C. (1998). The host environment promotes the development of primary and metastatic squamous cell carcinomas that constitutively express proinflammatory cytokines IL-1alpha, IL-6, GM-CSF, and KC. *Clin Exp Metastasis* 16(7):655–664.

Sombroek, C. C., Stam, A. G., Masterson, A. J., Lougheed, S. M., Schakel, M. J., Meijer, C. J., Pinedo, H. M., van den Eertwegh, A. J., Scheper, R. J., and de Gruijl, T. D. (2002). Prostanoids play a major role in the primary tumor-induced inhibition of dendritic cell differentiation. *J Immunol* 168(9):4333–4343.

Steinbrink, K., Graulich, E., Kubsch, S., Knop, J., and Enk, A. H. (2002). CD4(+) and CD8(+) anergic T cells induced by interleukin-10-treated human dendritic cells display antigen-specific suppressor activity. *Blood* 99(7):2468–2476.

Stoeck, M., Miescher, S., MacDonald, H. R., and Von Fliedner, V. (1989). Transforming growth factors beta slow down cell-cycle progression in a murine interleukin-2 dependent T-cell line. *J Cell Physiol* 141(1):65–73.

Strober, S. (1984). Natural suppressor (NS) cells, neonatal tolerance, and total lymphoid irradiation: exploring obscure relationships. *Annu Rev Immunol* 2:219–237.

Strober, S., Dejbachsh-Jones, S., Van Vlasselaer, P., Duwe, G., Salimi, S., and Allison, J. P. (1989). Cloned natural suppressor cell lines express the CD3+CD4-CD8- surface phenotype and the alpha, beta heterodimer of the T cell antigen receptor. *J Immunol* 143:1118–1122.

Subiza, J. L., Vinuela, J. E., Rodriguez, R., Gil, J., Figueredo, M. A., and De La Concha, E. G. (1989). Development of splenic natural suppressor (NS) cells in Ehrlich tumor-bearing mice. 44:307–314.

Suzuki, E., Kapoor, V., Jassar, A. S., Kaiser, L. R., and Albelda, S. M. (2005). Gemcitabine selectively eliminates splenic Gr-1+/CD11b+ myeloid suppressor cells in tumor-bearing animals and enhances antitumor immune activity. *Clin Cancer Res* 11(18):6713–6721.

Sykes, M., Sharabi, Y., and Sachs, D. H. (1990). Natural suppressor cells in spleens of irradiated, bone marrow- reconstituted mice and normal bone marrow: lack of Sca-1 expression and enrichment by depletion of Mac1-positive cells. *Cell Immunol* 127:260–274.

Takeda, K., Clausen, B. E., Kaisho, T., Tsujimura, T., Terada, N., Forster, I., and Akira, S. (1999). Enhanced Th1 activity and development of chronic enterocolitis in mice devoid of Stat3 in macrophages and neutrophils. *Immunity* 10(1):39–49.

Takeda, K., Hatakeyama, K., Tsuchiya, Y., Rikiishi, H., Kumagai, K., (1991). A correlation between GM-CSF gene expression and metastases in murine tumors. *Int J Cancer* 47:413–420.

Talmadge, J. E., Hood, K. C., Zobel, L. C., Shafer, L. R., Coles, M., and Toth, B. (2007). Chemoprevention by cyclooxygenase-2 inhibition reduces immature myeloid suppressor cell expansion. *Int Immunopharmacol* 7(2):140–151.

Terabe, M., and Berzofsky, J. A. (2004). Immunoregulatory T cells in tumor immunity. *Curr Opin Immunol* 16(2):157–162.

Terabe, M., Matsui, S., Noben-Trauth, N., Chen, H., Watson, C., Donaldson, D. D., Carbone, D. P., Paul, W. E., and Berzofsky, J. A. (2000). NKT cell-mediated repression of tumor immunosurveillance by IL-13 and the IL-4R-STAT6 pathway. *Nat Immunol* 1(6):515–520.

Terabe, M., Matsui, S., Park, J. M., Mamura, M., Noben-Trauth, N., Donaldson, D. D., Chen, W., Wahl, S. M., Ledbetter, S., Pratt, B., Letterio, J. J., Paul, W. E., and Berzofsky, J. A. (2003). Transforming growth factor-beta production and myeloid cells are an effector mechanism through which CD1d-restricted T cells block cytotoxic T lymphocyte-mediated tumor immunosurveillance: abrogation prevents tumor recurrence. *J Exp Med* 198(11): 1741–1752.

Terabe, M., Park, J. M., and Berzofsky, J. A. (2004). Role of IL-13 in regulation of anti-tumor immunity and tumor growth. *Cancer Immunol Immunother* 53(2):79–85.

Toi, M., Kondo, S., Suzuki, H., Yamamoto, Y., Inada, K., Imazawa, T., Taniguchi, T., and Tominaga, T. (1996). Quantitative analysis of vascular endothelial growth factor in primary breast cancer. *Cancer* 77(6):1101–1106.

Trikha, M., Corringham, R., Klein, B., and Rossi, J. F. (2003). Targeted anti-interleukin-6 monoclonal antibody therapy for cancer: a review of the rationale and clinical evidence. *Clin Cancer Res* 9(13):4653–4665.

Tsai, P., Weaver, J., Cao, G. L., Pou, S., Roman, L. J., Starkov, A. A., and Rosen, G. M. (2005). L-arginine regulates neuronal nitric oxide synthase production of superoxide and hydrogen peroxide. *Biochem Pharmacol* 69(6):971–979.

Van Ginderachter, J. A., Meerschaut, S., Liu, Y., Brys, L., De Groeve, K., Hassanzadeh Ghassabeh, G., Raes, G., and De Baetselier, P. (2006). Peroxisome proliferator-activated receptor gamma (PPARgamma) ligands reverse CTL suppression by alternatively activated (M2) macrophages in cancer. *Blood* 108(2):525–535.

Wang, S., Yang, J., Qian, J., Wezeman, M., Kwak, L. W., and Yi, Q. (2006). Tumor evasion of the immune system: inhibiting p38 MAPK signaling restores the function of dendritic cells in multiple myeloma. *Blood* 107(6):2432–2439.

Wang, T., Niu, G., Kortylewski, M., Burdelya, L., Shain, K., Zhang, S., Bhattacharya, R., Gabrilovich, D., Heller, R., Coppola, D., Dalton, W., Jove, R., Pardoll, D., and Yu, H. (2004). Regulation of the innate and adaptive immune responses by Stat-3 signaling in tumor cells. *Nat Med* 10(1):48–54.

Wing, E. J., Magee, D. M., Pearson, A. C., Waheed, A., and Shadduck, R. K. (1986). Peritoneal macrophages exposed to purified macrophage colony-stimulating factor (M-CSF) suppress mitogen- and antigen-stimulated lymphocyte proliferation. *J Immunol* 137(9):2768–2773.

Wolfraim, L. A., Walz, T. M., James, Z., Fernandez, T., and Letterio, J. J. (2004). p21Cip1 and p27Kip1 act in synergy to alter the sensitivity of naive T cells to TGF-beta-mediated G1 arrest through modulation of IL-2 responsiveness. *J Immunol* 173(5):3093–3102.

Wu, G., and Morris, S. M., Jr. (1998). Arginine metabolism:nitric oxide and beyond. *Biochem J* 336(Pt 1):1–17.

Xia, Y., Roman, L. J., Masters, B. S., and Zweier, J. L. (1998). Inducible nitric-oxide synthase generates superoxide from the reductase domain. *J Biol Chem* 273(35):22635–22639.

Xia, Y., and Zweier, J. L. (1997). Superoxide and peroxynitrite generation from inducible nitric oxide synthase in macrophages. *Proc Natl Acad Sci USA* 94(13):6954–6958.

Yang, L., DeBusk, L. M., Fukuda, K., Fingleton, B., Green-Jarvis, B., Shyr, Y., Matrisian, L. M., Carbone, D. P., and Lin, P. C. (2004). Expansion of myeloid immune suppressor Gr+CD11b+ cells in tumor-bearing host directly promotes tumor angiogenesis. *Cancer Cell* 6(4):409–421.

Yang, L., Yamagata, N., Yadav, R., Brandon, S., Courtney, R. L., Morrow, J. D., Shyr, Y., Boothby, M., Joyce, S., Carbone, D. P., and Breyer, R. M. (2003). Cancer-associated immunodeficiency and dendritic cell abnormalities mediated by the prostaglandin EP2 receptor. *J Clin Invest* 111(5):727–735.

Yang, R., Cai, Z., Zhang, Y., Yutzy, W. H. I. V., Roby, K. F., and Roden, R. B. S. (2006). CD80 in immune suppression by mouse ovarian carcinoma-associated Gr-1+CD11b+ myeloid cells. *Cancer Res* 66(13):6807–6815.

Yoshimura, A. (2006). Signal transduction of inflammatory cytokines and tumor development. *Cancer Sci* 97(6):439–447.

Young, M. R., and Cigal, M. (2006). Tumor skewing of CD34+ cell differentiation from a dendritic cell pathway into endothelial cells. *Cancer Immunol Immunother* 55(5):558–568.

Young, M. R., Lozano, Y., Ihm, J., Wright, M. A., and Prechel, M. M. (1996). Vitamin D3 treatment of tumor bearers can stimulate immune competence and reduce tumor growth when treatment coincides with a heightened presence of natural suppressor cells. *Cancer Lett* 104:153–161.

Young, M. R., Wright, M. A., Coogan, M., Young, M. E., and Bagash, J. (1992). Tumor-derived cytokines induce bone marrow suppressor cells that mediate immunosuppression through transforming growth factor beta. *Cancer Immunol Immunother* 35(1):14–18.

Young, M. R., Wright, M. A., Lozano, Y., Prechel, M. M., Benefield, J., Leonetti, J. P., Collins, S. L., and Petruzzelli, G. J. (1997). Increased recurrence and metastasis in patients whose primary head and neck squamous cell carcinomas secreted granulocyte-macrophage colony-stimulating factor and contained CD34+ natural suppressor cells. *Int J Cancer* 74(1):69–74.

Young, M. R., Wright, M. A., and Young, M. E. (1991). Antibodies to colony-stimulating factors block Lewis lung carcinoma cell stimulation of immune-suppressive bone marrow cells. *Cancer Immunol Immunother* 33:146–152.

Young, M. R., Young, M. E., and Wright, M. A. (1990a). Myelopoiesis-associated suppressor-cell activity in mice with Lewis lung carcinoma tumors: interferon-gamma plus tumor necrosis factor-alpha synergistically reduce suppressor cell activity. *Int J Cancer* 46(2):245–250.

Young, M. R., Young, M. E., and Wright, M. A. (1990b). Stimulation of immune-suppressive bone marrow cells by colony- stimulating factors. *Exp Hematol* 18(7):806–811.

Yu, X., Kennedy, R. H., and Liu, S. J. (2003). JAK2/STAT3, not ERK1/2, mediates interleukin-6-induced activation of inducible nitric-oxide synthase and decrease in contractility of adult ventricular myocytes. *J Biol Chem* 278(18):16304–16309.

Yu, Z., Zhang, W., and Kone, B. C. (2002). Signal transducers and activators of transcription 3 (STAT3) inhibits transcription of the inducible nitric oxide synthase gene by interacting with nuclear factor kappaB. *Biochem J* 367(Pt 1):97–105.

Zea, A. H., Rodriguez, P. C., Atkins, M. B., Hernandez, C., Signoretti, S., Zabaleta, J., McDermott, D., Quiceno, D., Youmans, A., O'Neill, A., Mier, J., and Ochoa, A. C. (2005). Arginase-producing myeloid suppressor cells in renal cell carcinoma patients: a mechanism of tumor evasion. *Cancer Res* 65(8):3044–3048.

Zhang, B., Bowerman, N. A., Salama, J. K., Schmidt, H., Spiotto, M. T., Schietinger, A., Yu, P., Fu, Y. X., Weichselbaum, R. R., Rowley, D. A., Kranz, D. M., and Schreiber, H. (2007). Induced sensitization of tumor stroma leads to eradication of established cancer by T cells. *J Exp Med* 204(1):49–55.

Signaling Pathways in Antigen-Presenting Cells Involved in the Induction of Antigen-Specific T-Cell Tolerance

Ildefonso Vicente-Suarez, Alejandro Villagra, and Eduardo M. Sotomayor

1 Introduction

In recent years, our view of activation of the immune system has changed dramatically given the identification of inhibitory signaling pathways in immune cells that, by counteracting positive/activating pathways, greatly influence the initiation, magnitude and duration of immune responses. These findings led immunologists to redefine the concept of immune activation as the net outcome of "turning on" activating genes and "downregulating" genes with inhibitory function (Ravetch and Lanier, 2000). By extension, it was proposed that these inhibitory regulatory pathways might also play a role in the induction and maintenance of peripheral tolerance to self-antigens. Experimental evidence supporting the role of negative regulatory pathways in immune tolerance was indeed provided by studies in mice with targeted disruption of specific inhibitory molecules in which unchecked inflammatory responses and autoimmunity were commonly observed (Hida et al., 2000; Tivol et al., 1995).

Bone marrow-derived APCs and in particular dendritic cells (DCs) play a central role in the generation of productive antigen-specific T-cell responses (Guermonprez et al., 2002). However, these same cells are also required for the induction of T-cell tolerance (Steinman et al., 2003). This seemingly dual function of APCs was attributed initially to the existence of specific APC subpopulation(s) that preferentially induce T-cell priming while other subpopulations are mainly involved in the induction of T-cell anergy (Huang et al., 2000; Munn et al., 2002; Scheinecker et al., 2002). The demonstration, however, that a single APC subpopulation can induce both T-cell outcomes (Belz et al. 2002) led to the alternative explanation that perhaps the functional status of the APC at the time of antigen presentation, rather than the phenotype of the APC, could be the central determinant of T-cell activation versusT-cell tolerance. Indeed, it has become increasingly clear now that while antigen

---

I.Vicente-Suarez
The Division of Immunology and Division of Malignant Hematology, Department of Interdisciplinary Oncology, H. Lee Moffitt Cancer Center & Research Institute at the University of South Florida, Tampa, FL 33612, USA

encounter by bone marrow-derived APCs in the presence of inflammatory mediators and/or microbial-derived molecules such as Toll-like receptor ligands triggers their maturation to a functional status capable of generating strong T-cell responses, antigen capture by these same APCs in the absence of inflammatory signals—or in the presence of inhibitory mediators—led instead to the development of antigen-specific T-cell tolerance (Steinman et al., 2003).

Given this plasticity of a defined APC population to induce divergent T-cell outcomes, it was then proposed that a delicate balance between activating and inhibitory pathways in the APC might play a role in influencing whether T cells would be activated or rendered tolerant following antigen recognition. As such, significant effort has been devoted in recent years to uncover those signaling pathways in APCs that by regulating the inflammatory properties of these cells might be central in the decision leading to T-cell activation versus T-cell unresponsiveness. In this chapter, we review those studies that provided some of the answers to these important questions. We discuss receptor–ligand interactions and novel intracellular signaling pathways that by limiting the ability of the APC to stimulate antigen-specific T-cells are important in preserving tolerance to self-antigens. Although these negative regulatory pathways in APC impose a significant barrier to our efforts to overcome immune tolerance to tumor antigens, their identification has provided novel molecular targets to potentially revert mechanisms of T-cell unresponsiveness in cancer.

2 Antigen-Presenting Cells and Tolerance to Tumor Antigens

An unexpected finding in the field of tumor immunology was the discovery that most of the antigens expressed by tumor cells were not necessarily neo-antigens uniquely present in cancer cells, but rather tissue-differentiation antigens shared between the tumor and normal tissues (Boon et al., 1996; Rosenberg, 1995). These surprising findings prompted some investigators to hypothesize that perhaps the greatest obstacle for harnessing the immune system against tumors is the immune system itself, and more specifically, its complex mechanisms for establishing T-cell tolerance against self- and by extension, to tumor antigens, most of them also "self" (Sotomayor et al., 1996). In the mid-1990s, the demonstration by the Bogen's and Levitsky's groups that antigen-specific $CD4^+$ T cells were rendered tolerant during tumor growth in vivo provided the first experimental evidence supporting the immune tolerance hypothesis (Bogen, 1996; Staveley-O'Carroll et al., 1998). Since then, several studies have confirmed that this state of T-cell unresponsiveness occurs during the progression of both hematologic and solid tumors expressing model or true tumor antigens (Cuenca et al., 2003; Morgan et al., 1998; Overwijk et al., 2003) and that also affects the CD8 T-cell compartment (Morgan et al., 1998; Ohlen et al., 2001; Overwijk et al., 2003; Shrikant et al., 1999). Furthermore, the demonstration that T-cell tolerance is seen during the progression of spontaneously arising tumors (Willimsky et al., 2005) and more importantly during the growth of human malignancies (Lee et al., 1999; Noonan et al., 2005) led to the undisputed realization that tolerance to tumor antigens, through mechanisms akin to those that regulate responses to self-antigens, represents an important immunosuppressive strategy by which tumor cells might escape T-cell-mediated antitumor responses.

This different view of tumor immunity has intimately linked the cancer immunology and autoimmunity fields. For instance, several principles learned from the better understanding of the cellular and molecular mechanisms by which tolerance to self is maintained in normal conditions, or broken in autoimmune diseases, have been applied to identify tolerogenic mechanisms in cancer patients (Pardoll, 2003). One such mechanism was provided by the identification of the central role that BM-derived APCs play in the induction of tolerance to self-antigens (Adler et al., 1998; Kurts et al., 1997), a concept that was then extended to the field of tumor immunology with the unambiguous demonstration by us and others that BM-derived APCs are also required for the induction of tolerance to antigens expressed by tumor cells (Cuenca et al., 2003; Sotomayor et al., 2001). These studies also provided evidence that the intrinsic antigen-presenting capacity of tumor cells has little influence over T-cell priming versus tolerance, a critical decision that is regulated at the level of the APC.

Dendritic cells (DCs), macrophages and B cells are all BM-derived cells that express MHC as well as costimulatory molecules and, as such, can potentially present tumor antigen to antigen-specific T cells. Although it is plausible that under particular conditions each subpopulation might induce T-cell tolerance (Fuchs and Matzinger, 1992; Lassila et al., 1988; Miyazaki et al., 1993; Ronchese and Hausmann, 1993; Ronchetti et al., 1999; Watson and Lopez, 1995), several lines of evidence have pointed to DCs as playing a central role in influencing the delicate balance between immunity and tolerance in vivo (Heath et al., 2004; Itano and Jenkins, 2003). For instance, it has been shown that in the steady state, DCs continually migrate between lymphoid and non-lymphoid tissues capturing self- and harmless environmental proteins through endocytic receptors such as DEC 205 (Steinman et al., 2003). Several studies tracking the fate of cellular antigens, particulate antigens and antigen-pulsed DCs at the site of injection and in draining lymphoid organs have now clearly established that antigen presentation by DCs in the steady state, which is characterized by the absence of inflammation, induced a modest T-cell proliferation but not polarization into Th1 or Th2 subsets. Instead, after several rounds of cell division almost all the antigen-specific T cells are deleted and those that remain are functionally anergic even to cognate antigen administered with strong immune adjuvant such as Complete Freund Adjuvant (CFA) (Bonifaz et al., 2002; Hawiger et al., 2001).

The above scenario also typifies how DCs would normally encounter tumor antigens in vivo and has been proposed as an explanation for how tolerance to tumor antigens is induced by these cells (Fig. 1). But unlike the steady state, in which the lack of inflammatory stimuli during antigen encounter by DCs is considered to be the major determinant of tolerance induction, in the tumor-bearing host the

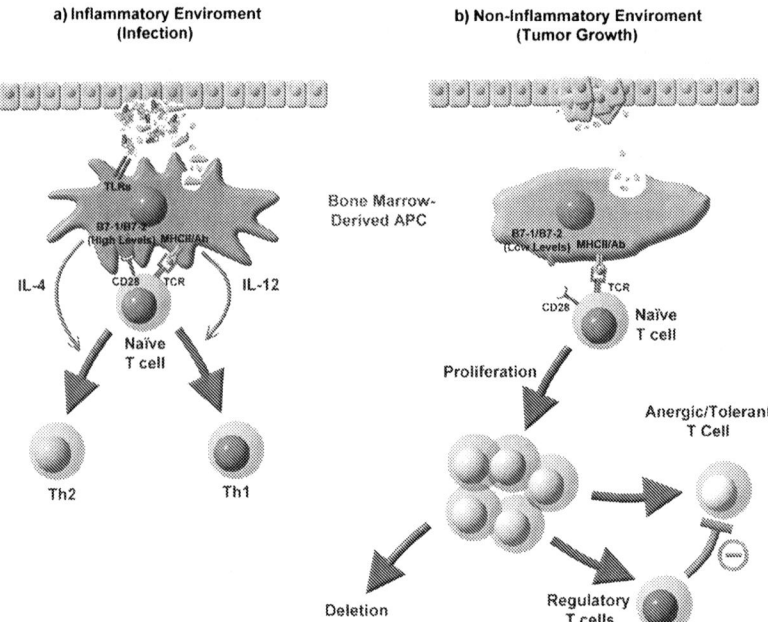

Fig. 1 The environment in which the antigen is encountered by BM-derived APCs influences T-cell priming versus tolerance. (**a**) In the course of an infection, microbial products are recognized by the APC through TLRs. TLR stimulation results in APC maturation, a process that increases the levels of costimulatory molecules and cytokine production. Antigen presentation by mature APCs leads to efficient priming of naïve T cells that proliferate and differentiate into effector cells. (**b**) Phagocytosis of self-antigens or tumor antigens in a non-inflammatory environment does not cause APC maturation. Immature APCs display low levels of costimulatory molecules and cytokine production. Antigen presentation by immature APCs to naïve T cells results in transient proliferation followed by induction of anergic and/or regulatory T cells

encounter of tumor antigens by DCs occurs not only in the absence of inflammatory signals needed for efficient maturation/activation of these cells, but also in the presence of inhibitory factors such as VEGF, IL-6, M-CSF, TGF-β, IL-10, PGE2 and gangliosides that further suppress DC maturation (Horna and Sotomayor, 2007). In this adverse environment, DCs will likely acquire "tolerogenic" properties that would in turn lead to the induction and maintenance of T-cell tolerance to tumor antigens (Munn et al., 2004). A better understanding of those ligand-receptors and/or intracellular pathways involved in the generation of "tolerogenic" APCs will provide novel molecular targets for therapeutic approaches that by converting APCs from "tolerizing" into "activating" in tumor-bearing hosts might ultimately result in the breaking of the remarkable barrier that tolerance to tumor antigens has imposed on cancer immunotherapy.

3 Signaling Pathways in APCs Influencing Antigen-Specific T-Cell Activation Versus Tolerance

3.1 Tyrosine Kinase Receptors

A wide spectrum of cellular functions such as cell proliferation, differentiation, survival and metabolism are regulated through tyrosine kinase receptors (TKRs). TKRs are characterized for their intrinsic tyrosine kinase activity. Ligand recognition by the TKR extra-cellular domain causes the receptor to dimerize or oligomerize which in turn activates its tyrosine kinase activity and initiates a specific signaling transduction cascade. Three TKR families have been identified in macrophages and other monocyte-derived cells: the receptor for macrophage colony-stimulating factor (M-CSF) involved in the survival of circulating monocytes and tissue macrophages and the closely related Tyro3 kinase and STK (mouse)/RON (human) receptors (Correll et al., 2004). Recently, a number of studies have demonstrated the important role of TKRs, especially those belonging to the Tyro3 kinase receptor family in limiting macrophage and dendritic cell activation (Lemke and Lu, 2003).

3.1.1 Tyro 3 Family Receptors

The Tyro3 family of tyrosine kinase receptors is composed of three members named Tyro3, Axl and Mer. This protein family was first identified in cells of the rat nervous system by using a homology-based cloning. This approach identifies novel RTK members because of the high similarity that exists on the TK domain of different RTK (Lai and Lemke, 1991). The central role of this receptor family in immune regulation was first highlighted by studies in triple mutant mice (TAM) lacking Tyro3, Axl as well as Mer (Lu and Lemke, 2001). Four-week-old TAM mice were found to have a progressive enlargement of the spleen and lymph nodes that was caused by aberrant T-cell and B-cell proliferation. Lymphocytes from these mice also show evidence of being activated as demonstrated by their increased expression of CD44 marker as well as by production of IFNγ. Eventually, these animals developed autoimmune disorders such as rheumatoid arthritis and systemic lupus erythematosus. Given that Tyro3 receptors are expressed in monocytes, macrophages and DCs, but not in B cells or T cells, it was concluded that the constitutive immune activation observed in these mutant mice was not due to an intrinsic defect in the lymphocyte compartment, but the result of lack of TKR in non-lymphocytic cells. Studies of APCs from TAM mice revealed that these cells are indeed functionally hyperactive and display higher levels of MHC class II and B7.2 costimulatory molecules relative to wild-type cells before and after activation with LPS. In addition, higher levels of TNF-α and IL-12 cytokines were produced by TAM $-/-$ macrophages in response to LPS stimulation. These results were reminiscent of previous studies in mice in which Mer kinase activity is suppressed (mer kinase deficient mice or mer^{kd}) (Camenisch et al., 1999). LPS-injected mer^{kd} mice showed an increase in TNF-α production in vivo that correlated with an enhanced susceptibility to endotoxic shock, which could be reverted when animals were treated with anti-TNFβ-blocking antibodies. Furthermore, macrophages from mer[kd] mice display increased NFκB translocation to the nucleus and TNF-α hyperproduction when stimulated with LPS in vitro. Finally, sera from mer[kd] mice show an increase in anti-DNA antibodies as compared to wild-type mice and later in life they develop autoimmunity features such as development of systemic lupus erythematosus (Cohen et al., 2002).

In addition to its role in regulation of APC function, recent evidence has demonstrated that Mer receptor is required for efficient phagocytosis of apoptotic bodies (Scott et al., 2001). Growth arrest-specific protein 6 (GAS6) is the common ligand for all the members of the Tyro3 kinase family (Stitt et al., 1995). The demonstration that GAS6 protein mediates the binding of phosphatidylserine displayed by cells that have initiated apoptosis suggested a potential role of the Tyro3 kinase family in the clearance of apoptotic cells. Indeed, injection of labeled apoptotic cells in animals with deficient mer receptor function resulted in excessive accumulation of apoptotic bodies (Scott et al., 2001). Furthermore, in vitro studies showed that macrophages isolated from mer kinase deficient mice displayed a marked decrease in their ability to phagocytate apoptotic cells but they were able to phagocytate bacteria, beads or opsonized cells. These data pointed to mer RTK as an important scavenger receptor in macrophages.

Deficient removal of cellular debris in animals with deficient mer receptor function resulted in persistence of self-antigens that could explain why these animals are prone to develop autoimmune disease (Casiano and Tan, 1996; Mevorach et al., 1998; Rosen and Casciola-Rosen, 1999; Taylor et al., 2000). Nonetheless, it does not explain why mer[kd] macrophages displayed overproduction of pro-inflammatory cytokines and costimulatory molecules following LPS stimulation. It has been proposed that phagocytosis of apoptotic bodies might inhibit the production of inflammatory cytokine in APCs through an enhancement in the production of anti-inflammatory mediators such as IL-10 and TGF-β (Fadok et al. 1998; Voll et al., 1997). In addition, it has been recently shown that uptake of apoptotic bodies by DCs prevents translocation of NFκB into the nucleus which leads to diminished production of pro-inflammatory mediators in response to LPS stimulation. Decreased NFκB nuclear translocation induced by apoptosis seems to be dependent on the activation of the phosphatidylinositol 3-kinase (PI3K)/AKT pathway since it was prevented by PI3K inhibitors (Sen et al. 2007). Given that mer-deficient macrophages have impaired phagocytosis of apoptotic bodies (Scott et al., 2001), the absence of this negative regulatory mechanism will be associated with increased NFκB translocation and enhanced pro-inflammatory response to LPS. Therefore, in mice in which signaling through the tyro3 kinase receptors and in particular mer has been abrogated, the accumulation of self-antigen due to deficient phagocytosis combined with the presence of APCs displaying enhanced pro-inflammatory features might result in aberrant activation of the lymphocytic compartment, breaking of tolerance to self and the subsequent development of autoimmunity (Fig. 2). In these mice, however, several questions remain unanswered, such as the potential contribution of the microbial flora to the autoimmune disease through TLR engagement. Important information will be obtained by crossing TAM mice or

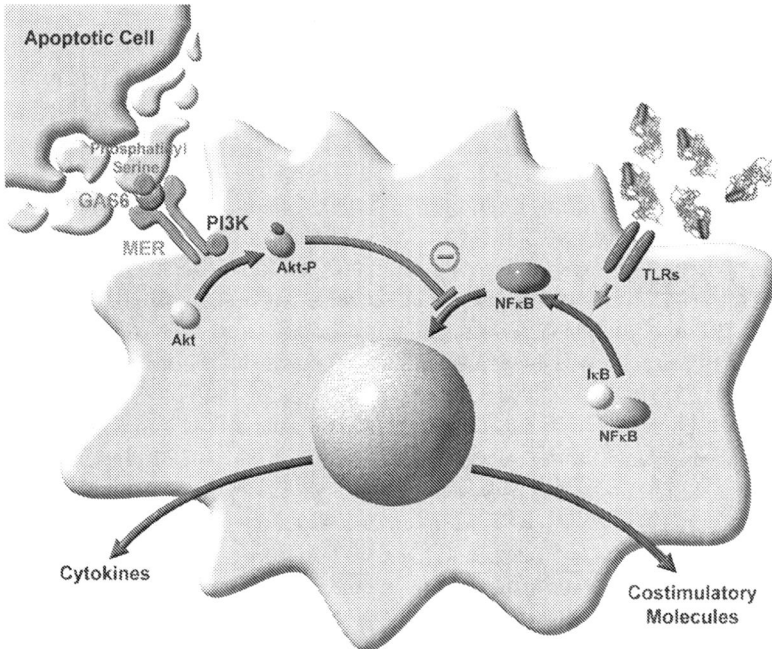

Fig. 2 Apoptosis, tyrosine kinase receptors and regulation of inflammatory responses in APCs. The Tyro3 kinase family receptors are involved in phagocytosis of apoptotic bodies through the recognition of GAS6-phosphatidyl serine complexes. Phagocytosis impairs the APC's ability to produce inflammatory cytokines by blocking NFκB translocation to the nucleus. This blockade seems to be dependent on PI3K pathway activation by Mer, one of the Tyro3 kinase family members (Sen et al., 2007)

mer^{kd} mice with MyD88$^{-/-}$ or mice lacking specific TLRs, specially TLR4$^{-/-}$ mice, given the increased sensitivity of TAM and mer^{kd} mice to LPS, the ligand for TLR4.

3.1.2 *c-kit* and Imatinib Mesylate

Given the above findings, we thought that APC function, and more specifically the ability to prime rather than tolerize tumor-specific T cells, might very well be amenable to modulation with pharmacologic agents targeting tyrosine phosphorylation of intracellular targets. One tyrosine kinase inhibitor that gained particular attention in recent years has been imatinib mesylate, a small molecule that strongly inhibits the *c-abl*, *c-kit* and PDGFR tyrosine kinases. Several studies in experimental tumor models as well as in clinical trials have shown that this drug is highly effective in blocking the tyrosine kinase activity of the *bcr/abl* fusion protein, leading to impressive hematologic as well as cytogenetic remissions in patients with chronic myelogenous leukemia (CML) (Druker et al., 1996; Kantarjian et al., 2002; Wang et al., 2004a). Because of its additional inhibitory activity upon *c-kit* (Heinrich et al., 2002; Wang et al., 2004a) this drug was also evaluated in patients with gastrointestinal sarcoma tumors (GIST), a malignancy characterized by *c-kit* over-expression and by its refractoriness to chemotherapy treatment. Treatment with imatinib mesylate was again associated with impressive tumor regression and induction of complete and sustained responses in patients with this otherwise lethal malignancy (Heinrich et al., 2002, 2003).

The potent tyrosine kinase inhibitory effect of imatinib mesylate prompted us to evaluate the in vitro and in vivo effects of this drug on APC's function and antigen-specific T-cell responses to cognate antigen. Reminiscent of the phenotype observed in TAM and mer kinase deficient mice, wild-type macrophages or DC treated with imatinib mesylate were found to be hyper-responsive to LPS stimulation as demonstrated by their enhanced production of the pro-inflammatory mediators IL-12, IL-1β, IL-6 and the chemokine RANTES (Wang et al., 2004a). Several studies have shown that persistent IL-12 production and signaling is linked to disruption of immune tolerance (Cheng et al., 2003; Evel-Kabler et al., 2006; Trinchieri, 2003). In addition, studies in our laboratory have demonstrated that production of RANTES by APCs is required for the in vitro reversion of antigen-specific T-cell tolerance (Cheng et al., 2003). Accordingly, we found that in vitro treatment of DCs or macrophages with imatinib mesylate was capable of restoring the responsiveness of tolerized T cells isolated from tumor-bearing hosts. In addition, in vivo treatment with this drug prevented tumor-induced antigen-specific T-cell tolerance and enhanced the efficacy of therapeutic vaccination. An interesting observation in our in vivo studies was the finding that imatinib mesylate consistently induced the expansion of NK 1.1^+ cells in response to vaccination. Similar results were reported by Borg et al. [2004] who found that in vivo treatment with this drug promotes a strong antitumor effect that was mediated by NK 1.1^+ cells (Borg et al. 2004). Further studies by Zitvogel's group showed that the combination of imatinib mesylate with IL-2 was also associated with an enhanced antitumor effect that was dependent on the expansion of a NK 1.1^+ population. Phenotypic and functional characterization of this cell population by two independent groups uncovered a novel DC subset expressing NK markers. Given their dual ability to uptake, process and present antigen (APC function) and to kill target cells (NK function), this novel population has been termed "interferon-producing killer dendritic cells" or IKDCs (Chan et al., 2006; Taieb et al., 2006). Whether IKDCs play a role in the prevention of tumor-induced antigen-specific T-cell tolerance observed in imatinib-treated tumor-bearing mice remains to be elucidated.

Molecular studies of imatinib-treated APCs showed that among all the known molecular targets of this drug, inhibition of *c-kit* phosphorylation seems to be the likely target in these cells (Wang et al., 2004a). First, *c-kit* but not *c-abl* was activated in LPS-treated APCs. Second, LPS-induced *c-kit* phosphorylation was inhibited in APCs that were simultaneously treated with imatinib. Third, inhibition of *c-kit* signaling was associated with enhanced production of pro-inflammatory mediators by APCs in response to LPS stimulation. These findings suggest that *c-kit* signaling might act as a negative regulator of inflammation, and its blockade might represent an appealing strategy to unleash inflammatory responses in APCs and tip the balance toward immune activation. These results have also opened the gate to novel lines of investigation focusing on the use of imatinib mesylate and other novel TKIs such as dasatinib and nilotinib as promising adjuvants in cancer immunotherapy. Furthermore, future studies will provide important insights into the signaling mechanisms by which these drugs augment APC function and trigger immune activation rather than tolerance.

3.2 STAT3

Signal transducers and activators of transcription (STATs) are cytoplasmic transcription factors that are key mediators of cytokine and growth factor signaling pathways (Darnell et al., 1997). Engagement of cell surface cytokine or growth factor receptors activates the Janus kinase (JAK) family of protein tyrosine kinases, which in turn phosphorylate and activate cytoplasmic STAT proteins. Activated STATs dimerize and translocate to the nucleus, where they bind to specific DNA response elements and induce expression of STAT-regulated genes (Darnell et al., 1997). One STAT family member, STAT3, has recently emerged as a negative regulator of inflammation. STAT3 is activated in response to various cytokines and growth factors including IL-6, VEGF and most importantly IL-10, an anti-inflammatory cytokine known to play a central role in limiting immune responses and establishing tolerance. Indeed, activation of STAT3 is essential for the anti-inflammatory properties of IL-10 signaling since blockade of STAT3 signaling abrogates IL-10-mediated suppression of LPS-induced cytokines and costimulatory molecules (Lang et al., 2002; Qin et al., 2006; Williams et al., 2004). Conversely, constitutively activated STAT3 resembles the anti-inflammatory effects induced by IL-10 (Williams et al., 2007).

Studies in mice with targeted disruption of STAT3 in different cellular compartments have highlighted the critical role of this pathway in the control of inflammatory responses and the development of autoimmunity. The earliest evidence was obtained in conditional STAT3 KO mice generated by utilizing the Cre-*loxP* recombination system. Mice carrying *loxP*-flanked STAT3 were crossed with mice expressing the Cre recombinase under the control of the lysozyme M promoter. Since lysozyme M is expressed in granulocytes, macrophages and in a small percent of myeloid DC, STAT3 is floxed out only in these cells (Takeda et al. 1999). LysMcre/Stat3$^{flox/-}$ mice develop severe inflammatory bowel disease as they age, seemingly as a result of aberrant immune responses to the enteric flora because breeding of these mice with TLR4$^{-/-}$ mice significantly reversed the autoimmune pathology (Kobayashi et al., 2003). These findings are quite reminiscent of those observed in IL-10 KO mice, in which inflammatory bowel disease always develops unless the animals are kept in a microbial-free environment.

Given the central role of STAT3 in regulating inflammatory responses, we explored whether disruption of this signaling pathway in APCs could influence their inflammatory status and their ability to prime antigen-specific CD4$^+$ T cells. First, inhibition of the JAK–STAT3 signaling pathway in macrophages and DCs using the tyrosine kinase inhibitor, tyrphostin AG490, resulted in enhanced priming of naïve antigen-specific T cells and the restoration of responsiveness of anergic $CD4^+$ T cells in vitro. Importantly, the ability of AG490-treated APCs to break T-cell tolerance correlated with a complete inhibition of STAT3 DNA-binding activity as determined by electromobility shift assay (EMSA) of nuclear extracts obtained from these APCs (Cheng et al., 2003). Second, studies in macrophages isolated from conditional KO mice (LysMcre/Stat3$^{flox/-}$ mice) revealed that LPS stimulation renders these APCs capable of effectively priming naïve antigen-specific T cells and to overcome the state of unresponsiveness of tolerized T cells in vitro. Third, the demonstration that in mice lacking functional STAT3 in macrophages and neutrophils the in vivo response to a tolerogenic stimuli is T-cell priming rather than T-cell tolerance uncovered a previously unknown role for STAT3 in the induction of immune tolerance (Cheng et al., 2003) (Fig. 3).

Phenotypic and functional analysis of macrophages isolated from LysMcre/Stat3$^{flox/-}$ mice provided important insights into the potential mechanism(s) by which these APCs can restore the responsiveness of tolerized T cells. Freshly isolated (non-stimulated) thioglycolate-elicited peritoneal macrophages (PEM from Stat3$^{-/-}$ mice) displayed an increased expression of MHC class II molecules as well as B7.1 and B7.2 costimulatory molecules relative to non-stimulated PEM from control mice. LPS stimulation of PEM from Stat3$^{-/-}$ mice resulted in significantly higher mRNA levels of the chemokines RANTES, MIP-1α, MIP-1β, MIP-2, IP-10 and the cytokine IL-6 as compared to LPS-stimulated PEM from control mice. In

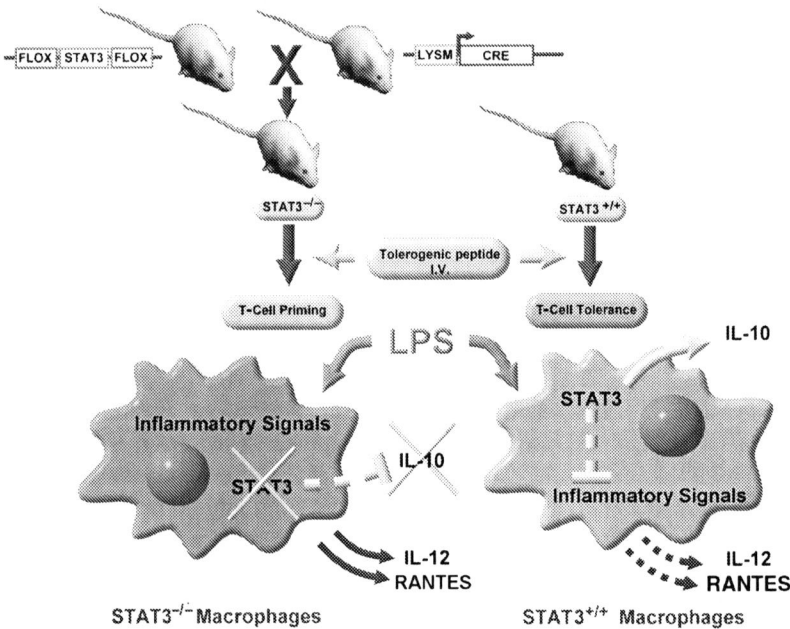

Fig. 3 Priming, not tolerance, of antigen-specific T cells is the functional outcome in STAT3 conditional KO in response to tolerogenic stimuli in vivo

addition, IL-12 mRNA as well as IL-12 protein production was detected in LPS-stimulated Stat3$^{-/-}$ PEM at the time that no IL-12 mRNA or protein could yet be detected in PEM from control mice. Importantly, no IL-10 was detected in the supernatants of LPS-stimulated Stat3$^{-/-}$ PEM, the cytokine that was present at significant levels in the supernatants of Stat3$^{+/+}$ PEM controls (Fig. 3, bottom). Further studies demonstrated that supernatants from LPS-stimulated STAT3$^{-/-}$ macrophages were sufficient to effectively break antigen-specific T-cell tolerance and that this ability was dependent on the combined effect of IL-12 and RANTES (Cheng et al., 2003).

It is important to note that macrophages devoid of STAT3 share phenotypic and functional characteristics displayed by macrophages lacking the inhibitory Tyro3 family of receptor tyrosine kinase, Tyro3, Axl and Mer (TAM mutant mice). As discussed previously, freshly isolated macrophages from these mice have increased expression of MHC class II molecules, produced elevated amounts of IL-12 in response to LPS and induced strong lymphocyte activation, findings quite similar to those observed in STAT3$^{-/-}$ macrophages. It is noteworthy, however, that while genetic disruption of all three inhibitory Tyr-3 receptors (triple mutant mice) is required to generate "inflammatory" macrophages, a similar outcome can be achieved by just disrupting STAT3 signaling pathway in these cells. The common findings in macrophages from STAT3$^{-/-}$ mice and in TAM triple mutant mice raise the interesting possibility that STAT3 may represent a common signaling pathway linking different inhibitory receptors with their downstream intracellular targets. It is plausible therefore that the activated phenotype of STAT3$^{-/-}$ PEM could be related to an enhanced activity of different pro-inflammatory pathways that are tightly regulated by an intact STAT3 signaling in these APCs.

Although significant advances have been made in recent years in the understanding of the anti-inflammatory effects associated with the activation of STAT3 in normal and malignant cells (Yu et al., 2007), the underlying molecular mechanism(s) through which this signaling pathway exerts its inhibitory effects in APCs are not fully elucidated but they will likely involve direct and/or indirect mechanism(s) and a vast array of molecules and intracellular pathways. STAT3 could act either *directly* by binding to other transcription factors (i.e., NFκB or STAT1) and repress the expression of inflammatory genes or *indirectly* by inducing the production of other proteins with anti-inflammatory properties.

In support of the direct mechanism of action, several studies have shown that STAT3 can interact directly with transcription factors involved in the maturation/activation of the APC. Maturation of antigen-presenting cells by classical inflammatory stimuli as microbial products or inflammatory cytokines such as TNF-α requires activation of the NFκB transcription factor family. The mammalian NFκB family includes RELA (p65), NFκB1 (p50; p105), NFκB2 (p52; p100), c-REL and RELB1 proteins. The main activated form of NFκB is a heterodimer composed of the p65 subunit associated with either the p50 or p52 subunit (Ghosh et al., 1998). By using NFκB reporter constructs, it has been shown that STAT3 antagonizes NFκB signaling and NFκB activated genes such as nitric oxide or IL-12 p40 (Hoentjen et al. 2005; Yu et al., 2002). This effect, which has been proposed, could be mediated by direct interaction between STAT3 and different NFκB subunits since STAT3 has been shown to co-immunoprecipitate with the p50 and p65 NFκB subunits in LPS-stimulated mesangial cells or with cRel in BM-derived DCs (Nefedova et al., 2005; Yu et al., 2002). Interestingly, STAT3 activation resulted in a decrease of cRel DNA binding activity in DCs while it did not affect NFκB DNA binding activity in mesangial cells, suggesting a mechanism of action that is cell-type-specific.

STAT1 is another transcription factor that has been shown to interact directly with STAT3. A common finding in APCs with genetic or pharmacologic disruption of STAT3 is the parallel increase in STAT1:1 homodimers, a functionally active complex that positively regulates genes encoding inflammatory factors (Ramana et al., 2000). Previous studies have demonstrated a frequent co-activation of STAT1 and STAT3 by the same ligand (Bromberg et al., 2000), leading to formation of STAT1:STAT3 heterodimers that may have a higher association constant than the STAT1:STAT1 homodimeric complex (Kotenko and Pestka, 2000). It is plausible therefore that STAT3 activation in APCs, by binding activated STAT1, may control the formation of functionally active STAT1:STAT1 homodimers, thus limiting the magnitude and/or intensity of an inflammatory response. Perhaps, only in those scenarios when the level of STAT1 exceeds the level of activated STAT3 or when STAT3 signaling is disrupted, STAT1 homodimers may trigger the downstream signals that APCs need to efficiently activate T cells and/or restore the responsiveness of tolerized T cells. Although we have identified IL-12 and RANTES as important inflammatory signals by which $Stat3^{-/-}$ APCs can overcome T-cell tolerance (Cheng et al., 2003), the potential role of "over-activated" pro-inflammatory pathways—i.e., STAT1 signaling among others—in the generation of these mediators remains to be elucidated.

Among the indirect mechanisms of action, one that has gained particular attention relates to the intimate link between STAT3 signaling and IL-10, a cytokine with well-known anti-inflammatory properties. It has been shown for example that binding of activated STAT3 to the IL-10 promoter is required for efficient expression of the IL-10 gene and protein production (Benkhart et al., 2000). In turn, IL-10 can enhance STAT3 activation in those cells expressing IL-10 receptor, suggesting a positive feedback mechanism to amplify/maintain the production of this cytokine. Supporting further the indirect mechanism of action of STAT3, recent studies have demonstrated that IL-10-mediated anti-inflammatory effects—which depend on STAT3 signaling—require synthesis of de novo proteins (Murray, 2005).

In addition to IL-10, other cytokines such as IL-6 can induce high levels of STAT3 phosphorylation in APCs. However, the anti-inflammatory effect associated with STAT3 activation is observed only in cells stimulated with IL-10 but not in response to IL-6. For instance, IL-10 but not IL-6 can significantly reduce the ability of APCs to produce IL-12 and TNF-α in response to LPS (El Kasmi et al., 2006; Yasukawa et al., 2003). Recent studies have demonstrated that SOCS3 could be responsible for the different inflammatory responses to IL-10 and IL-6 stimulation in spite of the fact that both cytokines induce STAT3 activation. SOCS3 exerts its regulatory role via binding to the phosphorylated subunit glycoprotein 130 (gp130) that is present in the IL-6 receptor but not in the IL-10 receptor (Yasukawa et al.,

2003). In animals with an intact SOCS3, IL-6 induction of STAT3 phosphorylation decreases faster relative to the levels of phosphorylated STAT3 induced by IL-10. In SOCS3$^{-/-}$ animals, however, the differences in the kinetics of STAT3 phosphorylation in response to IL-10 and Il-6 are no longer observed (Lang et al., 2003; Yasukawa et al., 2003). Furthermore, unlike wild-type mice, treatment of SOCS3$^{-/-}$ deficient animals with IL-6 reproduced the anti-inflammatory effect associated with IL-10 treatment (Yasukawa et al., 2003). These data suggest that persistent STAT3 signaling is required for this pathway to exert its anti-inflammatory effect. Given that IL-6 only transiently activates the STAT3 pathway—because of the negative regulation mediated by SOCS3—this mechanism has been proposed to explain why this cytokine is not as anti-inflammatory as IL-10, a factor that induces a more persistent STAT3 activation.

The anti-inflammatory role of STAT3 in APCs cannot be solely explained by its effect upon IL-10. Although APCs devoid of STAT3 share phenotypic characteristics with APCs from IL-10$^{-/-}$ mice, important differences among these cells still remain. Similar to our findings in PEM devoid of STAT3, alveolar macrophages from IL-10$^{-/-}$ mutant mice display increased expression of B7.1 and B7.2 costimulatory molecules (Soltys et al., 2002). However, while no change in the expression of MHC class II molecules was observed in macrophages devoid of IL-10, a significant increase in the expression of MHC class II molecules is a characteristic of macrophages devoid of STAT3. The ability of STAT3 to regulate MHC class II expression has been attributed, at least in part, to the regulatory role of STAT3 upon cathepsin S, a protease involved in cleavage of the invariant chain (Ii). Studies using the STAT3 inhibitor cucurbitacin I (JSI-124) have also pointed to an increase in translocation from intracellular compartments to the cell surface as an explanation for the increased expression of MHC class II molecules in APCs (Nefedova et al., 2005). Finally, recent studies have found that pathways other than IL-10 signaling, such as NFκB activation by Toll-like receptors, M-CSF signaling and NADPH oxidase function, which are negatively regulated by STAT3, might also play a role in the enhanced innate immunity observed in mice with disruption of this signaling pathway in APCs (Welte et al., 2003; Yu et al., 2007).

Further confirmation of the role that STAT3 plays in limiting immune responses was provided by the identification of other factors that depend on an intact STAT3 to inhibit inflammation. Cholinergic agonists (acetyl choline and nicotine) signaling through the alpha7 nicotinic acetylcholine receptor (alpha7nAChR) limit macrophage activation by reducing nuclear NFκB localization (Wang et al., 2004b). Recently it has been demonstrated that cholinergic agonist signaling results in STAT3 phosphorylation. Nicotine failed to block LPS-induced TNF-α production in cells transfected with a dominant negative form of STAT3 (de Jonge et al., 2005). Interestingly, the presence or absence of SOCS3 did not affect the anti-inflammatory role played by the alpha7 nicotinic acetylcholine receptor suggesting that this receptor could be not regulated by SOCS3. Taken together, it is plausible that factors that activate STAT3 but are not susceptible to SOCS3 regulation might block inflammatory responses while those factors that activate STAT3 only transiently because of SOCS3 effects will be unable to trigger anti-inflammatory responses (El Kasmi et al., 2006).

In summary, the identification of STAT3 signaling in APCs as a critical regulator of inflammatory responses has provided a novel molecular target for manipulation of immune activation/immune tolerance, a central decision with profound implications in cancer immunotherapy, autoimmunity and transplantation.

3.3 SOCS (Suppressors of Cytokine Signaling)

Communication among immune cells is fundamental in order to elicit a coordinated immune response. Cytokines released by immune cells orchestrate such a response by binding to specific cell surface receptors and activating proteins that will carry on the signal from the cell surface to the nucleus. Members of the Janus kinase (JAK) protein family bind constitutively in a specific manner to the cytoplasmic domains of cytokine receptor chains. After ligand engagement, dimerization or higher order oligomerization of receptor complexes occurs, allowing JAK phosphorylation. Activated JAK proteins will then recruit and phosphorylate specific signal transducers and activators of transcription (STATs) proteins. Activated STATs dimerize, dissociate from the receptor and translocate to the nucleus where they will induce gene expression. Among the STAT-activated genes there will be not only those that mediate the cytokine biological effect but also genes that are involved in turning off cytokine signal when their production is not longer needed. The suppressors of cytokine signaling (SOCS) are a protein family, consisting of eight members, the cytokine-inducible SH2 domain-containing protein (CIS) and SOCS1 through SOCS7, each of which has a central SH2 domain, and a C-terminal 40 amino acid sequence known as the SOCS box. While the SH2 domain binds phosphorylated tyrosine residues present in activated members of the cytokine signaling pathway, the SOCS box targets the complex for ubiquitination, proteosome degradation and as such termination of cytokine signaling and its biologic effect. Given that overexpression of certain SOCS proteins inhibits signaling by a variety of cytokines through the JAK/STAT pathway, it has been proposed that SOCS proteins are critical in providing a negative feedback loop for cytokine production.

3.3.1 SOCS1, a Fundamental Regulator of the Innate Immune Response

SOCS1 has been shown to play a central role in regulation of autoimmunity and in tumor rejection. This protein was first discovered using yeast two-hybrid assay to identify molecules that interact with Janus kinase 2 (JAK2) (Endo et al., 1997). In addition to the SH2 and SOCS box domain another region designated kinase inhibitory region (KIR) might play a role in SOCS1-mediated inhibition of JAK2. KIR might increase the binding strength of SOCS1 to JAK2 and block the access of substrates and/or ATP to the kinase catalytic pocket (Yasukawa et al., 1999). The importance of SOCS1 in controlling autoimmunity was unveiled in mice lacking functional $SOCS1$ ($SOCS1^{-/-}$ mice) (Alexander et al. 1999). These animals die within 2–3 weeks after birth because of a complex organ pathology that includes peripheral T-cell activation and massive infiltration of macrophages in the liver, spleen, lung and heart. These pathologic findings seem to be related to aberrant IFNγ production, since treatment with anti-IFNγ blocking antibodies or by crossing the SOCS1$^{-/-}$ strain with IFNγ $^{-/-}$ animals prevented the development of disease.

SOCS1 deficiency in the hematopoietic compartment is thought to be sufficient to cause disease since transfer of SOCS1$^{-/-}$ bone marrow into irradiated JAK3$^{-/-}$ recipients resulted in premature lethality (Marine et al., 1999). In addition, SOCS1$^{-/-}$ Rag2$^{-/-}$ mice do not develop pathologic abnormalities suggesting that lymphocyte subsets contribute to the SOCS1$^{-/-}$ pathology (Marine et al., 1999). However, SOCS1 deficiency in T/NKT cells alone is not sufficient to cause inflammatory pathology. Indeed, experimental studies in conditional KO mice in which SOCS1 was deleted in CD4$^+$ T cells, CD8$^+$ T cells and NKT cells but not in NK cells, B cells, monocytes or granulocytes (flox-SOCS1 mice crossed with mice expressing the Cre-recombinase protein under the control of the Lck promoter) did not show any abnormal activation and/or inflammatory changes (Chong et al., 2003). Conversely, in mice in which SOCS1 expression was specifically restituted in T and B cells (SOCS1$^{-/-}$ Tg) autoimmunity developed (Hanada et al., 2003). These animals die within 6 months when kept in pathogen-free conditions and within 3 months when kept in regular conditions. Splenomegaly and lymphadenopathy were observed as early as 10 weeks after birth and these findings coincided with the accumulation in the spleen of phenotypically mature DCs displaying high levels of costimulatory molecules. Studies of BM-derived DCs from SOCS1$^{-/-}$ Tg animals showed that these cells display enhanced responses to IL-4 and IFNγ stimulation. Finally, autoimmune disease in SOCS1$^{-/-}$ Tg animals resembles systemic lupus erythematosus (SLE) and the skin lesions, glomerulonephritis, hypergammaglobulinemia and autoantibody production seem to be the result of aberrant B-cell activation in these mice. Of note, SOCS1$^{-/-}$ DCs constitutively produce TNF-family B-cell growth factors and induce strong B-cell proliferation and antibody production (Hanada et al., 2003).

The studies above point toward the APC as being the critical cell in which disruption of the negative regulatory effect of SOCS1 results in dramatic pro-inflammatory changes and development of autoimmunity in vivo. Recent evidence supports the ability of SOCS1-deficient DCs to disrupt T-cell tolerance (Evel-Kabler et al., 2006). In this study the authors used lentiviral transfection to introduce SOCS1 siRNA into BM-derived DCs. SOCS1-deficient DCs were matured ex vivo with LPS and loaded with a self-antigen (TRP2) expressed by melanocytes. Adoptive transfer of LPS-matured SOCS1$^{-/-}$-deficient DCs, but not regular DCs, resulted in the development of autoimmune disease as evidenced by skin depigmentation or vitiligo. In addition, SOCS1-deficient cells induced strong antitumor responses against B16 melanoma tumors expressing the TRP2 antigen. Although the mechanism(s) involved in the generation of antitumor responses are not fully elucidated, IL-12 signaling seems to play an important role. Indeed, disruption of IL-12 signaling by using IL-12 receptor KO SOCS1-deficient DCs suppresses their ability to

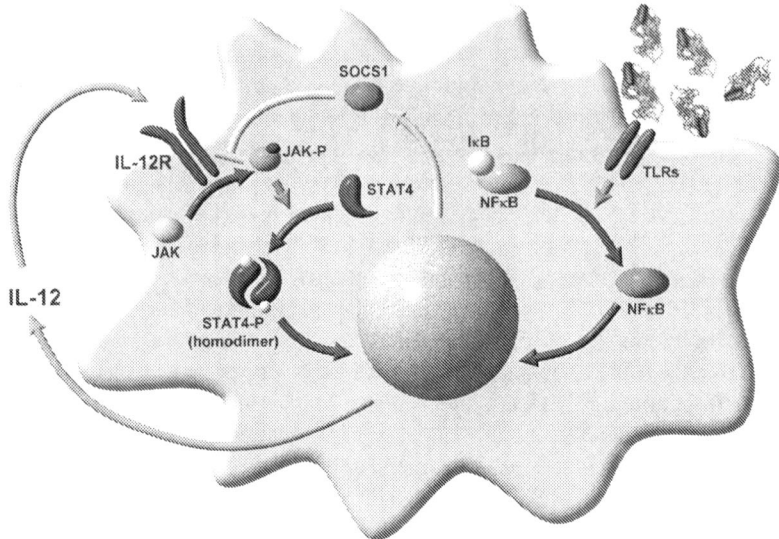

Fig. 4 Regulation of innate immune responses by SOCS1. IL-12 production is induced in macrophages and DCs by TLR ligands via NFκB activation. Through autocrine/paracrine mechanisms IL-12 signaling leads to the production of negative regulators needed to limit the cytokine's effects. Cells lacking SOCS1 are hyperactivated in response to microbial products and/or cytokines resulting in unrestrained inflammation and the development of autoimmune disease

induce autoimmunity or antitumor immune responses. Furthermore, IL-12 signaling in the absence of SOCS1 results in persistent STAT4 activation and increased production of IL-12. Interestingly, SOCS1-deficient DCs are characterized by an increased life-span since they were detected up to 4 days following their adoptive transfer into recipient animals. In sharp contrast, wild-type DCs have almost disappeared by 48 h after adoptive transfer. How much the immune response induced by SOCS1$^{-/-}$ DCs depends on their increased survival remains, however, to be elucidated.

In summary, SOCS1 has been unveiled as a master regulator of innate immune responses (Fig. 4). SOCS1-deficient DCs and macrophages stimulated with microbial products or cytokines are hyperactivated and as such prone to initiate pathological immune responses that might lead to autoimmunity. A better understanding of the mechanisms by which SOCS1 regulates inflammation would not only have a significant impact in the autoimmunity field but also provide novel molecular tools to overcome immune tolerance to tumor antigens.

4 Concluding Remarks

Since the initial description over a decade ago by the Bogen's and Levitsky's group of the phenomenon of tumor-induced antigen-specific T-cell tolerance, significant advances have been made in the understanding of the cellular and molecular mechanisms underlying this phenomenon. Bone marrow-derived antigen-presenting cells and specifically dendritic cells, through mechanisms akin to those that regulate responses to self-antigens, have been shown to be central in the induction of this state of T-cell unresponsiveness. More recently, studies of receptor ligands and intracellular signaling pathways in APCs have unveiled a complex network in which a delicate balance among stimulatory and inhibitory pathways critically influences the inflammatory status of these cells and as such their ability to induce priming versus tolerance of antigen-specific T cells. These studies have also shown the dominant role of inhibitory pathways in preserving tolerance toward self-, since their genetic and/or pharmacologic disruption in APCs was associated with over-activation of the immune system and development of autoimmunity. Inhibitory signaling pathways like those described here represent stringent safeguard mechanism to face the threat of autoimmunity. However, they have also imposed a significant barrier to our efforts to overcome tolerance to tumor antigens and effectively harness the immune system against malignant cells. Although therapeutic strategies targeting these molecular pathways in APCs have the inherent risks of inducing autoimmunity, a breeze of optimism has been provided by pre-clinical studies in which blockade of these inhibitory molecules resulted in strong antitumor effect and limited autoimmune damage. Future studies not only will provide answers to several questions that remain in our understanding of inhibitory signaling pathways in APCs but they will likely provide novel targets to augment antitumor immune responses while minimizing "collateral damage" to normal tissues.

Acknowledgments This work was supported by PHS grants CA87583 and CA100850 to Eduardo M. Sotomayor.

References

Adler, A. J., Marsh, D. W., Yochum, G. S., Guzzo, J. L., Nigam, A., Nelson, W. G. and Pardoll, D. M. (1998). CD4(+) T cell tolerance to parenchymal self-antigens requires presentation by bone marrow-derived antigen-presenting cells [In Process Citation]. *J Exp Med* 187(10):1555–1564.

Alexander, W. S., Starr, R., Fenner, J. E., Scott, C. L., Handman, E., Sprigg, N. S., Corbin, J. E., Cornish, A. L., Darwiche, R., Owczarek, C. M., Kay, T. W., Nicola, N. A., Hertzog, P. J., Metcalf, D. and Hilton, D. J. (1999). SOCS1 is a critical inhibitor of interferon gamma signaling and prevents the potentially fatal neonatal actions of this cytokine. *Cell* 98(5): 597–608.

Belz, G. T., Behrens, G. M., Smith, C. M., Miller, J. F., Jones, C., Lejon, K., Fathman, C. G., Mueller, S. N., Shortman, K., Carbone, F. R. and Heath, W. R. (2002). The CD8alpha(+) dendritic cell is responsible for inducing peripheral self-tolerance to tissue-associated antigens. *J Exp Med* 196(8):1099–1104.

Benkhart, E. M., Siedlar, M., Wedel, A., Werner, T. and Ziegler-Heitbrock, H. W. (2000). Role of Stat3 in lipopolysaccharide-induced IL-10 gene expression. *J Immunol* 165(3):1612–1617.

Bogen, B. (1996). Peripheral T cell tolerance as a tumor escape mechanism: deletion of CD4+ T cells specific for a monoclonal immunoglobulin idiotype secreted by a plasmacytoma. *Eur J Immunol* 26(11):2671–2679.

Bonifaz, L., Bonnyay, D., Mahnke, K., Rivera, M., Nussenzweig, M. C. and Steinman, R. M. (2002). Efficient targeting of protein antigen to the dendritic cell receptor DEC-205 in the steady state leads to antigen presentation on major histocompatibility complex class I products and peripheral CD8+ T cell tolerance. *J Exp Med* 196(12):1627–1638.

Boon, T. and van der Bruggen, P. (1996). Human tumor antigens recognized by T lymphocytes. *J Exp Med* 183(3):725–729.

Borg, C., Terme, M., Taieb, J., Menard, C., Flament, C., Robert, C., Maruyama, K., Wakasugi, H., Angevin, E., Thielemans, K., Le Cesne, A., Chung-Scott, V., Lazar, V., Tchou, I., Crepineau, F., Lemoine, F., Bernard, J., Fletcher, J. A., Turhan, A., Blay, J. Y., Spatz, A., Emile, J. F., Heinrich, M. C., Mecheri, S., Tursz, T. and Zitvogel, L. (2004). Novel mode of action of c-kit tyrosine kinase inhibitors leading to NK cell-dependent antitumor effects. *J Clin Invest* 114(3): 379–388.

Bromberg, J. and Darnell, J. E., Jr (2000). The role of STATs in transcriptional control and their impact on cellular function. *Oncogene* 19(21):2468–2473.

Camenisch, T. D., Koller, B. H., Earp, H. S. and Matsushima, G. K. (1999). A novel receptor tyrosine kinase, Mer, inhibits TNF-alpha production and lipopolysaccharide-induced endotoxic shock. *J Immunol* 162:3498–3503.

Casiano, C. A. and Tan, E. M. (1996). Recent developments in the understanding of antinuclear autoantibodies. *Int Arch Allergy Immunol* 111(4):308–313.

Chan, C. W., Crafton, E., Fan, H. N., Flook, J., Yoshimura, K., Skarica, M., Brockstedt, D., Dubensky, T. W., Stins, M. F., Lanier, L. L., Pardoll, D. M. and Housseau, F. (2006). Interferon-producing killer dendritic cells provide a link between innate and adaptive immunity. *Nat Med* 12(2):207–213.

Cheng, F., Wang, H. W., Cuenca, A., Huang, M., Ghansah, T., Brayer, J., Kerr, W. G., Takeda, K., Akira, S., Schoenberger, S. P., Yu, H., Jove, R. and Sotomayor, E. M. (2003). A critical role for Stat3 signaling in immune tolerance. *Immunity* 19(3):425–436.

Chong, M. M., Cornish, A. L., Darwiche, R., Stanley, E. G., Purton, J. F., Godfrey, D. I., Hilton, D. J., Starr, R., Alexander, W. S. and Kay, T. W. (2003). Suppressor of cytokine signaling-1 is a critical regulator of interleukin-7-dependent CD8+ T cell differentiation. *Immunity* 18(4):475–487.

Cohen, P. L., Caricchio, R., Abraham, V., Camenisch, T. D., Jennette, J. C., Roubey, R. A. S., Earp, H. S., Matsushima, G. and Reap, E. A. (2002). Delayed apoptotic cell clearance and lupus-like autoimmunity in mice lacking the c-mer membrane tyrosine kinase. *J Exp Med* 196:135–140.

Correll, P. H., Morrison, A. C. and Lutz, M. A. (2004). Receptor tyrosine kinases and the regulation of macrophage activation. *J Leukoc Biol* 75:731–737.

Cuenca, A., Cheng, F., Wang, H., Brayer, J., Horna, P., Gu, L., Bien, H., Borrello, I. M., Levitsky, H. I. and Sotomayor, E. M. (2003). Extra-lymphatic solid tumor growth is not immunologically ignored and results in early induction of antigen-specific T-cell anergy: dominant role of cross-tolerance to tumor antigens. *Cancer Res* 63(24):9007–9015.

Darnell, J. E., Jr (1997). STATs and gene regulation. *Science* 277(5332):1630–1635.

de Jonge, W. J., van der Zanden, E. P., The, F. O., Bijlsma, M. F., van Westerloo, D. J., Bennink, R. J., Berthoud, H. R., Uematsu, S., Akira, S., van den Wijngaard, R. M. and Boeckxstaens, G. E. (2005). Stimulation of the vagus nerve attenuates macrophage activation by activating the Jak2-STAT3 signaling pathway. *Nat Immunol* 6(8):844–851.

Druker, B. J., Tamura, S., Buchdunger, E., Ohno, S., Segal, G. M., Fanning, S., Zimmermann, J. and Lydon, N. B. (1996). Effects of a selective inhibitor of the Abl tyrosine kinase on the growth of Bcr-Abl positive cells. *Nat Med* 2(5):561–566.

El Kasmi, K. C., Holst, J., Coffre, M., Mielke, L., de Pauw, A., Lhocine, N., Smith, A. M., Rutschman, R., Kaushal, D., Shen, Y., Suda, T., Donnelly, R. P., Myers, M. G., Jr, Alexander, W., Vignali, D. A., Watowich, S. S., Ernst, M., Hilton, D. J. and Murray, P. J. (2006). General nature of the STAT3-activated anti-inflammatory response. *J Immunol* 177(11):7880–7888.

Endo, T. A., Masuhara, M., Yokouchi, M., Suzuki, R., Sakamoto, H., Mitsui, K., Matsumoto, A., Tanimura, S., Ohtsubo, M., Misawa, H., Miyazaki, T., Leonor, N., Taniguchi, T., Fujita, T., Kanakura, Y., Komiya, S. and Yoshimura, A. (1997). A new protein containing an SH2 domain that inhibits JAK kinases. *Nature* 387:921–924.

Evel-Kabler, K., Song, X. T., Aldrich, M., Huang, X. F. and Chen, S. Y. (2006). SOCS1 restricts dendritic cells' ability to break self tolerance and induce antitumor immunity by regulating IL-12 production and signaling. *J Clin Invest* 116(1):90–100.

Fadok, V. A., Bratton, D. L., Konowal, A., Freed, P. W., Westcott, J. Y. and Henson, P. M. (1998). Macrophages that have ingested apoptotic cells in vitro inhibit proinflammatory cytokine production through autocrine/paracrine mechanisms involving TGF-beta, PGE2, and PAF. *J Clin Invest* 101(4):890–898.

Fuchs, E. J. and Matzinger, P. (1992). B cells turn off virgin but not memory T cells. *Science* 258(5085):1156–1159.

Ghosh, S., May, M. J. and Kopp, E. B. (1998). NF-kappa B and Rel proteins: evolutionarily conserved mediators of immune responses. *Annu Rev Immunol* 16:225–260.

Guermonprez, P., Valladeau, J., Zitvogel, L., Thery, C. and Amigorena, S. (2002). Antigen presentation and T cell stimulation by dendritic cells. *Annu Rev Immunol* 20:621–667.

Hanada, T., Yoshida, H., Kato, S., Tanaka, K., Masutani, K., Tsukada, J., Nomura, Y., Mimata, H., Kubo, M. and Yoshimura, A. (2003). Suppressor of cytokine signaling-1 is essential for suppressing dendritic cell activation and systemic autoimmunity. *Immunity* 19:437–450.

Hawiger, D., Inaba, K., Dorsett, Y., Guo, M., Mahnke, K., Rivera, M., Ravetch, J. V., Steinman, R. M. and Nussenzweig, M. C. (2001). Dendritic cells induce peripheral T cell unresponsiveness under steady state conditions in vivo. *J Exp Med* 194(6):769–779.

Heath, W. R., Belz, G. T., Behrens, G. M., Smith, C. M., Forehan, S. P., Parish, I. A., Davey, G. M., Wilson, N. S., Carbone, F. R. and Villadangos, J. A. (2004). Cross-presentation, dendritic cell subsets, and the generation of immunity to cellular antigens. *Immunol Rev* 199:9–26.

Heinrich, M. C., Blanke, C. D., Druker, B. J. and Corless, C. L. (2002). Inhibition of KIT tyrosine kinase activity: a novel molecular approach to the treatment of KIT-positive malignancies. *J Clin Oncol* 20(6):1692–1703.

Heinrich, M. C., Corless, C. L., Demetri, G. D., Blanke, C. D., von Mehren, M., Joensuu, H., McGreevey, L. S., Chen, C. J., Van den Abbeele, A. D., Druker, B. J., Kiese, B., Eisenberg, B., Roberts, P. J., Singer, S., Fletcher, C. D., Silberman, S., Dimitrijevic, S. and Fletcher, J. A. (2003). Kinase mutations and imatinib response in patients with metastatic gastrointestinal stromal tumor. *J Clin Oncol* 21(23):4342–4349.

Hida, S., Ogasawara, K., Sato, K., Abe, M., Takayanagi, H., Yokochi, T., Sato, T., Hirose, S., Shirai, T., Taki, S. and Taniguchi, T. (2000). CD8(+) T cell-mediated skin disease in mice lacking IRF-2, the transcriptional attenuator of interferon-alpha/beta signaling [In Process Citation]. *Immunity* 13(5):643–655.

Hoentjen, F., Sartor, R. B., Ozaki, M. and Jobin, C. (2005). STAT3 regulates NF-kappaB recruitment to the IL-12p40 promoter in dendritic cells. *Blood* 105:689–696.

Horna, P. and Sotomayor, E. M. (2007). Cellular and molecular mechanisms of tumor-induced T-cell tolerance. *Curr Cancer Drug Targets* 7(1):41–53.

Huang, F. P., Platt, N., Wykes, M., Major, J. R., Powell, T. J., Jenkins, C. D. and MacPherson, G. G. (2000). A discrete subpopulation of dendritic cells transports apoptotic intestinal epithelial cells to T cell areas of mesenteric lymph nodes [see comments]. *J Exp Med* 191(3):435–444.

Itano, A. A. and Jenkins, M. K. (2003). Antigen presentation to naive CD4 T cells in the lymph node. *Nat Immunol* 4(8):733–739.

Kantarjian, H., Sawyers, C., Hochhaus, A., Guilhot, F., Schiffer, C., Gambacorti-Passerini, C., Niederwieser, D., Resta, D., Capdeville, R., Zoellner, U., Talpaz, M., Druker, B., Goldman, J., O'Brien, S. G., Russell, N., Fischer, T., Ottmann, O., Cony-Makhoul, P., Facon, T., Stone, R., Miller, C., Tallman, M., Brown, R., Schuster, M., Loughran, T., Gratwohl, A., Mandelli, F., Saglio, G., Lazzarino, M., Russo, D., Baccarani, M. and Morra, E. (2002). Hematologic and cytogenetic responses to imatinib mesylate in chronic myelogenous leukemia. *N Engl J Med* 346(9):645–652.

Kobayashi, M., Kweon, M.-N., Kuwata, H., Schreiber, R., Kiyono, H., Takeda, K. and Akira, S. (2003). Toll-like receptor-dependent production of IL-12p40 causes chronic enterocolitis in myeloid cell-specific Stat3-deficient mice. *J Clin Invest* 111:1297–1308.

Kotenko, S. V. and Pestka, S. (2000). Jak-Stat signal transduction pathway through the eyes of cytokine class II receptor complexes. *Oncogene* 19(21):2557–2565.

Kurts, C., Carbone, F. R., Barnden, M., Blanas, E., Allison, J., Heath, W. R. and Miller, J. F. (1997). CD4+ T cell help impairs CD8+ T cell deletion induced by cross- presentation of self-antigens and favors autoimmunity. *J Exp Med* 186(12):2057–2062.

Lai, C. and Lemke, G. (1991). An extended family of protein-tyrosine kinase genes differentially expressed in the vertebrate nervous system. *Neuron* 6(5):691–704.

Lang, R., Patel, D., Morris, J., Rutschman, R. and Murray, P. (2002). Shaping gene expression in activated and resting primary macrophages by IL-10. *J Immunol* 169:2253–2263.

Lang, R., Pauleau, A. L., Parganas, E., Takahashi, Y., Mages, J., Ihle, J. N., Rutschman, R. and Murray, P. J. (2003). SOCS3 regulates the plasticity of gp130 signaling. *Nat Immunol* 4(6): 546–550.

Lassila, O., Vainio, O. and Matzinger, P. (1988). Can B cells turn on virgin T cells? *Nature* 334(6179):253–255.

Lee, P. P., Yee, C., Savage, P. A., Fong, L., Brockstedt, D., Weber, J. S., Johnson, D., Swetter, S., Thompson, J., Greenberg, P. D., Roederer, M. and Davis, M. M. (1999). Characterization of circulating T cells specific for tumor-associated antigens in melanoma patients. *Nat Med* 5(6):677–685.

Lemke, G. and Lu, Q. (2003). Macrophage regulation by Tyro 3 family receptors. *Curr Opin Immunol* 15:31–36.

Lu, Q. and Lemke, G. (2001). Homeostatic regulation of the immune system by receptor tyrosine kinases of the Tyro 3 family. *Science* 293(5528):306–311.

Marine, J. C., Topham, D. J., McKay, C., Wang, D., Parganas, E., Stravopodis, D., Yoshimura, A. and Ihle, J. N. (1999). SOCS1 deficiency causes a lymphocyte-dependent perinatal lethality. *Cell* 98:609–616.

Mevorach, D., Zhou, J. L., Song, X. and Elkon, K. B. (1998). Systemic exposure to irradiated apoptotic cells induces autoantibody production. *J Exp Med* 188(2):387–392.

Miyazaki, T., Suzuki, G. and Yamamura, K. (1993). The role of macrophages in antigen presentation and T cell tolerance. *Int Immunol* 5(9):1023–1033.

Morgan, D. J., Kreuwel, H. T., Fleck, S., Levitsky, H. I., Pardoll, D. M. and Sherman, L. A. (1998). Activation of low avidity CTL specific for a self epitope results in tumor rejection but not autoimmunity. *J Immunol* 160(2):643–651.

Munn, D., Sharma, M., Hou, D., Baban, B., Lee, J., Antonia, S., Messina, J., Chandler, P., Koni, P. and Mellor, A. (2004). Expression of indoleamine 2,3-dioxygenase by plasmacytoid dendritic cells in tumor-draining lymph nodes. *J Clin Invest* 114:280–290.

Munn, D., Sharma, M., Lee, J., Jhaver, K., Johnson, T., Keskin, D., Marshall, B., Chandler, P., Antonia, S., Burgess, R., Slingluff, J. C. and Mellor, A. (2002). Potential regulatory function of human dendritic cells expressing indoleamine 2,3-dioxygenase. *Science* 297:1867–1870.

Murray, P. (2005). The primary mechanism of the IL-10-regulated antiinflammatory response is to selectively inhibit transcription. *Proc Natl Acad Sci USA* 102:8686–8691.

Nefedova, Y., Cheng, P., Gilkes, D., Blaskovich, M., Beg, A. A., Sebti, S. M. and Gabrilovich, D. I. (2005). Activation of dendritic cells via inhibition of Jak2/STAT3 signaling. *J Immunol* 175(7):4338–4346.

Noonan, K., Matsui, W., Serafini, P., Carbley, R., Tan, G., Khalili, J., Bonyhadi, M., Levitsky, H., Whartenby, K. and Borrello, I. (2005). Activated marrow-infiltrating lymphocytes effectively target plasma cells and their clonogenic precursors. *Cancer Res* 65(5):2026–2034.

Ohlen, C., Kalos, M., Hong, D. J., Shur, A. C. and Greenberg, P. D. (2001). Expression of a tolerizing tumor antigen in peripheral tissue does not preclude recovery of high-affinity CD8+ T cells or CTL immunotherapy of tumors expressing the antigen. *J Immunol* 166(4):2863–2870.

Overwijk, W. W., Theoret, M. R., Finkelstein, S. E., Surman, D. R., de Jong, L. A., Vyth-Dreese, F. A., Dellemijn, T. A., Antony, P. A., Spiess, P. J., Palmer, D. C., Heimann, D. M., Klebanoff, C. A., Yu, Z., Hwang, L. N., Feigenbaum, L., Kruisbeek, A. M., Rosenberg, S. A. and Restifo, N. P. (2003). Tumor regression and autoimmunity after reversal of a functionally tolerant state of self-reactive CD8+ T cells. *J Exp Med* 198(4):569–580.

Pardoll, D. (2003). Does the immune system see tumors as foreign or self? *Annu Rev Immunol* 21:807–839.
Qin, H., Wilson, C., Roberts, K., Baker, B., Zhao, X. and Benveniste, E. (2006). IL-10 inhibits lipopolysaccharide-induced CD40 gene expression through induction of suppressor of cytokine signaling-3. *J Immunol* 177:7761–7771.
Ramana, C. V., Chatterjee-Kishore, M., Nguyen, H. and Stark, G. R. (2000). Complex roles of Stat1 in regulating gene expression. *Oncogene* 19(21):2619–2627.
Ravetch, J. V. and Lanier, L. L. (2000). Immune inhibitory receptors [In Process Citation]. *Science* 290(5489):84–89.
Ronchese, F. and Hausmann, B. (1993). B lymphocytes in vivo fail to prime naive T cells but can stimulate antigen-experienced T lymphocytes. *J Exp Med* 177(3):679–690.
Ronchetti, A., Rovere, P., Iezzi, G., Galati, G., Heltai, S., Protti, M. P., Garancini, M. P., Manfredi, A. A., Rugarli, C. and Bellone, M. (1999). Immunogenicity of apoptotic cells in vivo: role of antigen load, antigen-presenting cells, and cytokines. *J Immunol* 163(1):130–136.
Rosen, A. and Casciola-Rosen, L. (1999). Autoantigens as substrates for apoptotic proteases: implications for the pathogenesis of systemic autoimmune disease. *Cell Death Differ* 6(1): 6–12.
Rosenberg, S. A. (1995). The development of new cancer therapies based on the molecular identification of cancer regression antigens. *Cancer J Sci Am* 1(2):90.
Scheinecker, C., McHugh, R., Shevach, E. M. and Germain, R. N. (2002). Constitutive presentation of a natural tissue autoantigen exclusively by dendritic cells in the draining lymph node. *J Exp Med* 196(8):1079–1090.
Scott, R. S., McMahon, E. J., Pop, S. M., Reap, E. A., Caricchio, R., Cohen, P. L., Earp, H. S. and Matsushima, G. K. (2001). Phagocytosis and clearance of apoptotic cells is mediated by MER. *Nature* 411(6834):207–211.
Sen, P., Wallet, M., Yi, Z., Huang, Y., Henderson, M., Mathews, C., Earp, H., Matsushima, G., Baldwin, J. A. and Tisch, R. (2007). Apoptotic cells induce Mer tyrosine kinase-dependent blockade of NF-kappaB activation in dendritic cells. *Blood* 109:653–660.
Shrikant, P. and Mescher, M. F. (1999). Control of syngeneic tumor growth by activation of CD8+ T cells: efficacy is limited by migration away from the site and induction of nonresponsiveness. *J Immunol* 162(5):2858–2866.
Soltys, J., Bonfield, T., Chmiel, J. and Berger, M. (2002). Functional IL-10 deficiency in the lung of cystic fibrosis (cftr(−/−)) and IL-10 knockout mice causes increased expression and function of B7 costimulatory molecules on alveolar macrophages. *J Immunol* 168(4):1903–1910.
Sotomayor, E. M., Borrello, I. and Levitsky, H. I. (1996). Tolerance and cancer: a critical issue in tumor immunology. *Crit Rev Oncog* 7(5–6):433–456.
Sotomayor, E. M., Borrello, I., Rattis, F. M., Cuenca, A. G., Abrams, J., Staveley-O'Carroll, K. and Levitsky, H. I. (2001). Cross-presentation of tumor antigens by bone marrow-derived antigen-presenting cells is the dominant mechanism in the induction of T-cell tolerance during B-cell lymphoma progression. *Blood* 98(4):1070–1077.
Staveley-O'Carroll, K., Sotomayor, E., Montgomery, J., Borrello, I., Hwang, L., Fein, S., Pardoll, D. and Levitsky, H. (1998). Induction of antigen-specific T cell anergy: an early event in the course of tumor progression. *Proc Natl Acad Sci USA* 95(3):1178–1183.
Steinman, R. M., Hawiger, D. and Nussenzweig, M. C. (2003). Tolerogenic dendritic cells. *Annu Rev Immunol* 21:685–711.
Stitt, T. N., Conn, G., Gore, M., Lai, C., Bruno, J., Radziejewski, C., Mattsson, K., Fisher, J., Gies, D. R., Jones, P. F. et al. (1995). The anticoagulation factor protein S and its relative, Gas6, are ligands for the Tyro 3/Axl family of receptor tyrosine kinases. *Cell* 80(4):661–670.
Taieb, J., Chaput, N., Menard, C., Apetoh, L., Ullrich, E., Bonmort, M., Pequignot, M., Casares, N., Terme, M., Flament, C., Opolon, P., Lecluse, Y., Metivier, D., Tomasello, E., Vivier, E., Ghiringhelli, F., Martin, F., Klatzmann, D., Poynard, T., Tursz, T., Raposo, G., Yagita, H., Ryffel, B., Kroemer, G. and Zitvogel, L. (2006). A novel dendritic cell subset involved in tumor immunosurveillance. *Nat Med* 12(2):214–219.

Takeda, K., Clausen, B. E., Kaisho, T., Tsujimura, T., Terada, N., Forster, I. and Akira, S. (1999). Enhanced Th1 activity and development of chronic enterocolitis in mice devoid of Stat3 in macrophages and neutrophils. *Immunity* 10:39–49.

Taylor, P. R., Carugati, A., Fadok, V. A., Cook, H. T., Andrews, M., Carroll, M. C., Savill, J. S., Henson, P. M., Botto, M. and Walport, M. J. (2000). A hierarchical role for classical pathway complement proteins in the clearance of apoptotic cells in vivo. *J Exp Med* 192(3):359–366.

Tivol, E. A., Borriello, F., Schweitzer, A. N., Lynch, W. P., Bluestone, J. A. and Sharpe, A. H. (1995). Loss of CTLA-4 leads to massive lymphoproliferation and fatal multiorgan tissue destruction, revealing a critical negative regulatory role of CTLA-4. *Immunity* 3(5):541–547.

Trinchieri, G. (2003). Interleukin-12 and the regulation of innate resistance and adaptive immunity. *Nat Rev Immunol* 3:133–146.

Voll, R. E., Herrmann, M., Roth, E. A., Stach, C., Kalden, J. R. and Girkontaite, I. (1997). Immunosuppressive effects of apoptotic cells [letter]. *Nature* 390(6658):350–351.

Wang, H., Cheng, F., Cuenca, A., Horna, P., Zheng, Z., Bhalla, K. and Sotomayor, E. M. (2004a). Imatinib mesylate (STI-571) enhances antigen presenting cell function and overcomes tumor-induced CD4+ T-cell tolerance. *Blood* 105(3):1135–43.

Wang, H., Liao, H., Ochani, M., Justiniani, M., Lin, X., Yang, L., Al-Abed, Y., Wang, H., Metz, C., Miller, E. J., Tracey, K. J. and Ulloa, L. (2004b). Cholinergic agonists inhibit HMGB1 release and improve survival in experimental sepsis. *Nat Med* 10(11):1216–1221.

Watson, G. A. and Lopez, D. M. (1995). Aberrant antigen presentation by macrophages from tumor-bearing mice is involved in the down-regulation of their T cell responses. *J Immunol* 155(6):3124–3134.

Welte, T., Zhang, S. S., Wang, T., Zhang, Z., Hesslein, D. G., Yin, Z., Kano, A., Iwamoto, Y., Li, E., Craft, J. E., Bothwell, A. L., Fikrig, E., Koni, P. A., Flavell, R. A. and Fu, X. Y. (2003). STAT3 deletion during hematopoiesis causes Crohn's disease-like pathogenesis and lethality: a critical role of STAT3 in innate immunity. *Proc Natl Acad Sci USA* 100(4):1879–1884.

Williams, L., Bradley, L., Smith, A. and Foxwell, B. (2004). Signal transducer and activator of Transcription 3 is the dominant mediator of the anti-inflammatory effects of IL-10 in human macrophages. *J Immunol* 172:567–576.

Williams, L., Sarma, U., Willets, K., Smallie, T. and Brennan, F. (2007). Expression of constitutively active STAT3 can replicate the cytokine-suppressive activity of interleukin-10 in human primary macrophages. *J Biol Chem* 282:6965–6975.

Willimsky, G. and Blankenstein, T. (2005). Sporadic immunogenic tumours avoid destruction by inducing T-cell tolerance. *Nature* 437:141–146.

Yasukawa, H., Misawa, H., Sakamoto, H., Masuhara, M., Sasaki, A., Wakioka, T., Ohtsuka, S., Imaizumi, T., Matsuda, T., Ihle, J. N. and Yoshimura, A. (1999). The JAK-binding protein JAB inhibits Janus tyrosine kinase activity through binding in the activation loop. *EMBO J* 18:1309–1320.

Yasukawa, H., Ohishi, M., Mori, H., Murakami, M., Chinen, T., Aki, D., Hanada, T., Takeda, K., Akira, S., Hoshijima, M., Hirano, T., Chien, K. and Yoshimura, A. (2003). IL-6 induces an anti-inflammatory response in the absence of SOCS3 in macrophages. *Nat Immunol* 4:551–556.

Yu, H., Kortylewski, M. and Pardoll, D. (2007). Crosstalk between cancer and immune cells: role of STAT3 in the tumour microenvironment. *Nat Rev Immunol* 7:41–51.

Yu, Z., Zhang, W. and Kone, B. (2002). Signal transducers and activators of transcription 3 (STAT3) inhibits transcription of the inducible nitric oxide synthase gene by interacting with nuclear factor kappaB. *Biochem J* 367:97–105.

Arginine Availability Regulates T-Cell Function in Cancer

Paulo C. Rodríguez and Augusto C. Ochoa

1 Introduction

The clinical experiments of William Coley in the 1890s and the work of Prehn and Main in the 1950s firmly demonstrated the presence of an immune response against tumor antigens, which could potentially be used in the treatment of cancer. More recent findings in cancer biology, including the viral etiology of some malignancies, the presence of mutated oncoproteins and the over-expression of certain normal proteins, further support the concept that an antigen-specific immune response can be generated to control tumor growth. However, it has also become evident that tumor cells have sophisticated mechanisms to induce a state of anergy or tolerance that allows malignant cells to evade the immune response, proliferate and metastasize. These mechanisms are still a matter of active research and debate, with possible explanations ranging from the inability of the immune system to recognize tumor antigens (tumor ignorance) to the gradual deterioration of the immune response caused directly by tumor-derived factors or indirectly by the stimulation of immune cells with suppressor function. Here we will discuss how tumors can induce T-cells anergy by modulating the availability of the amino acid arginine which results in the development of discrete molecular alterations that impair T-cell signal transduction and function. This mechanism is the result of an increased arginase activity in myeloid-derived suppressor cells (MDSC).

2 Dysfunctional Immune Response in Cancer: A Historic Perspective

The first known demonstration of the therapeutic potential of an anti-tumor immune response was shown by William Coley in the 1890s using bacterial extracts to vaccinate cancer patients (Coley, 1893) This concept was further supported by the

---

A.C. Ochoa
Stanley S. Scott Cancer Center, Louisiana State University Health Sciences Center, 533 Bolivar Street, New Orleans, LA 70112, USA

observations of Prehn and Main (1957) who demonstrated that mice with chemically induced tumors developed an immune response against specific tumor-associated antigens that was unique for each tumor. Studies in mice injected with oncogenic viruses confirmed that a protective immune response could arise during the early stages of tumor development, but disappeared with the progressive growth of the tumor (Eggers and Wunderlich, 1975; Fefer et al., 1968; Jaroslow et al., 1975). However, studies in cancer patients failed to consistently show a protective immune response to the progressively growing tumor and instead suggested that the cellular immune response was impaired. Hersh and Oppenheim (1965) found that patients with Hodgkin's disease (HD) had a decreased delayed-type hypersensitivity (DTH) response to PPD and DNBC (di-nitrochlorobenzene) and a diminished in vitro response to mitogen stimulation. This immune dysfunction persisted even in patients who had achieved a complete clinical response to chemotherapy (Fisher et al., 1980). Observations in melanoma patients also showed a decrease in the cellular immune response, but a marked increase in the levels of serum immunoglobulins. Similarly, patients with other solid tumors including renal cell carcinoma, prostate and bladder cancer (Catalona et al., 1975), lung cancer (Alberola, 1985), breast cancer (Jerrells et al., 1978) and gastric cancer (Iwahashi et al., 1992) consistently showed a decreased cellular response. Additional studies in tumor-bearing mice showed that T cells from animals infected with the murine sarcoma virus were unable to lyse tumor cells in vitro and were unresponsive to stimulation with mitogens and allo-antigens, suggesting a profound T-cell dysfunction (Bhatnagar et al., 1975; Bluestone and Lopez, 1979; Fernbach et al., 1976; Gorczynski and Knight, 1975; Kirchner et al., 1976).

Several mechanisms were proposed to explain how tumors inhibit the immune response. These included the presence of "blocking antibodies" that could interfere with antigen recognition, preventing the priming and activation of T cells (Hellstrom et al., 1971, 1983), the existence of suppressor T cells (Dye and North, 1984; Mills and North, 1985; North, 1985; North and Bursuker, 1984) that could be eliminated by low-dose cyclophosphamide and the existence of suppressor macrophages among others. Several approaches were tested in animal models to block these "suppressor" mechanisms, including cyclophosphamide and indomethacin, a potent prostaglandin inhibitor (Parhar and Lala, 1987). These observations demonstrated the dynamic interaction between the tumor and the immune system which could be manipulated to the benefit of the host.

The advent of immunotherapy trials in the 1980s using adoptive transfer of tumor-infiltrating lymphocytes (TIL) revealed to a greater extent the degree of T-cell dysfunction in patients with cancer. In vitro testing of freshly isolated TIL demonstrated a markedly decreased proliferation to mitogens and a diminished clonogenic potential (Miescher et al., 1986, 1988; Whiteside and Rabinowich, 1998; Whiteside et al., 1988). Using an adoptive transfer model, Loeffler and colleagues (Loeffler, 1991) showed that T cells from mice bearing subcutaneous MCA-38 colon carcinoma for more than 14 days lost their ability to induce any anti-tumor effect and had a decreased cytotoxic capacity and a diminished expression of the perforin family of cytolytic proteins.

In the early 1990s Mizoguchi et al. [1992] in mice and Finke and colleagues (Kolenko et al., 1999; Li et al., 1994) in humans described a series of discrete changes in T cells from tumor-bearing mice including a decreased expression of the CD3ζ chain of the T-cell receptor, an inability to upregulate JAK-3 and a decreased translocation of NFκB p65, which resulted in low proliferation and an impaired cytokine production. However, the mechanisms inducing these changes were unknown. Soon after, Gabrilovich and colleagues demonstrated that the production of VEGF by tumors resulted in the accumulation of immature myeloid cells that were able to suppress T-cell function (Gabrilovich, 2004; Gabrilovich et al., 1998). However, the possible association between the presence of immature dendritic cells and alterations in T-cell signal transduction molecules was unknown.

3 Arginine and T-Cell Function

L-Arginine (L-Arg) is a non-essential amino acid that plays a central role in several biological systems including immune response. The primary source of arginine is dietary, but it can also be synthesized de novo from citrulline. L-Arg levels are also regulated through its metabolism by L-Arg metabolizing enzymes, arginase I (ARG I), arginase II (ARG II) and the inducible nitric oxide synthase (NOS2) (Albina et al., 1989) (Fig. 1). ARG I and ARG II are encoded by two distinct genes and are located in the cytoplasm and mitochondria, respectively. Both enzymes hydrolyze L-Arg into urea and L-ornithine, the latter being the main substrate for the production of polyamines (putrescine, spermidine and spermine) that are required for cell proliferation. L-Arg can also be metabolized by NOS2 to produce citrulline and nitric oxide (NO), a highly reactive compound important in vascular homeostasis and the cytotoxic mechanism of macrophages (Hibbs et al., 1987). ARG I is constitutively expressed in the liver and human granulocytes and is inducible in murine macrophage cells, while ARG II is constitutively expressed in various tissues, especially the kidney. The expression of ARG I and NOS2 in murine macrophages is differentially regulated by Th1 and Th2 cytokines (Hesse et al., 2001; Munder et al., 1999). Stimulation of murine macrophages with IFN-γ or tumor necrosis factor-alpha (TNF-α) upregulates iNOS exclusively, while IL-4, IL-10, transforming growth factor-β (TGF-β) and IL-13 induce ARG I (Munder et al., 1998; Rutschman et al., 2001). The mitochondrial isoform ARG II is not significantly modulated by Th1 or Th2 cytokines (Louis et al., 1999; Munder et al., 1999).

The association of arginine metabolism and the immune system came initially from reports in the 1970s demonstrating that the injection of L-Arg in mice undergoing extensive surgery prevented a well-described phenomenon of post-surgical thymus involution and appeared to increase the number of T cells (Barbul et al., 1977). In the late 1980s Albina and Mills (Albina et al., 1989) studying wound

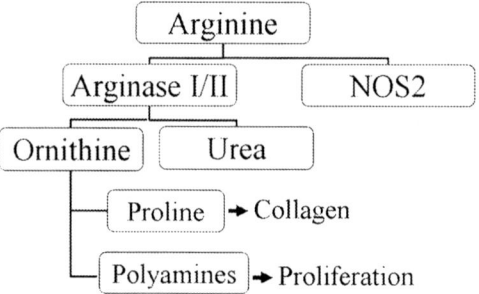

Fig. 1 Arginase induction in cancer. Malignant cells produce factors such as VEGF, GM-CSF, M-CSF or others which may stimulate an increase in production of MDSC from the bone marrow. The increased expression of COX-2 in tumor cells and the resulting production of PGE2 in turn induces the expression of arginase I and CAT-2 transporters in MDSC. This facilitates the rapid uptake of arginine and its subsequent depletion from the microenvironment. T cells stimulated in this arginine-depleted microenvironment are arrested in their cell cycle progression and develop specific alterations including the decreased expression of CD3ζ chain, an inability to translocate NFκB p65 and the inability to upregulate JAK-3. These changes result in the induction of T-cell anergy which is reversible by the supplementation of physiological levels of arginine or the inhibition of arginase

healing demonstrated that macrophages expressing arginase 1 infiltrated the site of surgical wounds rapidly incorporating and metabolizing arginine into ornithine, possibly as a means of increasing the production of proline and collagen by fibroblasts, facilitating the process of wound healing. A different but equally important association was suggested by reports showing the rapid depletion of plasma levels of arginine accompanied by a markedly decreased T-cell function in patients following liver transplantation, in trauma patients or in murine models of trauma (Makarenkova et al., 2006). This phenomenon was caused by a massive release of arginase from the transplanted liver caused by the hypoxia produced during the surgical procedure, while in trauma, arginase appeared to be released by immune cells in the peripheral blood (Bernard et al., 2001; Ochoa et al., 2000). Furthermore, the state of T-cell anergy caused by trauma could be rapidly reversed by the enteral or parenteral supplementation of L-Arg (Barbul, 1990).

4 Arginine, Arginase and Cancer

An increased arginase (ARG) activity has also been described in patients with different tumors (Singh et al., 2000; Suer et al., 1999). However, the initial reports were limited to the expression of ARG II in the malignant cells, which was thought to be necessary for the production of ornithine and polyamines to sustain the rapid cell division and growth of the malignant cells (Chang et al., 2001). More recent reports by Young et al. (1987, 1989) and later by Bronte et al. (2000) demonstrated the existence of immature myeloid cells in the spleen of mice bearing colon carcinoma, which expressed arginase I and the inducible nitric oxide (NOS2). These cells were able to suppress alloreactive T cells through mechanisms that were unclear at the time. The phenotype of these cells was reminiscent of the immature dendritic cells described initially by Gabrilovich (2004).

Almost simultaneously, Taheri et al. (2001) demonstrated that the stimulation of T cells in tissue culture media containing low levels of L-Arg (but not other nonessential amino acids) induced loss of the expression of the T-cell receptor ζ chain (CD3ζ), blocked T-cell proliferation and caused a markedly decreased IFN-γ production, similar to the alterations previously described in anergic cancer patients. These T cells, however, were not undergoing apoptosis, showed normal calcium flux and IL-2 production (but not IFN-γ) and were able to upregulate the IL-2 receptor alpha chain (CD25) (Zabaleta et al., 2004; Zea et al., 2004). The changes in CD3ζ chain expression alone, however, could not explain the complete inhibition of T-cell proliferation induced by the depletion of arginine. Cell cycle analysis using propidium iodide showed that stimulated T cells cultured in the absence of L-Arg were arrested in the G_0-G_1 phase, while those cultured in the presence of L-Arg progressed into S and G_2-M phases (Rodriguez et al. 2007.) The progression into S phase of the cell cycle is tightly regulated by cyclin D/cdk complexes, which phosphorylate retinoblastoma (Rb) protein inducing the release of the transcription factor E2F-1 and causing the progression into late G_1 phase. The expression of cyclin D3 and cdk4 was significantly impaired in T cells cultured in medium without L-Arg. Consequently, they had a significant decrease in Rb phosphorylation and were unable to progress in their cell cycle. Therefore arginine depletion impairs TCR signaling and cell-cycle progression.

More recently Rodriguez et al. (2002) published a series of elegant experiments describing the molecular mechanisms by which arginine depletion inhibits the expression of CD3ζ chain. The two main mechanisms a decreased stability of certain RNAs and a decreased translation of several proteins. Cells cultured in the absence of L-Arg had a shorter half-life of CD3ζ mRNA compared to cells cultured in the presence of L-Arg, suggesting that L-Arg starvation induced a post-transcriptional regulation of the CD3ζ mRNA. The available information on mRNA stability in states of amino acid deprivation is complex and often contradictory. Guerrini et al. (1993) described a differential post-transcriptional regulation of asparagine synthase mRNA induced by starvation of amino acids. In contrast, Bruhat et al. (1997, 1999) found that mRNA from HeLa cells cultured in the absence of leucine had a longer half-life than mRNA of cells cultured in the presence of leucine. However, the molecular mechanisms that affect the stability of mammalian genes in these settings remain to be characterized. Our data show that a decrease in CD3ζ mRNA half-life was also associated with de novo protein synthesis, suggesting a possible role for new proteins in the modification of the mRNA turnover (Ross, 1996; Sachs, 1993; Saini et al., 1990). Transfection experiments using the coding region of CD3ζ cDNA under the control of a CMV promoter showed that COS-7 L cells cultured in the absence of L-Arg displayed a decreased CD3ζ mRNA half-life, confirming the post-transcriptional regulation of CD3ζ gene in L-Arg starvation conditions. However, the *cis* acting sites in the 3'UTR mediating the instability of these mRNAs have not been identified.

Arginine starvation, however, does not always result in a decreased RNA half-life or in a diminished translation. The lack of L-Arg has also been associated with the induction of certain genes regulating L-Arg metabolism such as arginosuccinate synthase (Fafournoux et al., 2000; Quillard et al., 1996), probably as a pathway for the synthesis of L-Arg from citrulline. Gazzola et al. (1972) and Hyatt et al. (Aulak et al., 1996, 1999; Hyatt et al., 1997) have reported that the absence of L-Arg induces the transcription of genes encoding the cationic amino acid transport system 1 (CAT-1), which transports amino acids such as L-Arg from the extracellular space into the cytoplasm. This increase in the expression of CAT-1 is transient and if the absence of L-Arg persists, it is downregulated. The absence of leucine and L-Arg has also been associated with an increase in the amount and stability of mRNA encoding the *CHOP* gene (Bruhat et al., 1997, 1999; Jousse et al., 1999, 2000). This gene encodes a transcription factor that blocks the action of the CCAAT/enhancer-binding protein β, which in turn inhibits the normal proliferation of cells (Bruhat et al., 1999).

Amino acid control of translational mechanisms has also been suggested by reports showing that GCN2 acts as a central sensor to amino acid deprivation and thus controls the response to nutrient starvation in mammalian cells. T cells isolated from GCN2 knock-out mice were insensitive to arginine starvation, i.e., they were able to upregulate cyclin D3 and cdk4 and proliferate even in the absence of arginine (Rodriguez et al., 2007). These results are similar to those observed in cells cultured in the absence of tryptophan (Munn et al., 2005), suggesting that GCN2 serves as a general amino acid sensor in mammalian cells. How GCN2 activation leads to an arrest in protein synthesis in the absence of amino acids has been partially described. Amino acid deprivation and accumulation of empty tRNAs in eukaryotes activate GCN2 kinase, which results in the phosphorylation of eIF2α. In turn, phosphorylated eIF2α suppresses the translation initiation and stability of some cellular mRNAs and blocks the access of methionyl tRNA to the ribosome, impairing the initiation of translation (Lee et al., 2003). T cells cultured in the absence of L-Arg had an increased expression of the phosphorylated form of eIF2 alpha (P. Rodriguez, unpublished data). An alternative mechanism of inhibition of the translation by amino acids starvation has been described and is the result of inhibition of signaling through mTOR (mammalian target of rapamycin). However, its role in arginine starvation is unclear.

5 MDSC and the Production of Arginase

How can arginase modulate the availability of arginine to the immune system? Murine peritoneal macrophages stimulated with Th2 cytokines (IL-4 + IL-13) produce high levels of ARG I and have an increased uptake of radioactive L-Arg mediated through cationic amino acid transport system (CAT) (Rodriguez et al., 2003). This particular carrier system is characterized by its high affinity for basic amino acids, its independence from Na^+ and the ability of substrate on the opposite site (*trans*) side of the membrane to increase transport activity (White, 1985). CAT genes have been recently cloned and designated CAT-1, CAT-2A, CAT-2B and

CAT-3. Whereas CAT-1, CAT-2B and CAT-3 are high-affinity transporters (K_m 100 μmol/L) for L-Arg, CAT-2A is an alternatively spliced variant of CAT-2B that possesses low affinity for L-Arg (K_m 1–2 μmol/L) (Closs, 2002). Stimulation of murine macrophages with IL-4 + IL-13 induced the expression of CAT-2B. In contrast, stimulation of macrophages with IL-4 + IL-13 did not induce major changes in the expression of CAT-1 and CAT-2A. The rapid uptake and depletion of L-arginine from the microenvironment result in the development of severe T-cell dysfunction, a phenomenon that was reversed by the addition of excess arginine or by the use of an arginase inhibitor Nor-NOHA. In contrast, macrophages producing ARG II or NOS2 do not deplete L-Arg from the microenvironment and do not impair T-cell function. This in vitro model was also demonstrated in vivo in mice injected with 3LL lung carcinoma. Separation of cell subsets infiltrating subcutaneous tumors showed the presence of myeloid-derived suppressor cells (MDSC) that expressed ARG I and CAT-2B, were able to deplete arginine, and induced the loss of CD3ζ and impaired T-cell proliferation. Furthermore, the injection of the arginase inhibitor Nor-NOHA in tumor-bearing mice resulted in a dose-dependent T-cell-mediated anti-tumor response that inhibited tumor growth (Rodriguez et al., 2004).

Recent studies in patients with cancer have confirmed the expression of ARG I by immune cells infiltrating tumors and ARG II in certain tumor cells including prostate cancer and renal cell carcinoma. The phenotype of the cells infiltrating tumors and producing ARG I in the murine models ranges from immature dendritic cells expressing GR1 to mature macrophages. In humans they also include mature granulocytes. However, the common marker for all these cells is CD11b. Recently an agreement was reached by researchers in the field to call them myeloid-derived suppressor cells (MDSC). This wide variation in the phenotype of MDSC is probably caused by the factors being secreted by the different tumors. What factors activate MDSC is still a matter of much research. Gabrilovich (2004) has suggested that MDSC are generated from bone-marrow hematopoietic precursors in response to cytokines produced by the tumor, including granulocyte macrophage-colony stimulating factor (GM-CSF), interleukin-3 (IL-3) and vascular endothelial growth factor (VEGF). However, none of these cytokines alone induce ARG I. The mechanisms regulating the induction of ARG I in MDSC are still unclear. In mice, cytokines such as IL-4, IL-10 and IL-13 can induce ARG I expression in peritoneal macrophages in vitro (Pauleau et al., 2004; Rutschman et al., 2001). However, ARG I is also found in tumors that do not produce these cytokines. Murine tumor models have also recently shown that prostanoids, and in particular PGE2, stimulate the expression of ASE I in MDSC through the stimulation of prostaglandin receptors EP2 and EP4 on their cell membrane (Rodriguez et al., 2005). The inhibition of COX-2 by siRNA (in vitro) or COX-2 inhibitors (in vitro and in vivo) blocks the induction of ASE I and induces an immune-mediated anti-tumor response.

In humans, however, the phenotype of MDSC and the induction of ARG I are less clear. MDSC producing ARG I in humans not only are found in infiltrating tumors, but also are increased in peripheral blood of cancer patients. Schmielau and Finn (2001) first reported an unusually large number of myeloid cells with a granulocyte phenotype (CD14−, CD15+, CD11b+), which co-purified with low-density peripheral blood mononuclear cells in the peripheral blood of patients with pancreatic cancer. More recently, Zea et al. [2005] demonstrated a large increase in MDSC with a similar phenotype in a study with 117 patients with metastatic renal cell carcinoma (RCC). These cells expressed levels of arginase activity that were 8–10 times higher than normal age-matched controls. The presence of these high numbers of arginase expressing MDSC in peripheral blood not only impaired T-cell function, but also resulted in decreased levels of arginine and increased levels of ornithine in plasma. Analysis of RCC biopsies also demonstrated the presence of these MDSC infiltrating the tumor (P. Rodriguez unpublished data). Thus MDSC not only have a suppressive effect at the tumor site, but can also impair T-cell function in the periphery.

6 Nitric Oxide Synthase (NOS2) and T-Cell Function

Arginase I and NOS2 are differentially regulated in myeloid cells. Stimuli that increase arginase 1 such as IL-4 and IL-13 decrease NOS2 expression, and vice versa. However, only cells expressing arginase 1 deplete arginine from the microenvironment, which in turn blocks the induction of NOS2 expression by reducing the de novo synthesis of NOS2 protein (El-Gayar et al., 2003; Lee et al., 2003). In special circumstances arginase 1 and NOS2 are expressed in the same cell.

NOS2 and its product NO can also block T-cell function by interfering with the IL-2 pathway (Duhe et al., 1998). Induction of NOS2 alone in MDSC, with subsequent release of NO, is responsible for inhibition of T-cell responses in some experimental settings, as demonstrated by the complete reversal of immunosuppression with specific NOS2 inhibitors (Bronte et al., 2000, 2003a; Mazzoni et al., 2002). Moreover, NO added directly to cultures is an extremely potent inhibitor of T-cell proliferation (Wu and Morris, 1998). IFN-γ released by activated T lymphocytes and a yet-to-be-understood cell-to-cell contact between T lymphocytes and MDSC is necessary for NO production and inhibition of immune functions. In some models, the complete dependence on NOS2 and NO for suppression has been unequivocally established by experiments showing that inhibitory cells from NOS2-deficient mice are devoid of immunosuppressive properties.

NO does not impair the early events triggered by T-cell receptor (TCR) crosslinking, but acts instead at the level of IL-2 receptor signaling, blocking the phosphorylation and activation of several signaling molecules, including Janus kinases (JAKs) 1 and 3, STAT5 (signal transducer and activator of transcription 5), Erk and Akt (Bingisser et al., 1998). Previous studies demonstrated that JAK-3 is oxidized and inactivated following direct exposure to NO and the enzymatic activities of several other intracellular signaling proteins are also negatively regulated by NO either directly, by S-nitrosylation of crucial cysteine residues, or indirectly, through activation of guanylyl cyclase (Duhe et al., 1998). Concentrations of NO at the tumor site may also result in apoptosis of infiltrating T lymphocytes (Saio et al., 2001).

It has been suggested that NOS2 and ARG pathways operate in distinct macrophage subsets. However, studies with cloned cell lines showed that stimulation with LPS of either mouse or rat macrophages upregulated both NOS2 and ARG 1 (Salimuddin et al., 1999; Sonoki et al., 1997), suggesting that under some circumstances both enzymes can work in the same cellular environment. It has been established that under conditions of low L-Arg concentrations, the reductase domain of NOS2 generates superoxide ion (O_2-) (Xia and Zweier, 1997). This has been demonstrated in macrophages in which the targeted expression of ARG 1 increased the production of O_2, a process that was blocked by selective inhibitors of NOS2 reductase domain or by the scavenger superoxide dismutase. Furthermore, cloned MDSC lines require both ARG 1 and NOS2 to block allogeneic T-cell responses in tumor-bearing mice. These inhibitors also restored alloreactive T-cell responses in mice bearing a highly immunosuppressive colon carcinoma (Bronte et al., 2003a,b).

NO reacts with O_2, giving rise to peroxynitrites ($ONOO_2$), a highly reactive oxidizing agent that nitrates tyrosines on proteins. Peroxynitrites can induce apoptosis in T lymphocytes by inhibiting activation-induced protein tyrosine phosphorylation or by nitrating a component of the mitochondrial permeability transition pore, which causes release of death-promoting factors, such as cytochrome C (Brito et al., 1999). Thus, when ARG I and NOS2 enzymes are induced together, peroxynitrites, generated by NOS2 under conditions of limiting arginine, cause activated T lymphocytes to undergo apoptosis. Thus, NOS2 and ARG 1 may act separately or synergistically in vivo.

7 Summary

Immunological tolerance can be induced by tumors through various mechanisms, including the depletion of the non-essential amino acid L-arginine. Stimulation of T cells in an arginine-depleted medium induces the loss of several signaling proteins including CD3 chain, JAK-3 and NFκB p65 and also arrests T-cell cycle by inhibiting cyclin D3 and cdk4 expression. This process is mediated by GCN2 kinase which acts as a sensor of amino acid depletion. L-Arginine can be depleted in vivo by an increased uptake and metabolism through one of three major enzymatic pathways, namely arginase I and NOS2 in MDSC and arginase II in tumor cells. The most common mechanism leading to arginine depletion in vivo appears to be arginase I in MDSC (Fig. 2). MDSC are induced from bone-marrow precursors by tumor-derived factors such as VEGF, G-CSF and GM-CSF. Murine MDSC express CD11b and can be induced to express arginase I by stimulation with IL-4, IL-10, IL-13 and PGE2. In vivo or in vitro use of COX-2 inhibitors or the arginase inhibitor Nor-NOHA blocks the induction of arginase I in MDSC and its function, respectively, and causes an immune-mediated, dose-dependent anti-tumor response. MDSC can simultaneously produce NO which in turn can further impair T-cell function through the accumulation of peroxynitrites.

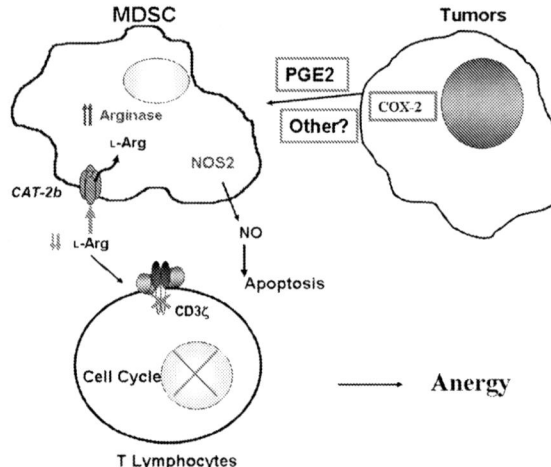

Fig. 2 Arginine metabolism. Arginine is preferentially metabolized in MDSC by arginase I or II or by nitric oxide synthase (NOS2). NOS2 converts arginine into NO, a major component of the cytotoxic mechanisms in phagocytic cells. Arginase I/II will convert arginine to ornithine and urea. The latter serves as a major detoxification mechanism (especially in the kidneys), while ornithine can then be utilized to form proline which is the substrate for the formation of collagen by fibroblasts. Alternatively, ornithine is used in the synthesis of polyamines, needed to sustain cell proliferation

In cancer patients MDSC are characterized by mature granulocyte morphology, express CD11b and constitutively express arginase I, which appears not to be regulated by cytokines. MDSC in cancer patients are present not only in the tumor, but also in peripheral blood where they can deplete normal levels of L-arginine.

In summary, tumors are "highjacking" a well-established cycle of wound healing where myeloid cells infiltrating a wound first produce NO to "cleanse" the site and then express arginase I which allows them to produce ornithine which is used to produce proline and collagen by fibroblasts, leading to the healing of the wound. In cancer, however, the cycle is never completed due to the continuous production of factors by the tumor cells which cause myeloid cells to continuously express arginase I and deplete arginine. Understanding the mechanisms by which tumors regulate MDSC function may provide novel therapeutic approaches to enhance the efficacy of novel immunotherapies.

References

Alberola, V., Gonzalez-Molina, A., Trenor, A., San Martin, B., Lluch, A., Palau, F., Marin, J., and Garcia-Conde, F. J. (1985). Mechanism of suppression of the depressed lymphocyte response in lung cancer patients. *Allergol Immunopathol (Madr)* 13, 213–219.

Albina, J. E., Caldwell, M. D., Henry, W. L., Jr, and Mills, C. D. (1989). Regulation of macrophage functions by L-arginine. *J Exp Med* 169, 1021–1029.

Aulak, K. S., Liu, J., Wu, J., Hyatt, S. L., Puppi, M., Henning, S. J., and Hatzoglou, M. (1996). Molecular sites of regulation of expression of the rat cationic amino acid transporter gene. *J Biol Chem* 271, 29799–29806.

Aulak, K. S., Mishra, R., Zhou, L., Hyatt, S. L., de Jonge, W., Lamers, W., Snider, M., and Hatzoglou, M. (1999). Post-transcriptional regulation of the arginine transporter Cat-1 by amino acid availability. *J Biol Chem* 274, 30424–30432.

Barbul, A. (1990). Arginine and immune function. *Nutrition* 6, 53–58.

Barbul, A., Rettura, G., Levenson, S. M., and Seifter, E. (1977). Arginine: a thymotropic and wound-healing promoting agent. *Surg Forum* 28, 101–103.

Bernard, A. C., Mistry, S. K., Morris, S. M., Jr, O'Brien, W. E., Tsuei, B. J., Maley, M. E., Shirley, L. A., Kearney, P. A., Boulanger, B. R., and Ochoa, J. B. (2001). Alterations in arginine metabolic enzymes in trauma. *Shock* 15, 215–219.

Bhatnagar, R. M., Zabriskie, J. B., and Rausen, A. R. (1975). Cellular immune responses to methylcholanthrene-induced fibrosarcoma in BALB/c mice. *J Exp Med* 142, 839–855.

Bingisser, R. M., Tilbrook, P. A., Holt, P. G., and Kees, U. R. (1998). Macrophage-derived nitric oxide regulates T cell activation via reversible disruption of the Jak3/STAT5 signaling pathway. *J Immunol* 160, 5729–5734.

Bluestone, J. A., and Lopez, C. (1979). Suppression of the immune response in tumor-bearing mice. II. Characterization of adherent suppressor cells. *J Natl Cancer Inst* 63, 1221–1227.

Brito, C., Naviliat, M., Tiscornia, A. C., Vuillier, F., Gualco, G., Dighiero, G., Radi, R., and Cayota, A. M. (1999). Peroxynitrite inhibits T lymphocyte activation and proliferation by promoting impairment of tyrosine phosphorylation and peroxynitrite-driven apoptotic death. *J Immunol* 162, 3356–3366.

Bronte, V., Apolloni, E., Cabrelle, A., Ronca, R., Serafini, P., Zamboni, P., Restifo, N. P., and Zanovello, P. (2000). Identification of a CD11b(+)/Gr-1(+)/CD31(+) myeloid progenitor capable of activating or suppressing CD8(+) T cells. *Blood* 96, 3838–3846.

Bronte, V., Serafini, P., De Santo, C., Marigo, I., Tosello, V., Mazzoni, A., Segal, D. M., Staib, C., Lowel, M., Sutter, G., Colombo, M. P., and Zanovello, P. (2003a). IL-4-induced arginase 1 suppresses alloreactive T cells in tumor-bearing mice. *J Immunol* 170, 270–278.

Bronte, V., Serafini, P., Mazzoni, A., Segal, D. M., and Zanovello, P. (2003b). L-Arginine metabolism in myeloid cells controls T-lymphocyte functions. *Trends Immunol* 24, 302–306.

Bronte, V., and Zanovello, P. (2005). Regulation of immune responses by L-arginine metabolism. *Nat Rev Immunol* 5, 641–654.

Bruhat, A., Jousse, C., and Fafournoux, P. (1999). Amino acid limitation regulates gene expression. *Proc Nutr Soc* 58, 625–632.

Bruhat, A., Jousse, C., Wang, X. Z., Ron, D., Ferrara, M., and Fafournoux, P. (1997). Amino acid limitation induces expression of CHOP, a CCAAT/enhancer binding protein-related gene, at both transcriptional and post-transcriptional levels. *J Biol Chem* 272, 17588–17593.

Catalona, W. J., Smolev, J. K., and Harty, J. I. (1975). Prognostic value of host immunocompetence in urologic cancer patients. *J Urol* 114, 922–926.

Chang, C. I., Liao, J. C., and Kuo, L. (2001). Macrophage arginase promotes tumor cell growth and suppresses nitric oxide-mediated tumor cytotoxicity. *Cancer Res* 61, 1100–1106.

Closs, E. I. (2002). Expression, regulation and function of carrier proteins for cationic amino acids. *Curr Opin Nephrol Hypertens* 11, 99–107.

Coley, W. B. (1893). The treatment of malignant tumors by repeated inoculations of erysipelas. With a report of ten original cases *Am J Med Sci* 1893 May, 105:487–511.

Duhe, R. J., Evans, G. A., Erwin, R. A., Kirken, R. A., Cox, G. W., and Farrar, W. L. (1998). Nitric oxide and thiol redox regulation of Janus kinase activity. *Proc Natl Acad Sci USA* 95, 126–131.

Dye, E. S., and North, R. J. (1984). Specificity of the T cells that mediate and suppress adoptive immunotherapy of established tumors. *J Leukoc Biol* 36, 27–37.

Eggers, A. E., and Wunderlich, J. R. (1975). Suppressor cells in tumor-bearing mice capable of nonspecific blocking of in vitro immunization against transplant antigens. *J Immunol* 114, 1554–1556.

El-Gayar, S., Thuring-Nahler, H., Pfeilschifter, J., Rollinghoff, M., and Bogdan, C. (2003). Translational control of inducible nitric oxide synthase by IL-13 and arginine availability in inflammatory macrophages. *J Immunol* 171, 4561–4568.

Fafournoux, P., Bruhat, A., and Jousse, C. (2000). Amino acid regulation of gene expression. *Biochem J* 351, 1–12.

Fefer, A., McCoy, J. L., Perk, K., and Glynn, J. P. (1968). Immunologic, virologic, and pathologic studies of regression of autochthonous Moloney sarcoma virus-induced tumors in mice. *Cancer Res* 28, 1577–1585.

Fernbach, B. R., Kirchner, H., Bonnard, G. D., and Herberman, R. B. (1976). Suppression of mixed lymphocyte response in mice bearing primary tumors induced by murine sarcoma virus. *Transplantation* 21, 381–386.

Fisher, R. I., DeVita, V. T., Jr, Bostick, F., Vanhaelen, C., Howser, D. M., Hubbard, S. M., and Young, R. C. (1980). Persistent immunologic abnormalities in long-term survivors of advanced Hodgkin's disease. *Ann Intern Med* 92, 595–599.

Gabrilovich, D. (2004). Mechanisms and functional significance of tumour-induced dendritic-cell defects. *Nat Rev Immunol* 4, 941–952.

Gabrilovich, D., Ishida, T., Oyama, T., Ran, S., Kravtsov, V., Nadaf, S., and Carbone, D. P. (1998). Vascular endothelial growth factor inhibits the development of dendritic cells and dramatically affects the differentiation of multiple hematopoietic lineages in vivo. *Blood* 92, 4150–4166.

Gazzola, G. C., Franchi, R., Saibene, V., Ronchi, P., and Guidotti, G. G. (1972). Regulation of amino acid transport in chick embryo heart cells. I. Adaptive system of mediation for neutral amino acids. *Biochim Biophys Acta* 266, 407–421.

Gorczynski, R. M., and Knight, R. A. (1975). Immunity to murine sarcoma virus induced tumours. IV. Direct cellular cytolysis of 51Cr labelled target cells in vitro and analysis of blocking factors which modulate cytotoxicity. *Br J Cancer* 31, 387–404.

Greenberg, P. D., Kern, D. E., and Cheever, M. A. (1985). Therapy of disseminated murine leukemia with cyclophosphamide and immune Lyt-1+,2− T cells. Tumor eradication does not require participation of cytotoxic T cells. *J Exp Med* 161, 1122–1134.

Guerrini, L., Gong, S. S., Mangasarian, K., and Basilico, C. (1993). Cis- and trans-acting elements involved in amino acid regulation of asparagine synthetase gene expression. *Mol Cell Biol* 13, 3202–3212.

Hellstrom, I., Sjogren, H. O., Warner, G., and Hellstrom, K. E. (1971). Blocking of cell-mediated tumor immunity by sera from patients with growing neoplasms. *Int J Cancer* 7, 226–237.

Hellstrom, K. E., Hellstrom, I., and Nelson, K. (1983). Antigen-specific suppressor ("blocking") factors in tumor immunity. *Biomembranes* 11, 365–388.

Hersh, E. M., and Oppenheim, J. J. (1965). Impaired in vitro lymphocyte transformation in Hodgkin's disease. *N Engl J Med* 273, 1006–1012.

Hesse, M., Modolell, M., La Flamme, A. C., Schito, M., Fuentes, J. M., Cheever, A. W., Pearce, E. J., and Wynn, T. A. (2001). Differential regulation of nitric oxide synthase-2 and arginase-1 by type 1/type 2 cytokines in vivo: granulomatous pathology is shaped by the pattern of L-arginine metabolism. *J Immunol* 167, 6533–6544.

Hibbs, J. B., Jr, Taintor, R. R., and Vavrin, Z. (1987). Macrophage cytotoxicity: role for L-arginine deiminase and imino nitrogen oxidation to nitrite. *Science* 235, 473–476.

Hyatt, S. L., Aulak, K. S., Malandro, M., Kilberg, M. S., and Hatzoglou, M. (1997). Adaptive regulation of the cationic amino acid transporter-1 (Cat-1) in Fao cells. *J Biol Chem* 272, 19951–19957.

Iwahashi, M., Tanimura, H., Yamaue, H., Tsunoda, T., Tani, M., Tamai, M., Noguchi, K., and Hotta, T. (1992). Defective autologous mixed lymphocyte reaction (AMLR) and killer activity generated in the AMLR in cancer patients. *Int J Cancer* 51, 67–71.

Jaroslow, B. N., Suhrbier, K. M., Fry, R. J., and Tyler, S. A. (1975). In vitro suppression of immunocompetent cells by lymphomas from aging mice. *J Natl Cancer Inst* 54, 1427–1432.

Jerrells, T. R., Dean, J. H., Richardson, G. L., McCoy, J. L., and Herberman, R. B. (1978). Role of suppressor cells in depression of in vitro lymphoproliferative responses of lung cancer and breast cancer patients. *J Natl Cancer Inst* 61, 1001–1009.

Jousse, C., Bruhat, A., Ferrara, M., and Fafournoux, P. (2000). Evidence for multiple signaling pathways in the regulation of gene expression by amino acids in human cell lines. *J Nutr* 130, 1555–1560.

Jousse, C., Bruhat, A., Harding, H. P., Ferrara, M., Ron, D., and Fafournoux, P. (1999). Amino acid limitation regulates CHOP expression through a specific pathway independent of the unfolded protein response. *FEBS Lett* 448, 211–216.

Kirchner, H., Glaser, M., Holden, H. T., Fernbach, B. R., and Herberman, R. B. (1976). Suppressor cells in tumor bearing mice and rats. *Biomedicine* 24, 371–374.

Kolenko, V., Rayman, P., Roy, B., Cathcart, M. K., O'Shea, J., Tubbs, R., Rybicki, L., Bukowski, R., and Finke, J. (1999). Downregulation of JAK3 protein levels in T lymphocytes by prostaglandin E2 and other cyclic adenosine monophosphate-elevating agents: impact on interleukin-2 receptor signaling pathway. *Blood* 93, 2308–2318.

Lee, J., Ryu, H., Ferrante, R. J., Morris, S. M., Jr, and Ratan, R. R. (2003). Translational control of inducible nitric oxide synthase expression by arginine can explain the arginine paradox. *Proc Natl Acad Sci USA* 100, 4843–4848.

Li, X., Liu, J., Park, J. K., Hamilton, T. A., Rayman, P., Klein, E., Edinger, M., Tubbs, R., Bukowski, R., and Finke, J. (1994). T cells from renal cell carcinoma patients exhibit an abnormal pattern of kappa B-specific DNA-binding activity: a preliminary report. *Cancer Res* 54, 5424–5429.

Loeffler, C. M. (1991). Antitumor effects of interleukin 2 liposomes and anti-CD3-stimulated T-cells against murine MCA-38 hepatic metastasis. *Cancer Res* 51, 2127–2132.

Louis, C. A., Mody, V., Henry, W. L., Jr, Reichner, J. S., and Albina, J. E. (1999). Regulation of arginase isoforms I and II by IL-4 in cultured murine peritoneal macrophages. *Am J Physiol* 276, R237–R242.

Makarenkova, V. P., Bansal, V., Matta, B. M., Perez, L. A., and Ochoa, J. B. (2006). CD11b+/Gr-1+ myeloid suppressor cells cause T cell dysfunction after traumatic stress. *J Immunol* 176, 2085–2094.

Mazzoni, A., Bronte, V., Visintin, A., Spitzer, J. H., Apolloni, E., Serafini, P., Zanovello, P., and Segal, D. M. (2002). Myeloid suppressor lines inhibit T cell responses by an NO-dependent mechanism. *J Immunol* 168, 689–695.

Miescher, S., Stoeck, M., Qiao, L., Barras, C., Barrelet, L., and von Fliedner, V. (1988). Preferential clonogenic deficit of CD8-positive T-lymphocytes infiltrating human solid tumors. *Cancer Res* 48, 6992–6998.

Miescher, S., Whiteside, T. L., Carrel, S., and von Fliedner, V. (1986). Functional properties of tumor-infiltrating and blood lymphocytes in patients with solid tumors: effects of tumor cells and their supernatants on proliferative responses of lymphocytes. *J Immunol* 136, 1899–1907.

Mills, C. D., and North, R. J. (1985). Ly-1+2− suppressor T cells inhibit the expression of passively transferred antitumor immunity by suppressing the generation of cytolytic T cells. *Transplantation* 39, 202–208.

Mizoguchi, H., O'Shea, J. J., Longo, D. L., Loeffler, C. M., McVicar, D. W., and Ochoa, A. C. (1992). Alterations in signal transduction molecules in T lymphocytes from tumor-bearing mice. *Science* 258, 1795–1798.

Munder, M., Eichmann, K., and Modolell, M. (1998). Alternative metabolic states in murine macrophages reflected by the nitric oxide synthase/arginase balance: competitive regulation by CD4+ T cells correlates with Th1/Th2 phenotype. *J Immunol* 160, 5347–5354.

Munder, M., Eichmann, K., Moran, J. M., Centeno, F., Soler, G., and Modolell, M. (1999). Th1/Th2-regulated expression of arginase isoforms in murine macrophages and dendritic cells. *J Immunol* 163, 3771–3777.

Munn, D. H., Sharma, M. D., Baban, B., Harding, H. P., Zhang, Y., Ron, D., and Mellor, A. L. (2005). GCN2 kinase in T cells mediates proliferative arrest and anergy induction in response to indoleamine 2,3-dioxygenase. *Immunity* 22, 633–642.

North, R. J. (1985). Down-regulation of the antitumor immune response. *Adv Cancer Res* 45, 1–43.

North, R. J., and Bursuker, I. (1984). Generation and decay of the immune response to a progressive fibrosarcoma. I. Ly-1+2− suppressor T cells down-regulate the generation of Ly-1−2+ effector T cells. *J Exp Med* 159, 1295–1311.

Ochoa, J. B., Bernard, A. C., Mistry, S. K., Morris, S. M., Jr, Figert, P. L., Maley, M. E., Tsuei, B. J., Boulanger, B. R., and Kearney, P. A. (2000). Trauma increases extrahepatic arginase activity. *Surgery* 127, 419–426.

Parhar, R. S., and Lala, P. K. (1987). Amelioration of B16F10 melanoma lung metastasis in mice by a combination therapy with indomethacin and interleukin 2. *J Exp Med* 165, 14–28.

Pauleau, A. L., Rutschman, R., Lang, R., Pernis, A., Watowich, S. S., and Murray, P. J. (2004). Enhancer-mediated control of macrophage-specific arginase I expression. *J Immunol* 172, 7565–7573.

Prehn, R. T., and Main, J. M. (1957). Immunity to methylcholanthrene-induced sarcomas. *J Natl Cancer Inst* 18, 759–778.

Quillard, M., Husson, A., and Lavoinne, A. (1996). Glutamine increases argininosuccinate synthetase mRNA levels in rat hepatocytes. The involvement of cell swelling. *Eur J Biochem* 236, 56–59.

Rodriguez, P. C., Hernandez, C. P., Quiceno, D., Dubinett, S. M., Zabaleta, J., Ochoa, J. B., Gilbert, J., and Ochoa, A. C. (2005). Arginase I in myeloid suppressor cells is induced by COX-2 in lung carcinoma. *J Exp Med* 202, 931–939.

Rodriguez, P. C., Quiceno, D. G., and Ochoa, A. C. (2007). L-Arginine availability regulates T-lymphocyte cell-cycle progression. *Blood* 109, 1568–1573.

Rodriguez, P. C., Quiceno, D. G., Zabaleta, J., Ortiz, B., Zea, A. H., Piazuelo, M. B., Delgado, A., Correa, P., Brayer, J., Sotomayor, E. M., Antonia, S., Ochoa, J. B., and Ochoa, A. C. (2004). Arginase I production in the tumor microenvironment by mature myeloid cells inhibits T-cell receptor expression and antigen-specific T-cell responses. *Cancer Res* 64, 5839–5849.

Rodriguez, P. C., Zea, A. H., Culotta, K. S., Zabaleta, J., Ochoa, J. B., and Ochoa, A. C. (2002). Regulation of T cell receptor CD3 zeta chain expression by L-arginine. *J Biol Chem* 277, 21123–21129.

Rodriguez, P. C., Zea, A. H., DeSalvo, J., Culotta, K. S., Zabaleta, J., Quiceno, D. G., Ochoa, J. B., and Ochoa, A. C. (2003). L-Arginine consumption by macrophages modulates the expression of CD3zeta chain in T lymphocytes. *J Immunol* 171, 1232–1239.

Ross, J. (1996). Control of messenger RNA stability in higher eukaryotes. *Trends Genet* 12, 171–175.

Rutschman, R., Lang, R., Hesse, M., Ihle, J. N., Wynn, T. A., and Murray, P. J. (2001). Cutting edge: Stat6-dependent substrate depletion regulates nitric oxide production. *J Immunol* 166, 2173–2177.

Sachs, A. B. (1993). Messenger RNA degradation in eukaryotes. *Cell* 74, 413–421.

Saini, K. S., Summerhayes, I. C., and Thomas, P. (1990). Molecular events regulating messenger RNA stability in eukaryotes. *Mol Cell Biochem* 96, 15–23.

Saio, M., Radoja, S., Marino, M., and Frey, A. B. (2001). Tumor-infiltrating macrophages induce apoptosis in activated CD8(+) T cells by a mechanism requiring cell contact and mediated by both the cell-associated form of TNF and nitric oxide. *J Immunol* 167, 5583–5593.

Salimuddin, Nagasaki, A., Gotoh, T., Isobe, H., and Mori, M. (1999). Regulation of the genes for arginase isoforms and related enzymes in mouse macrophages by lipopolysaccharide. *Am J Physiol* 277, E110–E117.

Schmielau, J., and Finn, O. J. (2001). Activated granulocytes and granulocyte-derived hydrogen peroxide are the underlying mechanism of suppression of t-cell function in advanced cancer patients. *Cancer Res* 61, 4756–4760.

Singh, R., Pervin, S., Karimi, A., Cederbaum, S., and Chaudhuri, G. (2000). Arginase activity in human breast cancer cell lines: N(omega)-hydroxy-L- arginine selectively inhibits cell proliferation and induces apoptosis in MDA-MB-468 cells. *Cancer Res* 60, 3305–3312.

Sonoki, T., Nagasaki, A., Gotoh, T., Takiguchi, M., Takeya, M., Matsuzaki, H., and Mori, M. (1997). Coinduction of nitric-oxide synthase and arginase I in cultured rat peritoneal macrophages and rat tissues in vivo by lipopolysaccharide. *J Biol Chem* 272, 3689–3693.

Suer, G. S., Yoruk, Y., Cakir, E., Yorulmaz, F., and Gulen, S. (1999). Arginase and ornithine, as markers in human non-small cell lung carcinoma. *Cancer Biochem Biophys* 17, 125–131.

Taheri, F., Ochoa, J. B., Faghiri, Z., Culotta, K., Park, H. J., Lan, M. S., Zea, A. H., and Ochoa, A. C. (2001). L-Arginine regulates the expression of the T-cell receptor zeta chain (CD3zeta) in Jurkat cells. *Clin Cancer Res* 7, 958s–965s.

White, M. F. (1985). The transport of cationic amino acids across the plasma membrane of mammalian cells. *Biochim Biophys Acta* 822, 355–374.

Whiteside, T. L., Miescher, S., Moretta, L., and von Fliedner, V. (1988). Cloning and proliferating precursor frequencies of tumor-infiltrating lymphocytes from human solid tumors. *Transplant Proc* 20, 342–343.

Whiteside, T. L., and Rabinowich, H. (1998). The role of Fas/FasL in immunosuppression induced by human tumors. *Cancer Immunol Immunother* 46, 175–184.

Wu, G., and Morris, S. M., Jr (1998). Arginine metabolism: nitric oxide and beyond. *Biochem J* 336 (Pt 1), 1–17.

Xia, Y., and Zweier, J. L. (1997). Superoxide and peroxynitrite generation from inducible nitric oxide synthase in macrophages. *Proc Natl Acad Sci USA* 94, 6954–6958.

Young, M. R., Aquino, S., and Young, M. E. (1989). Differential induction of hematopoiesis and immune suppressor cells in the bone marrow versus in the spleen by Lewis lung carcinoma variants. *J Leukoc Biol* 45, 262–273.

Young, M. R., Newby, M., and Wepsic, H. T. (1987). Hematopoiesis and suppressor bone marrow cells in mice bearing large metastatic Lewis lung carcinoma tumors. *Cancer Res* 47, 100–105.

Zabaleta, J., McGee, D. J., Zea, A. H., Hernandez, C. P., Rodriguez, P. C., Sierra, R. A., Correa, P., and Ochoa, A. C. (2004). *Helicobacter pylori* arginase inhibits T cell proliferation and reduces the expression of the TCR zeta-chain (CD3zeta). *J Immunol* 173, 586–593.

Zea, A. H., Rodriguez, P. C., Atkins, M. B., Hernandez, C., Signoretti, S., Zabaleta, J., McDermott, D., Quiceno, D., Youmans, A., O'Neill, A., Mier, J., and Ochoa, A. C. (2005). Arginase-producing myeloid suppressor cells in renal cell carcinoma patients: a mechanism of tumor evasion. *Cancer Res* 65, 3044–3048.

Zea, A. H., Rodriguez, P. C., Culotta, K. S., Hernandez, C. P., DeSalvo, J., Ochoa, J. B., Park, H. J., Zabaleta, J., and Ochoa, A. C. (2004). L-Arginine modulates CD3zeta expression and T cell function in activated human T lymphocytes. *Cell Immunol* 232, 21–31.

Protein–Glycan Interactions in the Regulation of Immune Cell Function in Cancer: Lessons from the Study of Galectins-1 and -3

Gabriel A. Rabinovich and Fu-Tong Liu

1 Galectins: Versatile Glycan-Binding Proteins Upregulated at Sites of Chronic Inflammation and Tumor Growth

Protein–glycan interactions are critical during tumor growth, metastasis and inflammation (Ohtsubo and Marth, 2006). Differential glycosylation of cell surface glycoconjugates can influence tumor growth and metastasis by modulating cell adhesion, cell migration, angiogenesis and inflammation (Liu and Rabinovich, 2005). Galectins, a family of evolutionarily conserved glycan-binding proteins, are defined by a conserved carbohydrate-recognition domain (CRD) with affinity for β-galactosides (Camby et al., 2006). To date, 15 mammalian galectins have been identified, which can be subdivided into three groups: single-CRD galectins (including galectins-1, -2, -5-, -7, -10, -11, -13, -14 and -15); two-CRD galectins (galectins-4, -6, -8-, -9 and -12) and the unique "chimera-type" galectin-3, which contains a single CRD fused to unusual tandem repeats of short amino acid stretches (Camby et al., 2006). Of these 15 members it should be highlighted that the inclusion of galectin-1, which was first characterized as a lens-specific protein called GRIFIN (*g*alectin-*r*elated *i*nter*f*iber prote*in*), remains controversial. In fact, galectin-1 lacks two of the seven key amino acid residues conserved in most galectin CRDs and does not display β-galactoside-binding activity (Cooper 2002).

Many galectins are either bivalent or multivalent with regard to their carbohydrate-binding activities: some one-CRD galectins (e.g. galectin-1) exists as dimers; two-CRD galectins have two carbohydrate-binding sites in tandem; and galectin-3 forms oligomers when it binds to multivalent carbohydrates (Brewer, 2002). Cross-linkage of cell surface receptors by galectins can trigger transmembrane signaling events through which diverse processes such as apoptosis, cell migration, cytokine secretion and angiogenesis are modulated (Liu and Rabinovich, 2005). Remarkably, the responsiveness of cells to individual members of the galectin family can fluctuate

---

G.A. Rabinovich
Instituto de Biología y Medicina Experimental, Consejo Nacional de Investigaciones Científicas y Técnicas de Argentina, Vuelta de Obligado 2490, Buenos Aires C1428ADN, Argentina
e-mail: gabyrabi@ciudad.com.ar

depending on the repertoire of potentially glycosylated molecules expressed on the cell surface and the activities of specific glycosyltransferases that are responsible for generating galectin ligands. These variables can dramatically change according to the differentiation and activation state of the cells (Rabinovich et al., 2002a). Here we will focus on the role of galectins-1 and -3, the best studied members of the galectin family, in inflammatory and immunoregulatory processes involved in tumor progression and tumor-immune escape.

2 Galectin-1

2.1 Galectin-1: Biochemistry and Cell Biology

Galectin-1, a β-galactoside-binding protein widely expressed in human tissues, may occur as a monomer as well as a non-covalent homodimer composed of 14.5 kDa subunits (Camby et al., 2006). One of the main properties of the homodimeric galectin-1 protein is that it spontaneously dissociates at low concentrations ($K_d \sim 7\mu M$) into a monomeric form that is still able to bind carbohydrates but with lower affinity. Interestingly, recent evidence indicates that the monomeric and dimeric forms of galectin-1 are associated with different, and in some cases contrasting, biological functions (Barrionuevo et al., 2007). Although this protein binds preferentially to glycoconjugates containing the ubiquitous disaccharide N-acetyllactosamine (Gal β 1-3/4 GlcNAc), binding to individual lactosamine units is of relatively low affinity and it is the arrangement of lactosamine disaccharides in repeating chains (polylactosamine) that increases the binding activity (Rabinovich et al., 2002a).

Galectin-1 lacks recognizable secretion signal sequences and does not pass through the standard ER/Golgi pathway (Camby et al., 2006). In addition, it shows other characteristics of typical cytoplasmic proteins, including acetylated N-terminus and lack of glycosylation. Nevertheless, it is well known that this protein, as well as other members of the galectin family, is secreted through a novel mechanism distinct from classical vesicle-mediated exocytosis. Compelling evidence indicates that the galectin secretion mechanism involves the formation of exovesicles generated by membrane blebbing (Hughes, 1999). Galectin-1 secretion was initially described by Cooper and Barondes (1990) in skeletal muscle cells, although the molecular mechanisms involved in this secretory pathway remained elusive for many years. Recently, Nickel's group has demonstrated that secretion of galectin-1 from mammalian cells depends on functional interactions between the lectin and its counter-receptors (Nickel, 2005). Thus, single-site mutations in galectin-1 caused deficiencies in both counter-receptor binding and export in CHO cells, suggesting that functional interactions with counter-receptors are essential for the overall galectin-1 export process.

2.2 Galectin-1 in the Tumor Microenvironment

The potential role of galectins in tumor progression was proposed in the middle 1980s reporting a differential expression of endogenous β-galactoside-binding proteins on the surface of non-tumorigenic, tumorigenic and metastatic cells (Gabius et al., 1986; Raz et al., 1986). Expression of galectin-1 has been well documented in cancer cells and cancer-associated stromal cells of different tumor types including astrocytoma, melanoma, prostate, thyroid, colon, breast and ovary carcinomas and also in hematological malignancies (Camby et al., 2006; Van den Brule et al., 2004; Lahm et al., 2004). In addition, galectin-1 is highly expressed in endothelial cells at sites of tumor growth (He and Baum, 2006). Interestingly, in most cases this expression correlates with the aggressiveness of these tumors and the acquisition of metastatic phenotype (Liu and Rabinvich, 2005).

How does galectin-1 contribute to tumor progression? Emerging evidence indicates that galectin-1 has diverse functions in several aspects of cancer biology including the regulation of tumor transformation, cell cycle regulation and apoptosis. Furthermore, this glycan-binding protein may also contribute to tumor metastasis through modulation of cell–cell and cell–matrix interactions, thus controlling cell adhesion, migration and invasiveness (Liu and Rabinovich, 2005). In addition, recent evidence highlights a critical role for galectin-1 as a negative regulator of T-cell immunity, thus contributing to tumor cell evasion of T-cell responses (Le et al., 2005; Rubinstein et al., 2004). We will focus in the following section on the role of galectin-1 as an immunosuppressive mediator employed by tumors to evade immune responses.

2.3 Role of Galectin-1 in the Regulation of Adaptive Immune Responses

Galectin-1 modulates different events involved in the physiology and dynamics of T- and B-cell responses, including activation, cytokine secretion, cell adhesion, migration and survival (Fig. 1).

2.3.1 Impact of Galectin-1 on Immune Cell Activation, Receptor-Mediated Signal Transduction and the Function of Regulatory T cells

By influencing TCR-mediated signal transduction, galectin-1 can modulate T-cell activation at sites of antigen presentation. Vespa et al. (1999) found that galectin-1 inhibits T-cell receptor (TCR) signals and antagonizes TCR-induced IL-2 production in a murine T-cell hybridoma clone. Interestingly, the same group further demonstrated that galectin-1 induces partial TCR-ζ chain phosphorylation and is able to antagonize full TCR responses including the production of IL-2 (Chung et al., 2000). These effects may influence T-cell effector functions in galectin-1-enriched tumor microenvironments.

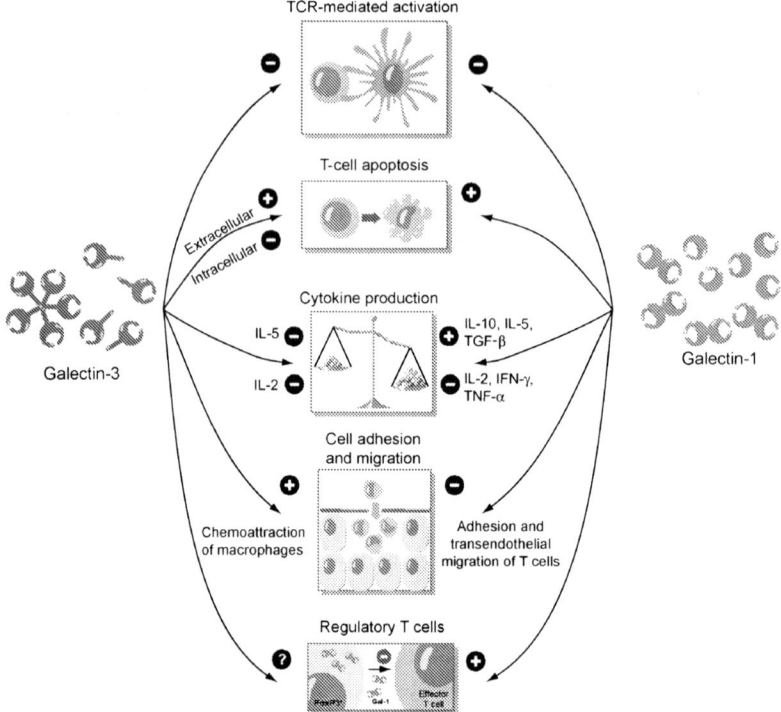

Fig. 1 Galectins-1 and -3 in the tumor microenvironment. Tumors elaborate a wide variety of soluble factors which can positively or negatively affect anti-tumor effector responses. Galectins-1 and -3 control T-cell-dependent anti-tumor immunity through different mechanisms including regulation of effector T-cell apoptosis, modulation of T-cell activation, inhibition of transendothelial T-cell migration and regulation of cytokine synthesis. In addition, galectins-1 and -3 can influence the physiology of other immune cell types in the tumor microenvironment, including B cells, dendritic cells, macrophages and polymorphonuclear leukocytes

Regarding the B-cell compartment, Gauthier et al. (2002). demonstrated that galectin-1 can act as a stromal ligand of the pre-B-cell receptor and contributes to synapse formation between pre-BCR and stromal cells. Thus, by acting at early events of TCR or BCR signaling, galectin-1 can control lymphocyte development and activation.

On the other hand, recent studies, using microarray analysis, highlighted the upregulated expression of galectin-1 in $CD4^+$ $CD25^+$ regulatory T cells and the ability of these cells to suppress T-cell effector functions through galectin-1-mediated mechanisms (Garin et al., 2007; Sugimoto et al., 2006). Further studies are warranted to investigate whether the upregulated expression of galectin-1 in regulatory T cells contributes to T-cell dysfunctions and tumor-immune escape. In addition, it would be of interest to investigate the essential signals and molecular mechanisms involved in the upregulated expression of galectin-1 in regulatory versus effector T-cell populations in tumor and chronic inflammatory microenvironments.

2.3.2 Role of Galectin-1 in the Control of Proinflammatory and Anti-inflammatory Cytokines

Compelling evidence indicates an essential role of galectin-1 in the control of cytokine synthesis and secretion, suggesting that this lectin may influence a variety of physiological processes. In addition, the ability to control cytokine production endows this sugar-binding protein with the capacity to interfere with pathological processes such as chronic inflammation and cancer.

Galectin-1 has been shown to selectively block secretion of Th1 and proinflammatory cytokines in vitro, including IL-2, IFN-γ and TNF-α (Rabinovich et al., 1999a; Vespa et al., 1999). In addition, studies in experimental models of chronic inflammation and autoimmunity (using gene and protein therapy strategies) showed the ability of galectin-1 to skew the balance toward a Th2-type cytokine profile in vivo, with decreased levels of IFN-γ and increased secretion of IL-5, IL-10 and TGF-β by pathogenic T cells (Baum et al., 2003; Perone et al., 2006; Rabinovich et al., 1999b; Santucci et al., 2000, 2003; Toscano et al., 2006). Furthermore, van der Leij et al. (2007) recently reported a marked increase in IL-10 secretion from non-activated and activated $CD4^+$ and $CD8^+$ T cells following exposure to galectin-1. Interestingly we have found that galectin-1 treatment at early phases of a pathogenic autoimmune response not only reduces the secretion of Th1-derived cytokines, but also promotes the expansion of IL-10-producing regulatory T cells in lymph nodes from treated mice (Toscano et al., 2006). These results underscore the ability of this endogenous lectin to counteract Th1-mediated responses through different, but potentially overlapping anti-inflammatory mechanisms. Furthermore, the ability of galectin-1 to promote the expansion of regulatory T cells which in turn are a major source of galectin-1 production could define a potential positive regulatory loop. This regulated mechanism may have critical implications in the promotion of tumor-immune escape, given the ability of regulatory T cells to restrain T-cell-mediated anti-tumor responses.

Taken together, these data suggest that under distinct physiological or pathological conditions, galectin-1 may provide inhibitory or stimulatory signals to control immune cell homeostasis and regulate inflammation following an antigenic challenge.

2.3.3 Influence of Galectin-1 in T-Cell Adhesion and Extravasation

In addition to its role in modulating T-cell activation and cytokine synthesis, previous studies demonstrated the ability of galectin-1 (at low concentrations $< 0.5\mu M$) to inhibit T-cell adhesion to extracellular matrix glycoproteins such as laminin and fibronectin without affecting T-cell survival (Rabinovich et al., 1999a). In addition, He and Baum (2006) recently found that galectin-1 specifically inhibits T-cell migration across endothelial cells. Since endothelial cells are a major source of galectin-1, inhibition of T-cell adhesion and migration may represent a novel anti-inflammatory effect of this protein, which might have profound implications in tumor-immune escape. Furthermore, recent studies highlight the ability of human galectin-1 to inhibit hematopoietic progenitor mobilization by obstructing the transendothelial migration of bone-marrow-derived stem cells (Kiss et al., 2007), suggesting the ability of galectin-1 to modulate immune cell adhesion and transendothelial migration at different levels of the lifespan of immunocytes with critical implications in immune surveillance.

2.3.4 Galectin-1-Mediated Modulation of T-Cell Proliferation and Apoptosis

Research over the past decade has demonstrated that galectin-1 plays a pivotal role in the regulation of T-cell homeostasis by regulating T-cell survival (Rabinovich et al., 2002a). Through interaction with specific carbohydrate ligands on T-cell surface glycoconjugates, this glycan-binding protein can induce cell cycle arrest and promote apoptosis of developing thymocytes and peripheral T lymphocytes (Blaser et al., 1998; Perillo et al., 1995, 1997; Rabinovich et al., 1997, 1998, 2002b). In addition, it has been demonstrated that galectin-1 is not expressed on resting T or B lymphocytes, but is upregulated upon T-cell activation and differentiation, suggesting a potential autocrine negative mechanism to kill effector T and B cells after the completion of an antigen-specific immune response (Blaser et al., 1998; Fuertes et al., 2004; Zuñiga et al., 2001).

Different glycoproteins on the surface of activated T cells appear to be primary receptors for galectin-1 including CD45, CD43, CD7, CD3 and CD2 (Pace et al., 1999; Stillman et al., 2006; Walzel et al., 2000). Susceptibility to galectin-1-induced cell death is tightly controlled by the regulated expression of specific glycosyltransferases which are regulated during T-cell development, activation and differentiation (Amano et al., 2003; Galvan et al., 2000). It has been shown that T cells lacking the core-2-β-1,6-N-acetylglucosaminyltransferase (C2GnT) are resistant to galectin-1-induced cell death. This enzyme is responsible for creating branched structures on O-glycans of T-cell surface glycoproteins, such as CD45 (Galvan et al., 2000). Consistent with these findings, recent studies showed that haploinsufficiency of C2GnT-I results in altered cellular glycosylation and resistance to galectin-1-induced cell death in T-lymphoma cells (Cabrera et al., 2006). In addition, other glycosyltransferases can also act to reduce galectin-1 binding by directly or indirectly masking galectin-1 saccharide ligands. In this context, addition of α 2,6-linked sialic acids to lactosamine units by the ST6Gal-I sialyltransferase has been shown to interrupt galectin-1 binding and cell death (Amano et al., 2003).

The signal transduction events triggered by galectin-1 are still controversial. It has been demonstrated that galectin-1-induced T-cell death requires the activation of specific transcription factors (i.e., AP-1) (Rabinovich et al., 2000b), cytochrome c release and modulation of caspases-8 and -3 (Matarrese et al., 2005). However, another report shows that apoptosis induced by galectin-1 in a T-cell line is not dependent on the activation of caspase-3 and does not involve cytochrome c release (Hahn et al., 2004). Interestingly, a recent study proposed a model in which galectin-1 induces the release of ceramide, which in turn promotes downstream events including decreased Bcl-2 protein expression, depolarization of mitochondria and activation of caspase-9 and -3 (Ion et al., 2006). All downstream events required the presence of two tyrosine kinases, p56^{lck} and ZAP-70 (Ion et al., 2005). On the other hand, other studies have shown that galectin-1 promotes phosphatidylserine exposure in different cell types with no apparent signs of apoptosis (Stowell et al., 2007). Therefore, it seems evident that galectin-1 may trigger distinct death pathways or different apoptosis end points in different cell types.

In contrast to the proapoptotic role of galectin-1 on activated T cells, Endharti et al. (2005) demonstrated that secretion of this protein by stromal cells is capable of supporting the survival of naïve T cells without promoting proliferation. In this regard, it has recently been found that murine dendritic cells transduced with an adenoviral vector expressing galectin-1 can elicit contrasting results when exposed to resting or activated T cells. While this protein stimulated activation of naïve T cells, it rapidly triggered apoptosis of activated T cells (Perone et al., 2006). Thus, galectin-1 might trigger different signals (i.e., apoptosis, proliferation or survival) depending on a number of factors including the activation state of the cells (resting vs activated or differentiated cells), the spatiotemporal expression of specific glycosyltransferases, the nature of the targeT cells (primary cell cultures or immortalized T-cell lines) and different biochemical or biophysical parameters (i.e., equilibrium between the monomeric or dimeric protein forms and the presence of reducing agents in the extracellular medium). Finally, the cross-talk between different mechanisms by which galectin-1 regulates T-cell fate and homeostasis (i.e., modulation of T-cell survival, activation, cytokine secretion and migration) still remains to be investigated. Does galectin-1 regulate cytokine production and cell trafficking through modulation of T-cell survival? Does galectin-1 impair cytokine secretion by influencing T-cell activation and differentiation? Is there any association between induction of regulatory T cells and the modulation of T-cell survival? These questions remain to be addressed in future studies.

2.4 Galectin-1 in the Establishment and Maintenance of Tumor-Immune Privilege

Different strategies are employed by tumors to foster a tolerant and privileged microenvironment including the tumor-induced impairment of antigen presentation, the activation of negative costimulatory signals and the elaboration of soluble immunosuppressive factors (Rabinovich et al., 2007). The ability of galectin-1 to impair T-cell effector functions and the expression of galectin-1 in highly aggressive tumors (Liu and Rabinovich, 2005) prompted us to investigate the ability of this glycan-binding protein to confer a status of immune privilege to tumor microenvironments. By a combination of in vitro and in vivo experiments, we established a link between galectin-1-mediated immunoregulation and its contribution to tumor-immune escape (Rubinstein et al., 2004). We found that human and mouse melanoma cells secrete substantial levels of galectin-1 which contribute to the immunosuppressive activity of these tumor cells. Blockade of the inhibitory effects of galectin-1 within the tumor tissue resulted in reduced tumor size and potentiation of tumor-specific T-cell responses in vivo. This effect required intact $CD4^+$ and $CD8^+$ T-cell compartments, since simultaneous depletion of these T-cell subpopulations in vivo completely abrogated the anti-tumor response induced by knockdown transfectants (Rubinstein et al., 2004). Furthermore, tumor-secreted galectin-1 induced in situ apoptosis of tumor-infiltrating lymphocytes and interruption of this proapoptotic pathway within the tumor microenvironment resulted in the generation of an otherwise repressed tumor-specific Th1 response in tumor-draining lymph nodes (Rubinstein et al., 2004).

In support of our findings, Le et al. (2005) have recently shown that galectin-1 can act as a molecular link between tumour hypoxia and tumor-immune privilege. Using proteomic analysis, the authors demonstrated that galectin-1 is overexpressed following exposure of tumor cells to hypoxic conditions. In addition, they found a strong inverse correlation between galectin-1 expression and T-cell recruitment in tumour sections corresponding to head and neck squamous cell carcinoma (HNSCC) patients (Le et al., 2005). Taken together, these results support the concept that galectin-1 contributes to immune privilege of tumors by modulating survival, cytokine production and differentiation of effector T cells. Nevertheless, the overall effects of galectin-1 in tumor progression could be multifactorial, including its ability to promote tumor escape, homotypic and heterotypic cell adhesion, tumor cell migration and cell cycle regulation (Liu and Rabinovich, 2005). Whether a connection exists between galectin-1-mediated immune and non-immune mechanisms related to tumor progression still remains to be elucidated. In addition, it remains to be determined whether galectin-1 may affect tumor-immune surveillance and escape by modulating the physiology of other immune cell types, including macrophages, dendritic cells, NK cells and NKT cells.

2.5 Impact of Galectin-1 on the Physiology of Other Immune Cell Types: Monocytes, Macrophages, Dendritic Cells and Neutrophils

Whereas compelling evidence has been accumulated regarding the effects of galectin-1 on T-cell fate, limited information is available on how galectin-1 may impact on other immune cell types, including monocytes and macrophage. In this regard, results from our laboratory showed that galectin-1, similarly to Th2 and Th3 cytokines, inhibits nitric oxide synthesis, favoring the expression of arginase (the alternative metabolic pathway of L-arginine) in activated peritoneal macrophages (Correa et al., 2003). In addition, we have recently demonstrated that galectin-1 can differentially regulate the expression and function of critical regulatory molecules (i.e., Fcγ RI and MHC-II) on human monocytes and mouse macrophages through a non-apoptotic ERK1/2-mediated pathway (Barrionuevo et al., 2007). This effect was clearly observed in macrophages recruited in response to inflammatory stimuli following treatment with recombinant galectin-1 and further confirmed in galectin-1-deficient ($gal\text{-}1^{-/-}$) mice (Barrionuevo et al., 2007). Altogether, these results suggest that this endogenous lectin might promote a state of "alternative activation" or "deactivation" in elicited macrophages. These observations may have profound implications in fostering tumor-immune escape, since alternatively activated macrophages play a critical role in promoting tumor progression and metastasis (Gordon, 2003).

Galectin-1 has also been shown to impact on dendritic cell function. Lee's group reported the ability of galectin-1 to augment the secretion of proinflammatory cytokines and to influence dendritic cell migration through the extracellular matrix (Fulcher et al., 2006). On the other hand, it has been demonstrated that dendritic cells engineered to overexpress galectin-1 are highly activated; these transgenic cells can stimulate naïve T cells, but induce apoptosis of activated T cells (Perone et al., 2006). Whether these effects influence the generation of immunogenic or tolerogenic dendritic cells in the tumor microenvironment still remains to be examined.

On the other hand, galectin-1 has also been shown to influence the functions of human and murine neutrophils. Treatment with galectin-1 inhibits chemotaxis and transendothelial migration of neutrophils in vivo (La et al., 2003; Rabinovich et al., 2000, 2000a; Gil et al., 2006). Furthermore, Stowell et al. (2007) reported that galectin-1 can induce phosphatidylserine exposure in activated human neutrophils thus compromising their further removal by phagocytes. In addition, Karlsson and colleagues showed that galectin-1 can activate NADPH oxidase in primed neutrophils (Almkvist et al., 2002), suggesting that galectin-1 may positively or negatively regulate neutrophil function during the development, progression and resolution of an inflammatory response by either inhibiting cell extravasation and chemotaxis, activating the release of proinflammatory mediators or regulating phagocytic removal. Further studies are warranted, using galectin-1-deficient mice, to dissect the precise role of galectin-1 in neutrophil-mediated immune functions in vivo.

2.6 Other Biological Functions Modulated by Galectin-1: Regulation of Cell Adhesion, Migration and Angiogenesis

In addition to the ability of galectin-1 to regulate immune processes related to innate or adaptive responses, this carbohydrate-binding protein can have important roles in cancer by contributing to neoplastic transformation, tumor cell survival, angiogenesis and tumor metastasis (extensively revised in Liu and Rabinovich, 2005). Therefore, the overall effect of galectin-1 on tumor progression may result not only from the modulation of immune-mediated mechanisms, but also from the control of other biological functions, including homotypic cell aggregation (Tinari et al., 2001), cell adhesion (Elola et al., 2007), migration (Camby et al., 2001), tumor cell proliferation (Kopitz et al., 2001) and angiogenesis (Thijssen et al., 2006). The functional interplay between immune and non-immune mechanisms by which galectin-1 may affect tumor progression, escape and metastasis remains to be further elucidated.

3 Galectin-3

3.1 Galectin-3: Biochemistry and Cellular Biology

Galectin-3 is the only member of the third form of galectins with a single CRD domain and an N-terminal region of approximately 120 amino acids composed of tandem repeats of short amino acid segments. Earlier studies suggested that galectin-3 can self-associate (Hsu et al., 1992; Massa et al., 1993) and a more recent quantitative study revealed that it forms pentamers on binding to multivalent carbohydrates (Ahmad,2004). Galectin-3 can be cross-linked by transglutaminase (Mehul et al., 1995), resulting in the formation of covalently linked oligomers (Mahoney et al., 2000; Van den Brule et al., 1998). Affinity of galectin-3 for galactose is low, but that for certain galactose-containing oligosaccharides can be significantly higher (Hirabayashi et al., 2002; Sparrow et al., 1987).

Like other galectins, galectin-3 lacks a classical signal peptide and transmembrane domain required for secretion through the classical secretory pathway and for display on the cell surface, respectively (reviewed in Hughes, 1999). Nevertheless, the protein is released to extracellular spaces and possible mechanisms have been investigated (reviewed in Hughes, 1999).

Secreted galectin-3 may affect cells through an autocrine or paracrine mechanism by binding to and cross-linking glycoproteins and glycolipids present on the cell surfaces and extracellular matrices that contain appropriate oligosaccharides. Galectin-3's effectiveness in this process is likely to be associated with its ability to form oligomers. Indeed, oligomerization of galectin-3 on cell surfaces was documented by the fluorescence resonance energy transfer method under the condition that the lectin mediates cell activation and adherence (Nieminen et al., 2007).

A number of intracellular functions have been reported for galectin-3 and, for some of these, intracellular proteins with which it interacts and possibly mediates these functions have been identified (reviewed in Liu 2005). Notably, galectin-3 binds to them through protein–protein interactions and not lectin–carbohydrate interactions (Liu et al., 2002; Wang et al., 2004). Galectin-3 has been observed to shuttle between the cytoplasm and the nucleus (Davidson et al., 2002) and thus can function in the nucleus. One such function relates to the process of pre-mRNA splicing (Dagher et al., 1995; Liu et al., 2002). A role for galectin-3 in glycoprotein trafficking was recently described by Delacour et al. [2006]. They found the presence of galectin-3 in vesicles containing glycoproteins destined for the plasma membrane at the apical side of the cell. The authors suggested that galectin-3 is responsible for sorting these glycoproteins into these vesicles. Thus, galectin-3 may play a role in controlling the composition of cell surface glycoproteins.

Galectin-3 exerts a variety of functions by acting extra- or intracellularly. Through the extracellular mode of action, the lectin can affect cell growth and survival, activate or inhibit cellular responses, modulate cell adhesion and induce cell migration. These are mostly demonstrated with recombinant galectin-3 in vitro. Through the intracellular mode of action, galectin-3 has been shown to regulate many of the above-mentioned processes. Some of these were revealed from studies involving gene transfection and antisense nucleotides to influence expression, while others were confirmed by the use of cells from mice deficient in galectin-3. The intracellular functions of galectin-3 are supported by experiments showing that the activity in question is not affected by lactose added to the culture medium, which would inhibit galectin-3's carbohydrate-dependent extracellular actions. They are also supported by the identification of intracellular proteins with which galectin-3 interacts.

The intracellular functions demonstrated for endogenous galectin-3 may not be the same as those exerted by exogenously added galectin-3. This is best exemplified by the opposite activities noted for the protein in regulation of apoptosis: The endogenous protein is anti-apoptotic presumably through its intracellular action, while exogenously added recombinant protein induces apoptosis, through an extracellular mechanism (for details see Sect. 3.3.1).

3.2 Expression of Galectin-3 in Tumors

Galectin-3 has wide tissue distribution and is expressed by many tumor cells (reviewed in Liu and Rabinovich, 2005). In general, in tissues and cell types that normally express galectin-3, the protein is downregulated when they become cancerous. Conversely, in those tissues and cells that do not normally express galectin-3, upregulation of the protein is observed in tumor transformation. In some cases, detection of galectin-3 in tumor cells is useful for diagnosis of the cancer. Additionally, in certain types of cancers, the levels of galectin-3 in tumor cells correlate with the disease progression (reviewed in Danguy et al., 2002).

As in the case of galectin-1, galectin-3 expressed in tumors can contribute to tumor transformation, cell cycle regulation and apoptosis, as well as tumor metastasis through modulation of cell adhesion, migration and invasiveness (Liu and Rabinovich, 2005). In addition, also like galectin-1, extracellular galectin-3 is a negative regulator of T-cell immunity and thus may contribute to tumor cell evasion of T-cell responses. Furthermore, galectin-3 can affect many aspects of the immune and inflammatory responses. We will concentrate here on the role of galectin-3 as an immunosuppressive mediator and inflammatory modulator employed by tumors to evade immune responses and fuel their own growth.

3.3 Role of Galectin-3 in Regulation of T-Cell Responses

3.3.1 Control of T-Cell Survival

Galectin-3 can induce apoptosis in T cells, like galectin-1. This has been shown by using recombinant protein and human T-leukemic cell lines, human peripheral blood mononuclear cells and activated mouse T cells (Fukumori et al., 2003). Another study confirmed the ability of galectin-3 to induce apoptosis in T cells and also noted differential sensitivities of cells to galectins-1 and -3 (Stillman et al., 2006).

Galectin-3 induces apoptosis preferentially in double-negative thymocytes, while galectin-1 kills both double-negative and double-positive thymocytes.

The cell surface glycoproteins that mediate the apoptosis-inducing activities also differ between the two galectins. One study showed that binding of galectin-3 to CD7 and CD29 (β 1 integrin) is involved, as neutralizing antibodies binding to these proteins inhibited galectin-3-induced apoptosis (Fukumori et al., 2003). Another study found that CD45 and CD71, but not CD29 and CD43, mediate galectin-3-induced death. Further, CD7, an essential receptor for galectin-1-induced T-cell death, is not required (Stillman et al., 2006). In addition, galectin-3 is more potent than galectin-1 in inducing T-cell death. About 50 % of cells underwent apoptosis when MOLT, CEM and Jurkat T-cell lines were treated with galectin-3 at 1–5 μM, while a comparable level of apoptosis required 20 μM galectin-1 (Stillman et al., 2006). The lower galectin-3 concentration required for cell death is likely due to the fact that galectin-3 pentamerizes when bound to multimeric glycans, including glycoproteins on the cell surface (Ahmad, 2004), as mentioned in the previous section. In contrast, dimeric galectin-1 is the effective molecule in cell death induction, and the K_d for galectin-1 dimerization is $\sim 7\mu M$ (Perillo et al., 1995). Thus, galectin-3 may be more effective in cross-linking and engaging cell surface apoptosis-inducing receptors.

It is important to point out that while exogenously added galectin-3 induces apoptosis, endogenous galectin-3 expressed in T cells is anti-apoptotic. This was first demonstrated in the human T-cell line Jurkat transfected to express galectin-3 (Yang et al., 1996). The transfectants were found to be more resistant to apoptosis induced by anti-Fas antibody and staurosporine, compared to control transfectants not expressing the protein. Galectin-3 expression in another human T-cell line CEM also resulted in resistance to apoptosis induced by galectin-1 (Hahn et al., 2004) and C(2)-ceramide (Fukumori et al., 2003).

The anti-apoptotic activity of galectin-3 has been demonstrated in a number of other cell types and diverse apoptotic stimuli (reviewed in Liu et al., 2002). Endogenous galectin-3 appears to confer resistance to apoptosis through an intracellular mechanism. The fact that exogenously added galectin-3 kills T cells further supports this notion, because if endogenous galectin-3 in T cells acts extracellularly, expression of galectin-3 would be associated with more apoptosis rather than resistance to apoptosis. Current information suggests that intracellular galectin-3 regulates apoptosis through engagement of apoptosis-regulation pathways (reviewed in Hsu et al., 2006; Liu et al., 2002). Translocation of galectin-3 from the cytosol or the nucleus to mitochondria following exposure to apoptotic stimuli was described (Yu et al., 2002). In addition, galectin-3 was observed to suppress the loss in the mitochondrial membrane potential, suggesting that its anti-apoptotic action may involve interactions with other regulators of apoptosis operating in mitochondria (Matarrese et al., 2000).

Galectin-3 has been found to bind to proteins implicated in regulation of apoptosis, including Bcl-2 (Yang et al., 1996) and synexin (Yu et al., 2002). With regard to the structural requirement, phosphorylation of serine at position 6 is critical (Yoshii et al., 2002), which is also required for export of the protein from the nucleus upon exposure to apoptotic stimuli (Takenaka et al., 2004). In addition, the activity is dependent on the presence of Asn-Trp-Gly-Arg (NWGR) motif in the C-terminal domain (Akahani et al., 1997; Yang et al., 1996). This motif is also contained in the well-characterized members of the Bcl-2 family of apoptosis regulators and is indispensable in their functions (Muchmore et al., 1996).

3.3.2 Control of Cytokine Synthesis

Galectin-3 has been shown to induce cell activation through its multivalent lectin activity, by cross-linking cell surface glycans. With regard to T cells, galectin-3 can induce human Jurkat T cells to produce IL-2 (Hsu et al., 1996) as well as uptake extracellular calcium ion (Dong et al., 1996). Galectin-3 was found to be associated with the T-cell receptor (TCR) complex (Demetriou et al., 2001) in a fashion that is dependent on the glycosylation of TCR controlled by β 1,6-N-acetylglucosaminyltransferase V (Mgat5), an enzyme in the N-glycosylation pathway (Demetriou et al., 2001). The carbohydrate dependence is also supported by the inhibition of the association by lactose. Galectin-3 is a negative regulator of TCR-initiated signal transduction, as inhibition of galectin-3 binding to TCR results in increased TCR clustering and enhanced signaling through TCR (Demetriou et al., 2001).

3.3.3 Regulation of T-Cell Adhesion and Migration

Galectin-3 has been shown to bind to extracellular matrix proteins, including laminin (Hughes, 2001; Kuwabara and Liu, 1996; Massa et al., 1993; Sato et al., 1992), fibronectin (Sato et al., 1992), elastin (Ochieng et al., 1999) and hensin (Hikita et al., 2001), in a carbohydrate-dependent manner. It also interacts with integrins $\alpha_1\beta_1$ (Ochieng et al., 1998) and $\alpha_M\beta_1$ (CD11b/18) (Dong, 1997). Evidence has been provided that galectin-3 can modulate cell adhesion through engaging these extracellular matrix proteins and integrins (reviewed in Hughes, 2001).

There is a report on the involvement of endogenous galectin-3 in adhesion of mouse T cells: adhesion of T cells to dendritic cells or macrophages triggered by ligation of L-selectin is inhibitable by a galectin-3 sugar ligand or anti-galectin-3 antibody (Swarte et al., 1998).

3.4 Modulation of Immune Cell Activation

Galectin-3 has been shown to activate other immune cells. This lectin (1) induces mediator release in both IgE-sensitized and non-sensitized mast cells, possibly by cross-linking FcεRI-bound IgE, FcεRI or both (Frigeri et al., 1993; Zuberi et al., 1994); (2) induces superoxide anion production (Liu et al., 1995) and potentiates

LPS-induced IL-1 production in human peripheral blood monocytes (Jeng et al., 1994); (3) promotes oxidative responses in human peripheral blood neutrophils primed with a variety of agents (Almkvist et al., 2001, 2004; Faldt et al., 2001; Karlsson et al., 1998; Yamaoka et al., 1995); and (4) triggers L-selectin shedding and IL-8 production in both naïve neutrophils and those primed with cytochalasin-B (Nieminen et al., 2005). Additionally, galectin-3 causes apoptosis as well as promotes phagocytic activity in neutrophils (Fernandez et al., 2005). Finally, it induces migration of human monocytes/macrophages and alveolar macrophages, both in vitro and in vivo (Sano et al., 2000). All these activities were demonstrated with exogenously added recombinant galectin-3.

Galectin-3 can also exert a suppressive effect on myeloid cells, as exemplified by inhibition of IL-5 production in human eosinophils, an eosinophilic cell line, human peripheral blood mononuclear cells (PBMC) and an antigen-specific T-cell line (Cortegano et al., 1998).

The above results were demonstrated with exogenously added galectin-3. However, endogenous galectin-3 can function intracellularly without being secreted. For example, galectin-3 has been shown to have a critical role in phagocytosis, as demonstrated by comparing macrophages from galectin-3-deficient ($gal3^{-/-}$) mice and wild-type mice (Sano et al., 2003). The intracellular mechanism is also consistent with the intracellular localization of the protein—galectin-3 is located in the phagosomes of macrophages (Garin et al., 2001; Sano et al., 2003). Studies of mast cells from $gal3^{-/-}$ mice have revealed a role for galectin-3 in regulation of mast cell mediator release and cytokine production, induced by cross-linkage of cell surface IgE receptor (Chen et al., 2006). The mechanisms by which galectin-3 regulates mediator release and cytokine production are not known, but could be in part due to the ability of the protein to control the expression of c-jun-N-terminal kinase-1 (JNK1) (Chen et al., 2006), which is known to play a central role in the signaling pathways leading to production of selected cytokines.

Thus, in vitro studies suggest that galectin-3 can promote immune and inflammatory responses through its functions on cell activation, cell migration or inhibition of apoptosis (thus prolonging the survival of inflammatory cells), although it can also suppress some responses and induce apoptosis. The proinflammatory role has been confirmed by in vivo studies of mouse models using $gal3^{-/-}$ mice. For example, $gal3^{-/-}$ mice exhibited attenuated peritoneal inflammation induced by peritoneal injection of thioglycollate broth (Colnot et al., 1998; Hsu et al., 2000) compared to wild-type mice. In addition, in a mouse model of asthma, ovalbumin-sensitized $gal3^{-/-}$ mice developed lower lung eosinophilia relative to similarly treated wild-type mice following airway ovalbumin challenge (Zuberi et al., 2004). The suppressive effect is best demonstrated by the observation of a reduced eosinophil infiltration in response to airway antigen challenge in rats and mice that received cDNA encoding galectin-3 delivered intranasally (Del Pozo et al., 2002; Lopez et al., 2006). These contrasting results may be explained by the potentiating role for endogenous galectin-3, but the suppressive effect of pharmacological concentrations of galectin-3.

3.5 Galectin-3 in Tumor-Immune Escape

Secretion of galectin-3 by tumor cells under certain conditions have been demonstrated, for example when breast carcinoma cells are treated with the glycoprotein, fetuin (Zhu and Ochieng, 2001), or when cultured adherent breast carcinoma cells are detached from the substratum (Baptiste et al., 2007). Galectin-3 is detectable in the sera and its serum levels are higher in patients with various cancers than in healthy individuals (Iruisci et al., 2000). Galectin-3 released by tumor cells can affect the immune cells in a number of ways. First, the protein can induce apoptosis in T cells and as reported for galectin-1 (Rubinstein et al., 2004), T-cell killing by galectin-3 may result in immune escape by tumor cells expressing the lectin. While this possibility has not been formally demonstrated, it was suggested by the observed correlation between galectin-3 expression in tumor cells and the degree of apoptosis in tumor-associated lymphocytes, in biopsied samples of human melanoma (Zubieta et al., 2006).

As mentioned above, intracellular galectin-3 is anti-apoptotic and this activity has been demonstrated in T cells. This activity has also been amply demonstrated with tumor cells expressing the protein. Thus, the current view is galectin-3 expressed in tumor cells may confer the cells resistance to killing by the host immune response or chemotherapeutic agents. Therefore, while galectin-3 released by tumor cells can kill T cells targeting the tumor, with galectin-3 being released, tumor cells would contain a lower amount of the protein and actually might become more sensitive to killing by cytotoxic T cells. The battle between the tumor cells and T cells may thus depend on the amount of galectin-3 in tumor cells and T cells, respectively, the amount released by tumor cells and the efficiency of cell killing by released galectin-3.

Galectin-3 released by tumor cells can also affect the inflammatory responses through other mechanisms. These include chemoattraction of macrophages, activation of various immune and inflammatory cells and modulation of the adhesive properties of these cells. On the basis of many studies, extracellular galectin-3 released by tumor is expected to potentiate the inflammatory response, especially from the point of view of galectin-3's ability to chemoattract macrophages and activate various leukocytes. Since the inflammatory response can potentially promote tumor development and progression, the overall effect of extracellular galectin-3 may benefit the tumors through mechanisms distinct from tumor-immune escape. It should be pointed out, however, that the effect of extracellular galectin-3 on cell adhesion can be either positive or negative, and whether galectin-3 released by tumor cells would enhance or suppress adhesion of different immune and inflammatory cells is yet to be elucidated. In the mean time, it is also to be recognized that galectin-3 can affect the tumor cell adhesion and migration and some through intracellular mechanisms (reviewed in Liu and Rabinovich, 2005). Thus, on one hand, galectin-3 released by tumor cells may affect how tumor cells adhere and migrate through the extracellular environment; on the other hand, it is possible the biological behavior of tumor cells might be affected by the decrease in the amount of galectin-3 inside the cells, as a result of release of the protein.

Finally, it should be pointed out that, similar to galectin-1, galectin-3 can also promote endothelial cell morphogenesis and angiogenesis (Nangia-Makker et al., 2000), suggesting another mechanism by which this protein can influence tumor progression.

4 Conclusions: Galectins as Targets for Anticancer Therapies: Current Status, Future Directions and Unresolved Issues

Given their contribution to tumor progression, galectins-1 and -3 have emerged as a promising molecular target for cancer therapy. Therefore, a current challenge is the development of potent and selective small inhibitors of galectins to delay tumor progression and stimulate anti-tumor T-cell responses. In fact, molecules with such properties have already been designed for galectins-1 and -3; these include synthetic glycoamines, glycodendrimers or natural products such as pectins (Andre et al., 2001; Glinsky et al., 1996; Nangia-Makker et al., 2002; Rabinovich et al., 2006; Tejler et al., 2006; Ingrassia et al., 2006). The emergence of novel high-throughput screening platforms, including glycan arrays (Blixt et al., 2004), will facilitate the characterization and profiling of the specificity of a diverse range of galectins and the identification of more specific and potent inhibitors of galectin–carbohydrate interactions. In addition, studies are being conducted in order to elucidate the anti-tumor effects of neutralizing anti-galectin antibodies and small molecule inhibitors in different pre-clinical models.

Although significant progress has been made on galectin research during the past decade, there are a number of aspects of galectin research remaining to be addressed. First, there is still scarce information regarding the molecular mechanisms and exogenous stimuli that regulate galectin expression in tumor cells. In this regard, evidence indicates that galectin-1 expression in tumor cells can be modulated by differentiating chemotherapeutic and antimetastatic agents including retinoic acid, transforming growth factor-β and cyclophosphamide (Daroqui et al., 2007; Lu et al., 2000; Zacarias Fluck et al., 2007). Galectin-3 expression is upregulated in a wide range of neoplasms, including those induced by virus, ultraviolet light or chemicals (Crittenden et al., 1984; Hébert and Monsigny, 1993, 1994; Raz et al., 1987). Galectin-3 expression is induced by transactivating proteins (such as HTLV-1 tax protein and hepatitis B virus X protein) (Hsu et al., 1996, 1999). Finally, future studies should be conducted to examine the cross-talk between galectins and other mechanisms of tumor-immune escape including PD-1/PD-L1, expansion and recruitment of regulatory T cells and indoleamine 2,3-deoxigenease expression.

In conclusion, although a great deal remains to be learned about the functional roles of galectins-1 and -3 in the tumor microenvironment, sufficient evidence has been accumulated demonstrating their importance, and new therapeutic approaches directed toward modulating their activities are being developed.

Acknowledgments We thank M.A. Toscano for her generous support and all the members of our laboratories for productive discussions. We apologize that we could not cite many excellent studies because of space limitations.

Work in G.A.R's laboratory is supported by grants from the Cancer Research Institute "Elaine Shepard Investigator Award" (USA), Mizutani Foundation for Glycoscience (Japan), Agencia de Promoción Científica y Tecnológica (PICT 2003-05-13787) (Argentina), Fundación Sales (Argentina), Fundación Bunge & Born (Argentina), Fundación Florencio Fiorini (Argentina) and the University of Buenos Aires (UBACYT-M091) (Argentina). G.A.R. is a member of the National Research Council (CONICET, Argentina). Work in F.-T.L laboratory is supported by grants from the NIH (RO1AI20958 and RO1AI39620).

References

Ahmad, N., Gabius, H. J., Andre, S., Kaltner, H., Sabesan, S., Roy, R., Liu, B., Macaluso, F., and Brewer, C. F. (2004). Galectin-3 precipitates as a pentamer with synthetic multivalent carbohydrates and forms heterogeneous cross-linked complexes. *J Biol Chem* 279:10841–10847.

Akahani, S., Nangia-Makker, P., Inohara, Kim, H. R. C., and Raz, A. (1997). Galectin-3: a novel antiapoptotic molecule with a functional BH1 (NWGR) domain of Bcl-2 family. *Cancer Res* 57:5272–5276.

Almkvist, J., Dahlgren, C., Leffler, H., and Karlsson, A. (2002). Activation of the neutrophil nicotinamide adenine dinucleotide phosphate oxidase by galectin-1, *J Immunol* 168:4034–4041.

Almkvist, J., Dahlgren, C., Leffler, H., and Karlsson, A. (2004). Newcastle disease virus neuraminidase primes neutrophils for stimulation by galectin-3 and formyl-Met-Leu-Phe. *Exp Cell Res* 298:74–82.

Almkvist, J., Faldt, J., Dahlgren, C., Leffler, H. and Karlsson, A. (2001). Lipopolysaccharide-induced gelatinase granule mobilization primes neutrophils for activation by galectin-3 and formylmethionyl-Leu-Phe. *Infect Immun* 69:832–837.

Amano, M., Galvan, M., He, J., Baum, L. G. (2003). The ST6Gal I sialyltransferase selectively modifies N-glycans on CD45 to negatively regulate galectin-1-induced CD45 clustering, phosphatase modulation and T cell death. *J Biol Chem* 278:7469–7475.

Andre, S., Pieters, R. J., Vrasidas, I., Kaltner, H., Kuwabara, I., Liu, F-T., Liskamp, R. M., and Gabius, H.-J. (2001). Wedgelike glycodendrimers as inhibitors of binding of mammalian galectins to glycoproteins, lactose, maxiclusters, and cell surface glycoconjugates. *Chembiochem* 2:822–830.

Baptiste, T. A., James, A., Saria, M. and Ochieng, J. (2007). Mechano-transduction mediated secretion and uptake of galectin-3 in breast carcinoma cells: Implications in the extracellular functions of the lectin. *Exp Cell Res* 313:652–664.

Barrionuevo, P., Beigier-Bompadre, M., Ilarregui, J. M., Toscano, M. A., Bianco, G. A., Isturiz, M. A., and Rabinovich, G. A. (2007). A novel function for galectin-1 at the crossroad of innate and adaptive immunity: Galectin-1 regulates monocyte/macrophage physiology through a nonapoptotic ERK-dependent pathway, *J Immunol* 178:436–445.

Baum, L. G., Blackall, D. P., Arias-Magallano, S., Nanigian, D., Uh, S. Y., Browne, J. M., Hoffmann, D., Emmanouilides, C. E., Territo, M. C., and Baldwin, G. C. (2003) Amelioration of graft versus host disease by galectin-1, *Clin Immunol* 109:295–307.

Blaser, C., Kaufmann, M., Muller, C., Zimmermann, C., Wells, V., Malluci, L., and Pircher, H. (1998). β-galactoside-binding protein secreted by activated T cells inhibits antigen-induced proliferation of T cells. *Eur J Immunol* 28:2311–2319.

Blixt, O., Head, S., Mondala, T., Scanlan, C., Huflejt, M. E., Alvarez, R., Bryan, M. C., Fazio, F., Calarese, D., Stevens, J., Razi, N., Stevens, D. J., Skehel, J. J., van Die, I., Burton, D. R., Wilson, I. A., Cummings, R., Bovin, N., Wong, C. H., and Paulson, J. C. (2004). Printed covalent glycan array for ligand profiling of diverse glycan-binding proteins. *Proc Natl Acad Sci USA* 101:17933–17938.

Brewer, C. (2002). Binding and cross-linking properties of galectins, *Biochim Biophys Acta* 1572:255–262.

Cabrera, P. V., Amano, M., Mitoma, J., Chan, J., Said, J., Fukuda, M., and Baum, L. G. (2006). Haploinsufficiency of C2GnT-1 glycosyltransferase renders T lymphoma cells resistant to cell death. *Blood* 108:2399–2406.

Camby, I., Mercier, M. L., Lefranc, F., and Kiss. R. (2006). Galectin-1: a small protein with major functions. *Glycobiology* 16:137–157.

Camby, I., Belot, N., Rorive, S., Lefranc, F., Maurage, C. A., Lahm, H., Kaltner, H., Hadari, Y., Ruchoux, M. M., Brotchi, J., Zick, Y., Salmon, I., Gabius, H. J., and Kiss, R. (2001) Galectins are differentially expressed in supratentorial pilocytic atrocytomas, astrocytomas, anaplastic astrocytomas and glioblastomas, and significantly modulate tumor astrocyte migration. *Brain Pathol* 11:12–26.

Chen, H. Y., Sharma, B. B., Yu, L., Zuberi, R., Weng, I. C., Kawakami, Y., Kawakami, T., Hsu, D. K., and Liu, F. T. (2006). Role of galectin-3 in mast cell functions: Galectin-3-deficient masT cells exhibit impaired mediator release and defective JNK expression. *J Immunol* 177:4991–4997.

Chung, C. D., Patel, V. P., Moran, M., Lewis, L. A., and Miceli, M. C. (2000). Galectin-1 induces partial TCR ζ-chain phosphorylation and antagonizes processive TCR signal transduction. *J Immunol* 165:3722–3729.

Colnot, C., Ripoche, M. A., Milon, G., Montagutelli, X., Crocker, P. R., and Poirier, F. (1998). Maintenance of granulocyte numbers during acute peritonitis is defective in galectin-3-null mutant mice. *Immunology* 94:290–296.

Cooper, D. N. (2002). Galectinomics: finding themes in complexity, *Biochim Biophys Acta* 1572:209–231.

Cooper, D. N., and Barondes, S. H. (1990). Evidence for export of a muscle lectin from cytosol to extracellular matrix and for a novel secretory mechanism. *J Cell Biol* 110:1681–1691.

Correa, S. G., Sotomayor, C. E., Aoki, M. P., Maldonado, C. A., and Rabinovich, G. A. (2003). Opposite effects of galectin-1 on alternative metabolic pathways of L-arginine in resident, inflammatory, and activated macrophages, *Glycobiology* 13:119–128.

Cortegano, I., del Pozo, V., Cardaba, B., de Andres, B., Gallardo, S., del Amo, A., Arrieta, I., Jurado, A., Palomino, P., Liu, F. T., and Lahoz, C. (1998). Galectin-3 down-regulates IL-5 gene expression on different cell types. *J Immunol* 161:385–389.

Crittenden, S. L., Roff, C. F., and Wang, J. L. (1984). Carbohydrate-binding protein 35: identification of the galactose- specific lectin in various tissues of mice. *Mol Cell Biol* 4: 1252–1259.

Dagher, S. F., Wang, J. L., and Patterson, R. J. (1995). Identification of galectin-3 as a factor in pre-mRNA splicing. *Proc Natl Acad Sci USA* 92:1213–1217.

Danguy, A., Camby, I., and Kiss, R. (2002). Galectins and cancer. *Biochim Biophys Acta* 1572: 285–293.

Daroqui, M. C., Ilarregui, J. M., Rubinstein, N., Salatino, M., Toscano, M. A., Vazquez, P., Bakin, A., Puricelli, L., Bal de Kier Joffe, E., and Rabinovich, G. A. (2007). Regulation of galectin-1 expression by transforming growth factor β 1: implications for tumor-immune escape. *Cancer Immunol Immunother* 56:491–499.

Davidson, P. J., Davis, M. J., Patterson, R. J., Ripoche, M. A., Poirier, F., and Wang, J. L. (2002). Shuttling of galectin-3 between the nucleus and cytoplasm. *Glycobiology* 12: 329–337.

Del Pozo, V., Rojo, M., Rubio, M. L., Cortegano, I., Cardaba, B., Gallardo, S., Ortega, M., Civantos, E., Lopez, E., Martin-Mosquero, C., Peces-Barba, G., Palomino, P., Gonzalez-Mangado, N., and Lahoz, C. (2002). Gene therapy with galectin-3 inhibits bronchial obstruction and inflammation in antigen-challenged rats through interleukin-5 gene downregulation. *Am J Respir Crit Care Med* 166:732–737.

Delacour, D., Cramm-Behrens, C. I., Drobecq, H., Le Bivic, A., Naim, H. Y., and Jacob, R. (2006). Requirement for galectin-3 in apical protein sorting. *Curr Biol* 16:408–414.

Demetriou, M., Granovsky, M., Quaggin, S., and Dennis, J. W. (2001). Negative regulation of T-cell activation and autoimmunity by Mgat5 N-glycosylation. *Nature* 409:733–779.

Dong, S. and Hughes, R. C. (1996). Galectin-3 stimulates uptake of extracellular Ca^{2+} in human Jurkat T-cells. *FEBS Lett* 395:165–169.
Dong, S. and Hughes, R. C. (1997). Macrophage surface glycoproteins binding to galectin-3 (Mac-2-antigen). *Glycoconjugate J* 14:267–274.
Elola, M. T., Wolfenstein-Todel, C., Troncoso, M. F., Vasta, G., and Rabinovich, G. A. (2007). Galectins: matricellular glycan-binding proteins linking cell adhesion, migration, and survival. *Cell Mol Life Sci* 64:1679–1700.
Endharti, A. T., Zhou, Y. W., Nakashima, I., and Suzuki, H. (2005). Galectin-1 supports survival of naive T cells without promoting cell proliferation, *Eur J Immunol* 35:86–97.
Faldt, J., Dahlgren, C., Ridell, M. and Karlsson, A. (2001). Priming of human neutrophils by mycobacterial lipoarabinomannans: role of granule mobilisation. *Microbes Infect* 3:1101–1109.
Fernandez, G. C., Ilarregui, J. M., Rubel, C. J., Toscano, M. A., Gomez, S. A., Beigier Bompadre, M., Isturiz, M. A., Rabinovich, G. A., and Palermo, M. S. (2005). Galectin-3 and soluble fibrinogen act in concert to modulate neutrophil activation and survival: involvement of alternative MAPK pathways. *Glycobiology* 15:519–527.
Frigeri, L. G., Zuberi, R. I., and Liu, F. T. (1993). εBP, a β-galactoside-binding animal lectin, recognizes IgE receptor (FceRI) and activates mast cells. *Biochemistry* 32:7644–7649.
Fuertes, M. B., Molinero, L. L., Toscano, M. A., Ilarregui, J. M., Rubinstein, N., Fainboim, L., Zwirner, N. W., and Rabinovich, G. A. (2004). Regulated expression of galectin-1 during T-cell activation involves Lck and Fyn kinases and signaling through MEK1/ERK, p38 MAP kinase and p70S6 kinase. *Mol Cell Biochem* 267:177–185.
Fukumori, T., Takenaka, Y., Yoshii, T., Kim, H. R., Hogan, V., Inohara, H., Kagawa, S., and Raz, A. (2003). CD29 and CD7 mediate galectin-3-induced type II T-cell apoptosis. *Cancer Res* 63:8302–8311.
Fulcher, J. A., Hashimi, S. T., Levroney, E. L., Pang, M., Gurney, K. B., Baum, L. G., and Lee, B. Galectin-1-matured human monocyte-derived dendritic cells have enhanced migration through extracellular matrix, *J Immunol* 177:216–226.
Gabius, H. J., Engelhardt, R., and Cramer, F. (1986). Endogenous tumor lectins: overview and perspectives. *Anticancer Res* 6:573–578.
Galvan, M., Tsuboi, S., Fukuda, M., and Baum, L. G. (2000). Expression of a specific glycosyltransferase enzyme regulates T cell death mediated by galectin-1. *J Biol Chem* 275:16730–16737.
Garin, M. I., Chu, C. C., Golshayan, D., Cernuda-Morollon, E., Wait, R., Lechler, R. I. (2007). Galectin-1: a key effector of regulation mediated by $CD4^+$ $CD25^+$ T cells. *Blood* 109:2058–2065.
Garin, J., Diez, R., Kieffer, S., Dermine, J. F., Duclos, S., Gagnon, E., Sadoul, R.,. Rondeau, C., and Desjardins, M. (2001). The phagosome proteome: insight into phagosome functions. *J Cell Biol* 152:165–180.
Gauthier, L., Rossi, B., Roux, F., Termine, E., and Schiff, C. (2002). Galectin-1 is a stromal cell ligand of the pre-B cell receptor (BCR) implicated in synapse formation between pre-B and stromal cells and in pre-BCR triggering. *Proc Natl Acad Sci USA* 99:13014–13019.
Gil, C. D., Cooper, D., Rosignoli, G., Perretti, M., and Oliani, S. M. (2006). Inflammation-induced modulation of cellular galectin-1 and -3 expression in a model of rat peritonitis. *Inflamm Res* 55:99–107.
Glinsky, G. V., Price, J. E., Glinsky, V. V., Mossine, V. V., Kiriakova, G. and Metcalf, J. B. (1996). Inhibition of human breast cancer metastasis in nude mice by synthetic glycoamines. *Cancer Res* 56:5319–5324.
Gordon, S. (2003). Alternative activation of macrophages. *Nat Rev Immunol* 3, 23–35.
Hahn, H. P., Pang, M., He, J., Hernandez, J. D., Yang, R. Y., Li, L. Y., Wang, X., Liu, F. T., and Baum, L. G. (2004). Galectin-1 induces nuclear translocation of endonuclease G in caspase- and cytochrome c-independent T cell death. *Cell Death Differ* 11:1277–1286.
He, J., and Baum, L. G. (2006). Endothelial cell expression of galectin-1 induced by prostate cancer cells inhibits T-cell transendothelial migration. *Lab Invest* 86:578–590.

Hébert, E. and Monsigny, M. (1993). Oncogenes and expression of endogenous lectins and glycoconjugates. *Biol Cell* 79:97–109.
Hébert, E. and Monsigny, M. (1994). Galectin-3 mRNA level depends on transformation phenotype in *ras*-transformed NIH 3T3 cells. *Biol Cell* 81:73–76.
Hikita, C., Vijayakumar, S., Takito, J., Erdjument-Bromage, H., Tempst, P., and Al-Awqati, Q. (2001). Induction of terminal differentiation in epithelial cells requires polymerization of hensin by galectin 3. *J Cell Biol* 151:1235–1146.
Hirabayashi, J., Hashidate, T., Arata, Y., Nishi, N., Nakamura, T., Hirashima, M., Urashima, T., Oka, T., Futai, M., Muller, W. E., Yagi, F., and Kasai, K. I. (2002). Oligosaccharide specificity of galectins: a search by frontal affinity chromatography. *Biochim Biophys Acta* 1572: 232–254.
Hsu, D. K., Dowling, C. A., Jeng, K.-C. G., Chen, J.-T., Yang, R.-Y. and Liu, F.-T. (1999). Galectin-3 expression is induced in cirrhotic liver and hepatocellular carcinoma. *Int J Cancer* 81: 519–526.
Hsu, D. K., Hammes, S. R., Kuwabara, I., Greene, W. C., and Liu, F. T. (1996). Human T lymphotropic virus-1 infection of human T lymphocytes induces expression of the β-galactose-binding lectin, galectin-3. *Am J Pathol* 148:1661–1670.
Hsu, D. K., Yang, R. Y., and Liu, F. T. (2006). Galectins in apoptosis. *Methods Enzymol* 417: 256–273.
Hsu, D. K., Yang, R. Y., Yu, L., Pan, Z., Salomon, D. R., Fung-Leung, W. P., and Liu, F. T. (2000). Targeted disruption of the galectin-3 gene results in attenuated peritoneal inflammatory responses. *Am J Pathol* 156:1073–1083.
Hsu, D. K., Zuberi, R. and Liu, F. T. (1992). Biochemical and biophysical characterization of human recombinant IgE-binding protein, an S-type animal lectin. *J Biol Chem* 267: 14167–14174.
Hughes, R. C. (1999). Secretion of the galectin family of mammalian carbohydrate-binding proteins. *Biochim Biophys Acta* 1473:172–185.
Hughes, R. C. (2001). Galectins as modulators of cell adhesion. *Biochimie* 83:667–676.
Ingrassia, L., Camby, I., Lefranc, F., Mathieu, V., Nshimyumukiza, P., Darro, F., and Kiss, R. (2006). Anti-galectin compounds as potential anti-cancer drugs. *Curr Med Chem* 13: 3513–3527.
Ion, G., Fajka-Boja, R., Kovacs, F., Szebeni, G., Gombos, I., Czibula, A., Matko, J., and Monostori, E. (2006). Acid sphingomyelinase-mediated release of ceramide is essential to trigger the mitochondrial pathway of apoptosis by galectin-1. *Cell Signal* 18:1887–1896.
Ion, G., Fajka-Boja, R., Toth, G. K., Caron, M., and Monostori, E. (2005). Role of p56lck and ZAP70-mediated tyrosine phosphorylation in galectin-1-induced cell death. *Cell Death Differ* 12:1145–1147.
Iruisci, I., Tinari, N., Natoli, C., Angelucci, D., Cianchetti, E., and Iacobelli, S. (2000). Concentrations of galectin-3 in the sera of normal controls and cancer patients. *Clin Cancer Res* 6: 1389–1393.
Jeng, K. C. G., Frigeri, L. G., and Liu, F. T. (1994). An endogenous lectin, galectin-3 (eBP/Mac-2), potentiates IL-1 production by human monocytes. *Immunol Lett* 42:113–116.
Karlsson, A., Follin, P., Leffler, H., and Dahlgren, C. (1998). Galectin-3 activates the NADPH-oxidase in exudated but not peripheral blood neutrophils. *Blood* 91:3430–3438.
Kiss, J. Kunstar, A., Fajka-Boja, R., Dudics, V., Tovari, J., Legradi, A., Monostori, E., and Uher, F. (2007). A novel anti-inflammatory function of human galectin-1: inhibition of hematopoietic progenitor cell mobilization. *Exp Hematol* 35:305–313.
Kopitz, J., von Reitzenstein, C., Andre, S., Kaltner, H., Uhl, J., Ehemann, V., Cantz, M., and Gabius, H. J. (2001). Negative regulation of neuroblastoma cell growth by carbohydrate-dependent surface binding of galectin-1 and functional divergence from galectin-3. *J Biol Chem* 276:35917–35923.
Kuwabara, I., and Liu, F. T. (1996). Galectin-3 promotes adhesion of human neutrophils to laminin. *J Immunol* 156:3939–3944.

La, M., Cao, T. V., Cerchiaro, G., Chilton, K., Hirabayashi, J., Kasai, K., Oliani, S. M., Chernajovsky, Y., and Perretti, M. (2003). A novel biological activity for galectin-1: inhibition of leukocyte–endothelial cell interactions in experimental inflammation. *Am J Pathol* 163:1505–1515.

Lahm, H., Andre, S., Hoeflich, A., Kaltner, H., Siebert, H. C., Sordat, B., von der Lieth, C. W., Wolf, E., and Gabius, H.-J. (2004). Tumor galectinology: insights into the complex network of a family of endogenous lectins. *Glycoconj J* 20:227–238.

Le, Q. T., Shi, G., Cao, H., Nelson, D. W., Wang, Y., Chen, E. Y., Zhao, S., Kong, C., Richardson, D., O'Byrne, K. J., Giaccia, A. J., and Koong, A. C. (2005). Galectin-1: a link between tumor hypoxia and tumor immune privilege. *J Clin Oncol* 23:8932–8941.

Liu, F. T. (2005). Regulatory roles of galectins in the immune response. *Int Arch Allergy Immunol* 136:385–400.

Liu, F. T., Hsu, D. K., Zuberi, R. I., Kuwabara, I., Chi, E. Y., and Henderson, W. R., Jr (1995). Expression and function of galectin-3, a β-galactoside-binding lectin, in human monocytes and macrophages. *Am J Pathol* 147:1016–1029.

Liu, F. T., Patterson, R. J., and Wang, J. L. (2002). Intracellular functions of galectins. *Biochim Biophys Acta* 1572:263–273.

Liu, F. T., and Rabinovich, G. A. (2005). Galectins as modulators of tumour progression. *Nat Rev Cancer* 5:29–41.

Lopez, E., Del Pozo, V., Miguel, T., Sastre, B., Seoane, C., Civantos, E., Llanes, E., Baeza, M. L., Palomino, P., Cardaba, B., Gallardo, S., Manzarbeitia, F., Zubeldia, J. M., and Lahoz, C. (2006). Inhibition of chronic airway inflammation and remodeling by galectin-3 gene therapy in a murine model. *J Immunol* 176:1943–1950.

Lu, Y., Lotan, D., and Lotan, R. (2000). Differential regulation of constitutive and retinoic acid-induced galectin-1 gene transcription in murine embryonal carcinoma and myoblastic cells. *Biochim Biophys Acta* 1491:13–19.

Mahoney, S., Wilkinson, M., Smith, S., and Haynes, L. W. (2000). Stabilization of neurites in cerebellar granule cells by transglutaminase activity: identification of midkine and galectin-3 as substrates. *Neuroscience* 101:141–155.

Massa, S. M., Cooper, D. N. W., Leffler, H., and Barondes, S. H. (1993). L-29, an endogenous lectin, binds to glycoconjugate ligands with positive cooperativity. *Biochemistry* 32:260–267.

Matarrese, P., Tinari, A., Mormone, E., Bianco, G. A., Toscano, M. A., Ascione, B., Rabinovich, G. A., and Malorni, W. (2005). Galectin-1 sensitizes resting human T lymphocytes to Fas (CD95)-mediated cell death via mitochondrial hyperpolarization, budding and fission. *J Biol Chem* 280:6969–6985.

Matarrese, P., Tinari, N., Semeraro, M. L., Natoli, C., Iacobelli, S., and Malorni, W. (2000). Galectin-3 overexpression protects from cell damage and death by influencing mitochondrial homeostasis. *FEBS Lett* 473:311–315.

Mehul, B., Bawumia, S., and Hughes, R. C. (1995). Cross-linking of galectin 3, a galactose-binding protein of mammalian cells, by tissue-type transglutaminase. *FEBS Lett* 360:160–164.

Muchmore, S. W., Sattler, M., Liang, H., Meadows, R. P., Harlan, J. E., Yoon, H. S., Nettesheim, D., Chang, B. S., Thompson, C. B., Wong, C. L., Ng, S. L., and Fesik, S. W. (1996). X-ray and NMR structure of human Bcl-xL, an inhibitor of programmed cell death. *Nature* 381:335–341.

Nangia-Makker, P., Hogan, V., Honjo, Y., Baccarini, S., Tait, L., Breaslier, R., and Raz, A. (2002). Inhibition of human cancer cell growth and metastasis in nude mice by oral intake of modified citrus pectin. *J Natl Cancer Inst* 94:854–1862.

Nangia-Makker, P., Honjo, Y., Sarvis, R., Akahani, S., Hogan, V., Pienta, K. J., and Raz, A. (2000). Galectin-3 induces endothelial cell morphogenesis and angiogenesis. *Am J Pathol* 156:899–909.

Nickel, W. (2005). Unconventional secretory routes: direct protein export across the plasma membrane of mammalian cells. *Traffic* 6:607–614.

Nieminen, J., Kuno, A., Hirabayashi, J., and Sato, S. (2007). Visualization of galectin-3 oligomerization on the surface of neutrophils and endothelial cells using fluorescence resonance energy transfer. *J Biol Chem* 282:1374–1383.

Nieminen, J., St-Pierre, C., and Sato, S. (2005). Galectin-3 interacts with naive and primed neutrophils, inducing innate immune responses. *J Leukoc Biol* 78:1127–1135.

Ochieng, J., Leite-Browning, M. L., and Warfield, P. (1998). Regulation of cellular adhesion to extracellular matrix proteins by galectin-3. *Biochem Biophys Res Commun* 246:788–791.

Ochieng, J., Warfield, P., Green-Jarvis, B., and Fentie, I. (1999). Galectin-3 regulates the adhesive interaction between breast carcinoma cells and elastin. *J Cell Biochem* 75:505–514.

Ohtsubo, K., and Marth, J. D. (2006). Glycosylation in cellular mechanisms of health and disease. *Cell* 126:855–867.

Pace, K. E., Lee, C., Stewart, P. L., and Baum, L. G. (1999). Restricted receptor segregation into membrane microdomains occurs on human T cells during apoptosis induced by galectin-1. *J Immunol* 163:3801–3811.

Perillo, N. L., Pace, K. E., Seilhamer, J. J., and Baum, L. G. (1995). Apoptosis of T cells mediated by galectin-1. *Nature* 378:736–739.

Perillo, N. L., Uittenbogaart, C. H., Nguyen, J. T., and Baum, L. G. (1997). Galectin-1, an endogenous lectin produced by thymic epithelial cells, induces apoptosis of human thymocytes. *J Exp Med* 97:1851–1858.

Perone, M. J., Larregina, A. T., Shufesky, W. J., Papworth, G. D., Sullivan, M. L., Zahorchak, A. F., Stolz, D. B., Baum, L. G., Watkins, S. C., Thomson, A. W., and Morelli, A. E. (2006). Transgenic galectin-1 induces maturation of dendritic cells that elicit contrasting responses in naïve and activated T cells. *J Immunol* 176:7207–7220.

Rabinovich, G. A., Alonso, C. R., Sotomayor, C. E., Durand, S., Bocco, J. L., and Riera, C. M. (2000b). Molecular mechanisms implicated in galectin-1-induced apoptosis: activation of the AP-1 transcription factor and downregulation of Bcl-2. *Cell Death Differ* 7:747–753.

Rabinovich, G. A., Ariel, A., Hershkoviz, R., Hirabayashi, J., Kasai, K. I., and Lider, O. (1999a). Specific inhibition of T-cell adhesion to extracellular matrix and pro-inflammatory cytokine secretion by human recombinant galectin-1. *Immunology* 97:100–106.

Rabinovich, G. A., Baum, L. G., Tinari, N., Paganelli, R., Natoli, C., Liu, F. T., and Iacobelli, S. (2002a). Galectins and their ligands: amplifiers, silencers or tuners of the inflammatory response. *Trends Immunol* 23:313–320.

Rabinovich, G. A., Correa, S. G., Bianco, I., Riera, C. M., and Sotomayor, C. E. (2000a). Evidence of a role for galectin-1 in acute inflammation. *Eur J Immunol* 30:1331–1339.

Rabinovich, G. A., Cumashi, A., Bianco, G. A., Ciavardelli, D., Iurisci, I., D'Egidio, M., Piccolo, E., Tinari, N., Nifantiev, N., and Iacobelli, S. (2006). Synthetic lactulose amines: novel class of anticancer agents that induce tumor-cell apoptosis and inhibit galectin-mediated homotypic cell aggregation and endothelial cell morphogenesis. *Glycobiology* 16:210–220.

Rabinovich, G. A., Daly, G., Dreja, H., Tailor, H., Riera, C. M, Hirabayashi, J., and Chernajovsky, Y. (1999b). Recombinant galectin-1 and its genetic delivery suppress collagen-induced arthritis via T cell apoptosis. *J Exp Med* 190:385–397.

Rabinovich, G. A., Gabrilovich, D. and Sotomayor, E. M. (2007). Immunosuppressive strategies that are mediated by tumor cells. *Annu Rev Immunol* 25:267–296.

Rabinovich, G. A., Modesti, N. M, Castagna, L., Landa, C. A, Riera, C. M, Sotomayor, C. E. (1997). Specific inhibition of lymphocyte proliferation and induction of apoptosis by CLL-I, a beta-galactoside-binding lectin. *J Biochem* 122:365–373.

Rabinovich, G. A., Modesti, N. M, Castagna, L., Todel, C. W, Riera, C. M, Sotomayor, C. E. (1998). Activated rat macrophages produce a galectin-1-like protein that induces apoptosis of T cells: biochemical and functional characterization. *J Immunol* 160:4831–4840.

Rabinovich, G. A., Ramhorst, R. E., Rubinstein, N., Corigliano, A., Daroqui, M. C., Kier-Joffe, E. B., and Fainboim, L. (2002b). Induction of allogeneic T-cell hyporesponsiveness by galectin-1-mediated apoptotic and non-apoptotic mechanisms. *Cell Death Differ* 9:661–670.

Raz, A., Meromsky, L., and Lotan, R. (1986). Differential expression of endogenous lectins on the surface of nontumorigenic, tumorigenic and metastatic cells. *Cancer Res* 46:3667–3672.

Raz, A., Meromsky, L., Zvibel, I., and Lotan, R. (1987). Transformation-related changes in the expression of endogenous cell lectins. *Int J Cancer* 39:353–360.

Rubinstein, N., Alvarez, M., Zwirner, N. W., Toscano, M. A., Ilarregui, J. M., Bravo, A., Mordoh, J., Fainboim, L., Podhajcer, O. L., and Rabinovich, G. A. (2004). Targeted inhibition of galectin-1 gene expression in tumor cells results in heightened T cell-mediated rejection; a potential mechanism of tumor-immune privilege. *Cancer Cell* 5:241–251.

Sano, H., Hsu, D. K., Apgar, J. R., Yu, L., Sharma, B. B., Kuwabara, I., Izui, S., and Liu, F. T. (2003). Critical role of galectin-3 in phagocytosis by macrophages. *J Clin Invest* 112: 389–397.

Sano, H., Hsu, D. K., Yu, L., Apgar, J. R., Kuwabara, I., Yamanaka, T., Hirashima, M., and Liu, F. T. (2000). Human galectin-3 is a novel chemoattractant for monocytes and macrophages. *J Immunol* 165:2156–2164.

Santucci, L., Fiorucci, S., Cammilleri, F., Servillo, G., Federici, B., and Morelli, A. (2000). Galectin-1 exerts immunomodulatory and protective effects on concanavalin A-induced hepatitis in mice. *Hepatology* 31:399–406.

Santucci, L., Fiorucci, S., Rubinstein, N., Mencarelli, A., Palazetti, B., Federici, B., Rabinovich, G. A., and Morelli, A.. (2003). Galectin-1 suppresses experimental colitis in mice. *Gastroenterology* 124:1381–1394.

Sato, S. and Hughes, R. C. (1992). Binding specificity of a baby hamster kidney lectin for H type I and II chains, polylactosamine glycans, and appropriately glycosylated forms of laminin and fibronectin. *J Biol Chem* 267:6983–6990.

Sparrow, C. P., Leffler, H. and Barondes, S. H. (1987). Multiple soluble b-galactoside-binding lectins from human lung. *J Biol Chem* 262:7383–7390.

Stillman, B. N., Hsu, D. K., Pang, M., Brewer, C. F., Johnson, P., Liu, F. T., and Baum, L. G. (2006). Galectin-3 and galectin-1 bind distinct cell surface glycoprotein receptors to induce T cell death. *J Immunol* 176:778–789.

Stowell, S. R., Karmakar, S., Stowell, C. J., Dias-Baruffi, M., McEver, R. P., and Cummings, R. D. (2007). Human galectin-1, -2, and -4 induce surface exposure of phosphatidylserine in activated human neutrophils but not in activated T cells. *Blood* 109:219–227.

Sugimoto, N., Oida, T., Hirota, K., Nakamura, K., Nomura, T., Uchiyama, T., and Sakaguchi, S. (2006). Foxp3-dependent and -independent molecules specific for $CD25^+$ $CD4^+$ natural regulatory T cells revealed by DNA microarray analysis. *Int Immunol* 18:1197–1209.

Swarte, V. V., Mebius, R. E., Joziasse, D. H., Van den Eijnden, D. H., and Kraal, G. (1998). Lymphocyte triggering via L-selectin leads to enhanced galectin-3-mediated binding to dendritic cells. *Eur J Immunol* 28:2864–2871.

Takenaka, Y., Fukumori, T., Yoshii, T., Oka, N., Inohara, H., Kim, H. R., Bresalier, R. S., and Raz, A. (2004). Nuclear export of phosphorylated galectin-3 regulates its antiapoptotic activity in response to chemotherapeutic drugs. *Mol Cell Biol* 24:4395–4406.

Tejler, J., Tullberg, E., Frejd, T., Leffler, H., and Nilsson, U. J. (2006). Synthesis of multivalent lactose derivatives by 1,2-dipolar cycloadditions: selective galectin-1 inhibition. *Carbohydr Res* 341:1353–1362.

Thijssen, V. L., Postel, R., Brandwijk, R. J., Dings, R. P., Nesmelova, I., Satijn, S., Verhofstad, N., Nakabeppu, Y., Baum, L. G., Bakkers, J, Mayo, K. H., Poirier, F., and Griffioen, A. W. (2006). Galectin-1 is essential in tumor angiogenesis and is a target for antiangiogenesis therapy, *Proc Natl Acad Sci USA*, 103:15975–15980.

Tinari, N., Kuwabara, I., Huflejt, M. E., Shen, P. F., Iacobelli, S., and Liu, F. T. (2001). Glycoprotein 90K/Mac-2BP interacts with galectin-1 and mediates galectin-1-induced cell aggregation. *Int J Cancer* 91:167–172.

Toscano, M. A., Commodaro, A. G., Bianco, G. A., Ilarregui, J. M., Liberman, A., Serra, H. M., Hirabayashi, J., Rizzo, L. V., and Rabinovich, G. A. (2006). Galectin-1 suppresses autoimmune retinal disease by promoting concomitant T helper (Th)2- and T regulatory mediated anti-inflammatory responses. *J Immunol* 176:6323–6332.

Van den Brule, F., Califice, S., and Castronovo, V. (2004). Expression of galectins in cancer: a critical review. *Glycoconj J* 19:537–542.

Van den Brule, F. A., Liu, F. T., and Castronovo, V. (1998). Transglutaminase-mediated oligomerization of galectin-3 modulates human melanoma cell interactions with laminin. *Cell Adhes Commun* 5:425–435.

Van der Leij, J., van den Berg, A., Harms, G., Eschbach, H., Vos, H., Zwiers, P., van Weeghel, R., Groen, H., Poppema, S., Visser, L. (2007). Strongly enhanced IL-10 production using stable galectin-1 homodimers. *Mol Immunol* 44:506–513.

Vespa, G. N., Lewis, L. A., Kozak, K. R., Moran, M., Nguyen, J. T., Baum, L. G., and Miceli, M. C. (1999). Galectin-1 specifically modulates TCR signals to enhance TCR apoptosis but inhibits IL-2 production and proliferation. *J Immunol* 162:799–806.

Walzel, H., Blach, M., Hirabayashi, J., Kasai, K. I., and Brock, J. (2000). Involvement of CD2 and CD3 in galectin-1 induced signaling in human Jurkat T-cells. *Glycobiology* 10:131–150.

Wang, J. L., Gray, R. M., Haudek, K. C., and Patterson, R. J. (2004). Nucleocytoplasmic lectins. *Biochim Biophys Acta* 1673:75–93.

Yamaoka, A., Kuwabara, I., Frigeri, L. G., and Liu, F. T. (1995). A human lectin, galectin-3 (εBP/Mac-2), stimulates superoxide production by neutrophils. *J Immunol* 154:3479–3487.

Yang, R. Y., Hsu, D. K., and Liu, F. T. (1996). Expression of galectin-3 modulates T cell growth and apoptosis. *Proc Natl Acad Sci USA* 93:6737–6742.

Yoshii, T., Fukumori, T., Honjo, Y., Inohara, H., Kim, H. R., and Raz, A. (2002). Galectin-3 phosphorylation is required for its anti-apoptotic function and cell cycle arrest. *J Biol Chem* 277:6852–6857.

Yu, F., Finley, R. L., Jr, Raz, A., and Kim, H. R. (2002). Galectin-3 translocates to the perinuclear membranes and inhibits cytochrome c release from the mitochondria. A role for synexin in galectin-3 translocation. *J Biol Chem* 277:15819–15827.

Zacarias Fluck, M. F., Rico, M. J., Gervasoni, S. I., Ilarregui, J. M., Toscano, M. A., Rabinovich, G. A., and Scharovsky, O. G. (2007). Low-dose cyclophosphamide modulates galectin-1 expression and function in an experimental rat lymphoma model. *Cancer Immunol Immunother* 56:237–248.

Zhu, W. Q., and Ochieng, J. (2001). Rapid release of intracellular galectin-3 from breast carcinoma cells by fetuin. *Cancer Res* 61:1869–1873.

Zuberi, R. I., Frigeri, L. G., and Liu, F. T. (1994). Activation of rat basophilic leukemia cells by εBP, an IgE- binding endogenous lectin. *Cell Immunol* 156:1–12.

Zuberi, R. I., Hsu, D. K., Kalayci, O., Chen, H. Y., Sheldon, H. K., Yu, L., Apgar, J. R., Kawakami, T., Lilly, C. M., and Liu, F. T. (2004). Critical role for galectin-3 in airway inflammation and bronchial hyperresponsiveness in a murine model of asthma. *Am J Pathol* 165:2045–2053.

Zubieta, M. R., Furman, D., Barrio, M., Bravo, A. I., Domenichini, E., and Mordoh, J. (2006). Galectin-3 expression correlates with apoptosis of tumor-associated lymphocytes in human melanoma biopsies. *Am J Pathol* 168:1666–1675.

Zuñiga, E., Rabinovich, G. A, Iglesias, M. M, Gruppi, A. (2001). Regulated expression of galectin-1 during B-cell activation and implications for T-cell apoptosis. *J Leukoc Biol* 70:73–79.

Role of Reactive Oxygen Species in T-Cell Defects in Cancer

Alex Corzo, Srinivas Nagaraj, and Dmitry I. Gabrilovich

1 Introduction

Nitrogen and oxygen are essential for sustaining life on earth. They are not only part of the amino acids and the nucleotides, but also critical for energy transfer in cells. The majority of living organisms have taken advantage of the availability of oxygen, and their existence has become dependent on the presence of this element at all times. At the cellular level, oxygen is necessary for the generation of energy during the aerobic respiration process which occurs in mitochondria. Oxygen is invariably used as the terminal electron acceptor. High-energy electrons from nicotinamide adenine dinucleotide (NADH) and flavin adenine dinucleotide (FADH) are ultimately transferred to oxygen via various electron carriers in the inner mitochondrial membrane and, in the process, a proton gradient is created that drives the synthesis of ATP. The high electronegativity of O_2 enables aerobic organisms to produce much higher amounts of ATP than anaerobic organisms. However, this efficiency comes with the price of the generation of toxic-free radicals via the reduction of O_2 to free radical superoxide (O_2^-), which happens primarily within the mitochondria. In fact, it has been speculated that 1–2 % of daily consumed oxygen is converted to O_2^- by mitochondrial respiration.

The generation of O_2^- if uncontrolled can become a predicament because of up to nine of the various mitochondrial enzymes and redox carriers having been reported as possible O_2^- producing sites (Andreyev et al., 2005). Reactive oxygen species (ROS) are produced in response to multiple stimuli. Even immunoglobulins and the T-cell receptor contain a catalytic site that promotes the formation of ROS (Datta et al., 2002). ROS are a crucial component of many immunological reactions and their significant role in biology and cancer is just now being fully acknowledged. In this review we will focus on the importance of oxygen for normal leukocyte biology and how oxygen metabolites are manipulated by cancer to inhibit T-cell functions and allow the progression of the disease.

Dmitry I. Gabrilovich
H. Lee Moffitt Cancer Center, MRC 2067, 12902 Magnolia Dr., Tampa, FL 33612, USA.
e-mail: dmitry.gabrilovich@moffitt.org

2 Types of Reactive Oxygen Species and Their Effects of Biological Function of Cells

A free radical is any molecule that has an unpaired electron capable of existing independently (Winterbourn, 1993). When a molecule loses an electron it becomes a radical starting an ongoing chain reaction that finally reaches an end when the free radical encounters a second radical and they neutralize each other. By nature, a free radical seeks to pair its unpaired electron. Due to this necessity, most free radicals are highly reactive and will react with another molecule to obtain the missing electron.

Molecular oxygen available in air is stable and long-lived. The origins of almost all reactive oxygen species begin with the monovalent reduction of oxygen to the radical superoxide (O_2^-). O_2^- can act as both oxidant and reductant (Valko et al., 2004). Superoxide performs most of its damage by serving as the building block to other reactive molecules listed below. A second mechanism exists that allows oxygen to be a precursor of other ROS. The two outermost pairs of electrons of O_2 have parallel spins, a characteristic that prevents oxygen from reacting with most molecules. However, if O_2 absorbs sufficient energy, the spin of one electron can be reversed creating antiparallel spins. This state of oxygen with a pair of electrons with opposite spins is called singlet oxygen ($^1O_2^*$) (Ryter and Tyrrell, 1998). Singlet oxygen, though not a free radical, is a more active oxygen state; it also has a longer lifetime unless it transfers its excess energy to another molecule. The list of ROS originated by O_2^- include:

Hydrogen peroxide (H_2O_2). At physiologic pH O_2^- is rapidly converted to H_2O_2 and oxygen (Rodriguez et al., 2000). The reaction is catalyzed by superoxide dismutases (SOD), a family of enzymes abundant in all cells. H_2O_2 is a weak acid with strong oxidizing properties and is one of the most powerful oxidizers known. One of the most important properties of H_2O_2 is its ability to diffuse freely through the cell membrane, thereby playing a role in many biological processes including antimicrobial defense and cell signaling. In fact, H_2O_2 is viewed as one of the most important oxidant species because of its known interactions with multiple signaling pathways.

Hydroxyl radical ($\cdot OH$). H_2O_2 in combination with some metals, particularly iron, can also give rise to one of the most reactive free radicals, the hydroxyl radical ($\cdot OH$). Under physiologic condition, iron exists in the Fe^{3+} form. When iron is reduced to Fe^{2+}, a reaction accomplished by O_2^- (Leonard et al., 2004), it is capable of reacting together with H_2O_2 culminating in the production of $\cdot OH$.

Hypochlorous acid (HOCl). Alternatively, H_2O_2 can also be used as a substrate by the enzyme myeloperoxidase (MPO) for the formation of another type of ROS: hypochlorous acid (HOCl). MPO is abundant in neutrophil granules and it produces HOCl from H_2O_2 and chloride anions during the respiratory burst, a process that will be explained in detail later. HOCl possesses potent antimicrobial properties. MPO is unique in its ability to generate reactive chlorinating species such as HOCl (Weiss et al., 1986). MPO has emerged as a potential mediator of atherosclerosis by promoting endothelial dysfunction, generating atherogenic lipoproteins, consumption of nitric oxide, all of which result in the promotion and/or propagation of the disease (Nicholls and Hazen, 2005). HOCl carries out a wider variety of oxidative reactions, including chlorination of tyrosines and the oxidative modification (and often inactivation) of enzymes (Fu et al., 2003). Hypochlorite can interact with ammonia or amines to produce chloramines.

Peroxynitrite ($ONOO^-$). Another molecule that reacts with O_2^- is nitric oxide (NO), a key biological messenger in mammals and a toxic reagent functioning as a tumoricidal and antimicrobial molecule. NO can quickly react with superoxide leading to the formation of the dangerous free radical peroxynitrite ($ONOO^-$), a potent oxidant that can attack a wide range of biological molecules. These chain reactions can be extremely detrimental when the main biological macromolecules (lipids, nucleic acids and proteins) become involved in the reaction, as all three are highly susceptible to free radical attack.

The vast majority of oxidative stress-induced DNA damage affects the primary structure of the double helix. Nucleic acids undergo single- and double-strand breaks in the backbone. Additionally, bulky adducts that do not fit in the standard double helix are introduced. Lipids are major targets during oxidative stress. Free radicals can attack polyunsaturated fatty acids directly in membranes and initiate lipid peroxidation, which decreases membrane fluidity, altering membrane properties and disrupting membrane-bound proteins (Eze, 1992). Reactive unsaturated aldehydes are also produced, which can act as either mutagens (Marnett et al., 1985) or inactivate enzymes (Chen and Yu, 1994; Szweda et al., 1993) or react with proteins and nucleic acids to form heterogeneous cross-links (Chio and Tappel, 1969). Unlike reactive free radicals, aldehydes are long-lived molecules that can reach and attack targets far from the initial free radical event.

Oxidative damages to proteins are just as severe because they ultimately lead to the inactivation of the protein. Damages include oxidation of sulfhydryl groups, reduction of disulfides and peptide fragmentation (Stadtman and Oliver, 1991; Starke-Reed and Oliver, 1989). Peroxynitrite can induce protein modifications via several mechanisms. Nitration of tyrosine residues has been long recognized as a marker of peroxynitrite activity. Peroxynitrite-dependent tyrosine nitration is likely to occur through the initial reaction of peroxynitrite with carbon dioxide or metal centers leading to secondary nitrating species as nitrogen dioxide radicals (Alvarez and Radi, 2003). In addition, peroxynitrite can react directly with cysteine, methionine and tryptophan residues (Alvarez and Radi, 2003). Furthermore, the degradation of proteins becomes compromised by oxidative stress. Oxidized proteins are poor substrates for degradation and can also inhibit proteases to degrade other oxidized proteins to prevent their accumulation (Dean et al., 1997; Grune et al., 1997). The removal of these proteins is a necessity, as accumulation leads to altered cellular metabolism.

Thus, ROS are harmful metabolites with the potential to induce drastic biological modifications to cells. Neutralization of ROS has become an important therapeutic strategy in a number of pathological conditions (Bonnefoy et al., 2002; St Clair et al., 2005).

3 Cellular Defenses Against ROS

In light of the critical cellular damage that uncontrolled ROS could create, cells have devised a proficient antioxidant system to scavenge the radicals before damage is irreparable and maintain the cellular redox in the reduced state. This antioxidant system consists of specialized enzymes and non-enzymatic molecules.

Superoxide dismutases. SOD catalyzes the conversion of O_2^- into H_2O_2. SOD requires a redox active metal ion in the active site: Mn, Fe, Cu or Ni. In mammals, only Cu and Mn binding forms are found. While the copper-binding SOD is primarily available in the cytosol, the manganese form of SOD is generally present in high levels in the mitochondrial matrix. The importance of SOD in oxidative stress protection cannot be overstated, as complete loss of the enzyme in mice results in neonatal lethality or mutations in SOD may result in amyotrophic lateral sclerosis.

Catalase. Found in peroxisomes, catalase detoxifies H_2O_2 by converting it to water and molecular oxygen. Catalase is an efficient enzyme; it has one of the highest turnover rates for all known enzymes: one molecule of catalase can convert 6 million molecules of hydrogen peroxide to water and oxygen each minute.

Glutathione peroxidase. It reduces free H_2O_2 to water during the simultaneous oxidation of glutathione, a cofactor of the enzyme. Glutathione itself is an extremely important antioxidant, being present in cells at very high concentrations. There are several isozymes encoded by different genes, which vary in cellular location and substrate specificity. Glutathione peroxidase 1 is the most abundant version and is found in the cytoplasm of nearly all mammalian tissues (Mates et al., 1999)

Glutathione (GSH). A thiol antioxidant, GSH reduces any disulfide bonds formed within cytoplasmic proteins to cysteines by acting as an electron donor. It also aids in the regeneration of the active forms of vitamin C and E from their oxidized forms. GSH is recycled back to its reduced form by the enzyme glutathione reductase, which is constitutively active and inducible upon oxidative stress. GSH is important not only for its ability to scavenge ROS directly, but also because it serves as a cofactor of various ROS-scavenging enzymes such as glutathione peroxidase and glutathione-S-transferase (GST). GST detoxifies some of the carbonyl-, peroxide- and epoxide-containing metabolites produced within the cell by oxidative stress, and mutations of GST have been reported in cancer patients. Unusually high levels of oxidized glutathione have been described in the circulation in patients with advanced stages of colon and breast cancers (Pastore et al., 2003).

Vitamin C or L-ascorbic acid is an essential nutrient and a very powerful antioxidant. Vitamin C has been found to protect against cell death through its antioxidant ability. It has also been implicated in cell signaling. Vitamin C regulates the AP-1 signaling pathway by preventing JNK phosphorylation (Kyaw et al., 2002).

Vitamin E or tocopherol is a powerful membrane-bound antioxidant and its main function is the prevention of lipid peroxidation. Natural vitamin E exists in eight different forms. A clinical trial reported that vitamin E induced the cell cycle inhibitor $p21^{wafi/cip1}$ and reduced the incidence of colon cancer (White et al., 1997).

Other ROS scavengers include carotenoids and selenium. Epidemiological studies have shown that deficiency of these compounds increases the chances of cancer occurrence (Valko et al., 2006).

4 Nicotinamide Adenine Dinucleotide Phosphate (NADPH) as the Main Source of ROS in Myeloid Cells

The notion that ROS are merely byproducts of cellular metabolism is only partially correct. While indeed mitochondria are organelles whose primary function is energy production and where ROS generation occurs accidentally, cells do possess an oxidative enzyme complex exclusively dedicated to the generation of oxygen radicals. In leukocytes, the primary producer of ROS is NADPH oxidase. Since its original description in neutrophils, various isoforms have been found in different cell types ranging from endothelial cells to microglia. NADPH oxidase catalyzes the one-electron reduction of oxygen to superoxide anion using electrons supplied by NADPH. The importance of this enzyme can be observed in the severity of hereditary chronic granulomatous disease (CGD). CGD is caused by mutations in any of the genes that encode the subunits of the oxidase. Over 410 possible CGD mutations have been identified, and all result in the complete or partial loss of the protein and its activity. Patients with CGD experience frequent life-threatening infections during their lifetime (Assari, 2006).

The oxidase is a multicomponent enzyme consisting of two membrane proteins, gp91 and p22, that together form a unique membrane-bound flavocytochrome and at least four cytosolic components: $p47^{phox}$, $p67^{phox}$, $p40^{phox}$ (phox is short for phagocyte oxidase) and a small G protein Rac (Groemping and Rittinger, 2005).

4.1 Components of NADPH Oxidase

4.1.1 Membrane-Bound Subunits

$gp91^{phox}$: Also known as NOX2, the $gp91^{phox}$ subunit consists of 570 amino acids and contains two heme groups as well as a flavin adenine dinucleotide (FAD), which serves as the binding site for NADPH. Over the last years, six homologs of this subunit have been described: NOX1, NOX3, NOX4, NOX5, DUOX1 and DUOX2.
$p22^{phox}$: A 195 amino acid protein that helps stabilize $gp91^{phox}$ for the docking of the cytosolic subunits. The cytoplasmic portion contains a proline-rich region that contains a consensus PxxP motif around Pro^{156}.

4.1.2 Cytosolic Subunits

$p47^{phox}$: The p47phox subunit consists of 390 amino acids. $p47^{phox}$ is primarily expressed in myeloid cells (Rodaway et al., 1990). The homolog is referred to as NOXO1. Although NOXO1 is highly expressed in the colon, it can also be found in other tissues such as the intestine, liver, kidney and pancreas (Banfi et al., 2003). It controls and facilitates translocation of the p67–p47–p40 complex to the membrane. It contains two SH3 domains that recognize the PxxP motif in the cytoplasmic tail of $p22^{phox}$ (Finan et al., 1994; Leto et al., 1994). *$p67^{phox}$*: It is a 526 amino acid protein (Francke et al., 1990). Similar to $p47^{phox}$, the $p67^{phox}$ subunit is expressed primarily in myeloid cells. It has a homolog referred

to as NOXA1. NOXA is expressed in the stomach, colon and small intestine (Banfi et al., 2004; Geiszt et al., 2003). p67^{phox} contains an activation domain (aa 199–210), which is absolutely required for O_2^- production (Han and Lee, 2000; Han et al., 1998). *p40^{phox}*: It consists of 339 amino acids. It has been suggested that p40^{phox} functions as a gp91^{phox} inhibitor (Sathyamoorthy et al., 1997). Regardless of its true function, the fact that p40^{phox} co-immunoprecipitates with p47^{phox} and p67^{phox} confirms its involvement in complex assembly. *Rac GTPase*: Three homologous Rac proteins have been described in mammalian cells. Rac1 is ubiquitously expressed and regulates cell adhesion, migration and differentiation in various cell types (Moll et al., 1991). Rac2 is expressed in hematopoietic cells only and is the most relevant Rac GTPase for activation of NOX2 in human neutrophils (Haataja et al., 1997). Rac3 is predominantly expressed in neuronal cells and although it shows high homology to Rac1, its function remains unknown (Bolis et al., 2003). Rac proteins are critical for many cellular functions, including actin polymerization and the formation of lamellipodia and membrane ruffles, as described first for Rac1 (Ridley et al., 1992). *S100A8/A9*: The S100 proteins comprise a family of 23 different members, which are characterized by high homology, low molecular weight (9–14 kDa) and calcium binding. They possess important regulatory functions including calcium buffering, regulation of kinases and phosphatases, cell proliferation, differentiation, energy metabolism, cytoskeletal–membrane interactions, embryogenesis, cell migration and inflammation (Kerkhoff et al., 1998). Two of its members, S100A8/A9, are expressed in large amounts in neutrophils and monocytes (Edgeworth et al., 1991), playing a role in cell activation (Kerkhoff et al., 1998) and adhesion (Newton and Hogg, 1998). S100A8/A9 expression is strongly upregulated in the advanced stages of multiple cancers in mice and humans (Gebhardt et al., 2006).

In the resting state the p47^{phox}, p67^{phox} and p40^{phox} subunits are believed to exist in a complex in cytoplasm. During activation, the cytosolic proteins translocate to the plasma membrane and associate with cytochrome b_{558}. In the resting state, p47^{phox} exists in an auto-inhibitory state that is only relieved after extensive phosphorylation of various serine residues. Once the complex is assembled, electrons are removed from NADPH in the cytoplasm and transferred through the gp91^{phox} component (which includes FAD and two hemes) across the membrane, where the reduction of O_2 begins. Sustained NADPH oxidase activity requires continuous renewal of the enzyme complex; without that, it is rapidly deactivated (Decoursey and Ligeti, 2005).

5 Biological Roles of NADPH Oxidase

NADPH oxidase-derived ROS have two specific purposes: the well-characterized destruction of pathogens through the respiratory burst (Hampton et al., 1998) and the more recently acknowledged intracellular signaling (Forman and Torres, 2002).

5.1 Microbicidal Function

The role of NADPH oxidase in phagocytosis has been described extensively. When a phagocyte encounters a bacterium, this interaction results in an activation of cell signaling via different receptors (primarily toll-like receptors) that initiate the assembly and activation of the NADPH oxidase (Babior, 1984). The rate of oxygen consumption increased 50–100 times over normal levels leading to the production of vast amounts of superoxide. In parallel with oxygen consumption, glucose metabolism is sharply increased. Because of the dramatic increase in oxygen intake, this series of events is commonly referred to as the respiratory burst (Babior, 1984).

The significance of the respiratory burst in innate immunity was traditionally believed to be due to the generation of O_2-derived antimicrobial molecules. The extremely high ROS concentrations found inside the phagosome were thought to be sufficient in destroying bacteria because of the high reactivity of ROS with the bacterial cell membrane. However, in recent years the importance of superoxide generation during phagocytosis in microbial killing has been questioned. A new school of thought has developed claiming that the transfer of electrons across the membrane, not the oxygen radicals, is the true mediator of antimicrobial defense. The transfer of electrons across the membrane generates superoxide in the vacuole. The superoxide molecules then consume protons in the vacuole elevating the pH turning the vacuole alkaline. The granule protein contents, which include the proteases elastase and cathepsin G, released in the vacuole, become activated by the elevation in pH (Segal, 2005). The proteases, not ROS, therefore are toxic of pathogens. Support for this hypothesis relies on the observation that cathepsin G and elastase-deficient mice cannot kill bacteria, despite normal ROS generation (Reeves et al., 2002). The conclusion from this study questions whether ROS themselves contribute much to microbial killing and calls for re-evaluation of treatment for diseases whose pathogenesis is thought to be mediated by ROS.

5.2 Cell Signaling

During the past two decades, an additional characteristic of ROS has become apparent: their ability to interact with multiple signaling cascades. Under physiological conditions, ROS are present within cells at very low concentrations: approximately in the picomolar to low nanomolar range. At these low concentrations, ROS can regulate the redox state of transcription factors and enzymes involved in signal transduction without inducing excessive oxidative damage. The activities of two transcription factors in particular have long been known to depend on the redox state of the cell: NF-κB and AP-1. Both NF-κB and AP-1 play a central role in immune responses and inflammation through regulation of expression of a large number of cytokines and other immune response genes. Although still debatable, recent studies seem to agree that ROS increase the translocation of cytoplasmic NF-κB to the nucleus through the activation of IKK, the kinase responsible for phosphorylation and degradation of the NF-κB inhibitory protein IκB (Kamata et al., 2002; Takada et al., 2003). The regulation of AP-1 is controlled by the c-Jun N-terminal kinase (JNK) cascade. JNK is a part of the mitogen-activated protein kinase (MAPK) superfamily of serine/threonine kinases. ROS can activate the upstream kinases in the cascade, ending in increased transcription of AP-1-controlled genes (Gotoh and Cooper, 1998; Tobiume et al., 2002).

The mechanism of ROS-mediated signaling revolves around post-translational modifications of proteins. These modifications include oxidations of cysteines, which can affect the activity of proteins. Cysteine residues are susceptible to oxidative damage, producing sulfinic acid intermediates, which can further react with thiols to form catalytically inactive protein disulfides (Valko et al., 2006). The best examples of ROS-modified enzymes are protein tyrosine phosphatases (PTPs), whose activity is extremely susceptible to ROS. PTPs are pivotal regulators of cell signaling: they negatively regulate protein tyrosine kinase receptors and also deactivate kinases involved in signal transduction and transcription factors such as JNK, ERK, p38 and STATs (Mustelin et al., 2005). All PTPS contain an oxidizable residue within their catalytic region and oxidation of this cysteine residue by ROS renders the phosphatase completely inactive.

Although there are countless cysteine residues that could potentially undergo this reaction, ROS are only capable of oxidizing cysteine residues existing in a deprotonated state (–S–) at physiological pH. Deprotonation of a cysteine residue requires a positively charged amino acid present in the vicinity of the cysteine residue. Cysteine has a pK_a around 8.5, significantly higher than the physiologic range. The surrounding positive residues provide a positively charged electrostatic field that allows dissociation to the deprotonated state and stabilization of the structure (Denu and Tanner, 1998).

6 Role of ROS in Myeloid Cell-Mediated Immune Suppression in Cancer

Elevated levels of ROS observed in many cancer cells contribute to tumorigenesis and metastasis (Mantovani et al., 2003; Waris and Ahsan, 2006). ROS is known to trigger signaling related to angiogenesis (Agostinelli and Seiler, 2006). Studies implicate ROS mediates activation of c-met, the HGF receptor which is involved in metastases (Ferraro et al., 2006). Abnormal NF-κB activity, documented in a number of cancers, can be mediated by ROS. Furthermore, NADPH oxidase-derived ROS from macrophages and neutrophils contribute to the metastasis of tumor cells through upregulation of genes, such as thymosin-beta 4 vital for tumor cell motility and invasiveness (Okada et al., 2006). The important role of ROS in tumor cell survival was confirmed in a recent study that demonstrated that antioxidants may synergize with cytotoxic drugs in the treatment of colon or liver cancers. When CT26 tumor-bearing mice were treated by oxaliplatin and one of the three SOD mimetics manganese(III)tetrakis(4-benzoic acid) porphyrin, copper(II)(3,5-

diisopropylsalicylate)2 or manganese dipyridoxyl diphosphate, the tumor volumes were comparatively smaller than oxaliplatin treatment (Laurent et al., 2005).

The major contributors to ROS pool in cancer are myeloid cells. They include granulocytes, macrophages and the recently described group of cells accumulated in tumor-bearing hosts—myeloid-derived suppressor cells (MDSC).

6.1 Characteristics of Myeloid-Derived Suppressor Cells

Tumors devised various strategies to escape immune system control (Rabinovich et al., 2007). MDSC are one of the major components of these strategies. These cells are generated in response to multiple tumor derived factors and represent a group of cells of myeloid origin comprised of immature macrophages, granulocytes, DCs and other myeloid cells at early stages of differentiation (Gabrilovich, 2004; Gabrilovich et al., 2007; Kusmartsev and Gabrilovich, 2006; Serafini et al., 2006). In mice, MDSC are defined as Gr-1^+CD11b$^+$ cells. Cells with this phenotype are present in the bone marrow and spleen of healthy mice and differentiate into mature myeloid cells—granulocytes, macrophages and DCs—in vitro in the presence of GM-CSF or in vivo after adoptive transfer into naïve healthy recipients (Kusmartsev and Gabrilovich, 2003). However, they accumulate in the spleen and, to some extent, in the lymph nodes of mice bearing many different tumors (Bronte et al., 1998; Fu et al., 1990; Gabrilovich et al., 2001; Kusmartsev et al., 2000; Li et al., 2004; Melani et al., 2003; Nefedova et al., 2004; Ruiz de Morales et al., 1999; Salvadori et al., 2000; Terabe et al., 2003; Young et al., 1987).

The hallmark of MDSC is their ability to suppress immune responses. This effect is partially mediated by (1) loss or significant decrease of the expression of the T-cell receptor ζ chain (CD3ζ), which is the principal part of TCR complex (Otsuji et al., 1996); (2) inhibition of CD3/CD28-induced T-cell activation/proliferation by production of reactive nitrogen and oxygen intermediates (Kusmartsev et al., 2000); (3) inhibition of interferon-γ (IFN-γ) production by CD8$^+$ T cells in response to the specific peptide presented by MHC class I molecules (Gabrilovich et al., 2001); (4) prevention of the development of CTL in vitro (Bronte et al., 2003).

Most MDSC effects on T cells require close cell–cell contact and depend on MHC expression by MDSC. Recent studies in vivo have demonstrated that MDSC, but not immature myeloid cells from control mice, were able to take up soluble protein in vivo, process it, present antigenic epitopes on their surface and induce antigen-specific T-cell anergy (Kusmartsev, 2005). The subset of MDSC, Gr-1^+CD115$^+$ cells in addition to being able to suppress T-cell proliferation in vitro, could induce the development of Foxp3$^+$ Tregs in vivo (Huang et al., 2006). The development of Tregs required antigen-associated activation of tumor-specific T cells, was dependent on the presence of IFN-γ and IL-10 and was independent of nitric oxide (Huang et al., 2006). Interestingly, while nitric oxide was required for suppression of mitogen-activated T cells by MDSC, inhibition of allogenic T-cell responses by MDSC was mediated by the enzyme arginase-1 (Bronte et al., 2003).

Gr-1$^+$ MDSC could be differentiated in vitro and in vivo into F4/80$^+$ macrophages. Inside the tumor these macrophages produce high levels of nitric oxide and can directly induce T-cell apoptosis. It appears that the Stat1 transcription factor is responsible for this effect (Kusmartsev and Gabrilovich, 2005).

In humans, MDSC are defined as cells that express the common myeloid marker CD33, but lack expression of markers of mature myeloid and lymphoid cells and HLA-DR (Almand et al., 2001) or CD14$^-$CD11b$^+$ cells (Zea et al., 2005). Advanced-stage cancer promoted the accumulation of these cells in blood, whereas surgical resection of the tumor decreased the number of immature cells. Consistent with data obtained in tumor-bearing mice, MDSC derived from patients with advanced tumors inhibited the production of IFN-γ by autologous CD8$^+$ T cells stimulated with specific peptide-pulsed DCs (Almand et al., 2001). Using blood samples from cancer patients, Schmielau and Finn (2001) observed an unusually large number of myeloid cells with granulocyte phenotype. These cells, if activated, can inhibit cytokine production by T cells. A more detailed description of the MDSC's role in tumor-associated immune suppression can be found in "Myeloid-Derived Suppressor Cells in Cancer" of this monograph.

6.2 MDSC and ROS

MDSC isolated from tumor-bearing mice produced high levels of ROS, which were dramatically reduced by catalase. This suggested a major contribution of H_2O_2 to this ROS pool (Kusmartsev and Gabrilovich, 2003). Inhibition of ROS in MDSC completely abrogated the negative effect of these cells on T cells (Kusmartsev, 2005). Thus, MDSC generated in tumor-bearing hosts could suppress CD8$^+$ T-cell responses via release of ROS. Immunosuppressive effect of myeloid cells with granulocyte phenotype described in cancer patients was also abrogated by addition of catalase, further implicating H_2O_2 as a critical effector molecule (Schmielau and Finn, 2001). The production of ROS by neutrophils isolated from the blood of 16 patients with larynx carcinoma was compared with that of neutrophils obtained from 15 healthy individuals. The levels of ROS, especially spontaneous and PMA-inducible superoxide, were substantially higher in cancer patients than in healthy volunteers and that increase was associated with the tumor stage. After partial or total laryngectomy, a significant decrease in ROS production and the serum activity of catalase and peroxidase was observed (Szuster-Ciesielska et al., 2004). Interaction of MDSC with antigen-specific T cells in the presence of specific but not control antigens resulted in a significant increase of ROS production. That increase was independent of IFN-γ production by T cells, but was mediated by integrins CD11b, CD18 and CD29 (Kusmartsev et al., 2004). Blockade of these integrins abrogated ROS production and MDSC-mediated suppression of CD8$^+$ T-cell responses. Importantly, no T-cell apoptosis or T-cell deletion has been observed. This data was consistent with reports implicating adhesion molecules, and integrins in particular, in ROS production by macrophages and fibroblasts (Chiarugi et al., 2003; Husemann et al., 2001; Werner and Werb, 2002).

Studies demonstrate that MDSC freshly isolated from tumor-bearing mice but not their control counterparts are able to inhibit antigen-specific response of CD8+ T cells (Kusmartsev et al., 2004, 2005). MDSC obtained from tumor-bearing mice had significantly higher levels of ROS than MDSC isolated from tumor-free animals. Since ROS production by MDSC can be blocked by arginase inhibitors, it appears that arginase activity may play an important role in ROS accumulation in these cells (Kusmartsev et al., 2004). MDSC upregulate arginase expression upon stimulation with IL-4. L-Arginine is the common substrate for arginase and iNOS. In contrast to arginase, iNOS is induced by IFN-γ. Hence L-arginine metabolism in myeloid cells is a potential target for selective intervention in reversing myeloid-induced dysfunction in tumor-bearing hosts (Bronte et al., 2003).

Is there a link between arginase activity and ROS production? Arginase catalyzes the hydrolysis of L-arginine to urea and L-ornithine. L-arginine is used by NO synthase as a substrate for generation of NO (Wu and Morris, 1998). However, low concentrations of L-arginine result in low NO formation and high generation of superoxide (O_2^-) (reviewed in Boucher et al., 1999). Thus, it is possible that high arginase activity in MDSCs may have lowered the level of L-arginine and resulted in increased production of O_2^- (Kusmartsev et al., 2004). Superoxide itself is very unstable and is converted to H_2O_2 and oxygen. This is consistent with the data showing that in MDSC ROS accumulates primarily in the form of H_2O_2, but not O_2^-.

The nature of factors responsible for upregulation of ROS production in myeloid cells as well the molecular mechanism of this phenomenon is currently not clear. It is likely that combination of many different factors as well as inflammation in tumor site may contribute to this process. Several known tumor-derived factors like TGFβ and IL-10 are able to increase arginase activity in macrophages (Boutard et al., 1995; Modolell et al., 1995), and a number of other cytokines and growth factors produced by tumor can induce ROS production, including IL-6, IL-3, PDGF, GM-CSF, FGF (reviewed in Sauer et al., 2001). Constant production of these factors in tumor-bearing mice could lead to the different levels of ROS observed in MDSC from tumor-bearing and tumor-free mice. It was shown that granulocyte colony-stimulating factor (G-CSF) induced activation of Lyn and Akt in a time- and dose-dependent manner. Lyn-deficient mice produced lower levels of ROS when the neutrophils were stimulated with G-CSF. Diphenyleneiodonium I, a specific inhibitor of ROS, could inhibit Lyn and Akt (Zhu et al., 2006). Patients with acute myeloid leukemia have a truncated G-CSF receptor, which leads to ROS dysregulation. Hence, targeting the Lyn–Akt cascade may provide a potential therapeutic benefit (Zhu et al., 2006).

6.3 ROS Involvement in T-Cell Suppression in Cancer

A number of studies have linked hyperproduction of ROS in cancer with T-cell abnormalities. Splenic T cells from tumor-bearing mice have been observed to have a reduction in the CD3ζ chain despite normal expression of other components of the

TCR complex. This loss is associated with the accumulation of macrophages and hydrogen peroxide (Kono et al., 1996). In another study examining patients with ovarian cancer, tumor-infiltrating lymphocytes show loss of CD3ζ, which could be due to macrophage secretion of ROS (Lockhart et al., 2001). Myeloid cells derived from tumor-bearing mice cause oxidative stress and inhibit ζ-chain expression in T cells and antigen-induced cell proliferation (Aoe et al., 1995; Otsuji et al., 1996). Similar results were obtained by Schmielau and Finn (2001), studying myeloid cells from cancer patients. T cell had a reduction in CD3ζ expression and decrease in cytokine production and this correlated with the presence of activated myeloid cells in the PBMC population. Freshly isolated myeloid cells from healthy donors if activated could inhibit cytokine production of T cells. This action was abrogated by addition of catalase, implicating H_2O_2 as an effector molecule. CD45RO$^+$ memory cells or CD45RO$^-$ naïve T cells were tested for the capacity to produce cytokines under conditions of oxidative stress. CD45RO$^+$ cells did not produce IFN-γ, TNF-α and IL-2 following exposure to H_2O_2, whereas CD45RO$^-$ cells producing these cytokines were not affected by oxidative stress. It was found that unresponsiveness of the memory T cells was due to block in NF-κB activation (Malmberg et al., 2001).

In studies of mice with ataxia-telangiectasia caused by mutations in Atm gene, stimulation of T cells resulted in ROS production leading to the induction of apoptosis. The data suggested that Atm plays a critical role in T-cell activation by regulating the cellular response to ROS following stimulation through the TCR (Bagley et al., 2007).

6.4 Mechanism of T-Cell Suppression Mediated by Peroxynitrite

Peroxynitrite is one of the most powerful and reactive oxidant species, which is responsible for most of the adverse effects linked with ROS. As we described above, peroxynitrite is a product of reaction of NO and superoxide. The combination occurs at a diffusion-limited rate and NO outcompetes SOD and the consequence of which is peroxynitrite, occurring at sites characterized by the presence of MDSC, inflammatory cells or immune reactions. Nitric oxide itself produced by macrophages could inhibit T cells via a variety of different mechanisms involving inhibition of the phosphorylation and activation of Janus kinase 3 (Jak3) and STAT5 transcription factor (Bingisser et al., 1998), inhibition of MHC class II gene expression (Harari and Liao, 2004) and induction of T-cell apoptosis (Rivoltini et al., 2002). Peroxynitrite can inhibit T-cell activation and proliferation via impairment of tyrosine phosphorylation and apoptotic death. We demonstrated that ONOO$^-$ was involved in T-cell inhibition by Gr-1$^+$ MDSC derived from tumor-bearing mice (Gabrilovich, 2004; Kusmartsev et al., 2000). Peroxynitrite could also cause a nitration of proteins involved in antigen processing. In normal conditions proteasome or immunoproteasome are involved in the removal of oxidatively modified proteins. Under oxidative stress proteasomal machinery could be impaired (Amici et al., 2003).

(Bronte et al. 2005) reported that human prostatic adenocarcinomas are infiltrated by terminally differentiated cytotoxic T lymphocytes. These lymphocytes, however, were in an unresponsive state. The authors demonstrated the presence of high levels of nitrotyrosine in prostatic tumor-infiltrating lymphocytes (TIL), suggesting a local production of peroxynitrites. Restoration of TIL responsiveness to tumor could be achieved by simultaneous inhibition of iNOS and arginase activity. Thus, local peroxynitrite production could represent one of the important mechanisms by which tumor escape immune response.

Peroxynitrite produced during inflammatory conditions can modify and inactivate proteins, especially zinc finger transcription factors such as p53. Due to oxidant stress the development and progression of pancreatic cancer involved higher levels of protein tyrosine nitration (Vickers et al., 1999). Malignant gliomas have evidence of peroxynitrite-mediated tyrosine nitration in vivo and that physiologically relevant concentrations of peroxynitrite could inhibit the specific DNA binding ability of purified wild-type p53 as well as wild-type p53 protein in D54MG cells in culture (Cobbs, 2003). Increased nitrotyrosine staining has been observed in tumor tissues from patients with head and neck cancer (Bentz et al., 2000). Peroxynitrite or nitrotyrosine plays a role in the etiology, progression and outcome of cancer treatment. Nitrotyrosylation is evident in cancer tissues. In MCF7 breast cancer cells peroxynitrite has been found to irreversibly inactivate arylamine N-acetyltransferases (NATs), which are required for detoxification and metabolic activation of a variety of aromatic xenobiotics, including numerous carcinogens. NAT1 is highly expressed in normal and cancerous human breast tissue. This indicated the role played by peroxynitrite in carcinogenesis and tumor progression (Dairou et al., 2005). High levels of nitrotyrosine expression by melanoma metastases may be predictive of a poor outcome (Ekmekcioglu et al., 2000). Patients with mesothelioma under oxidative stress express high levels of a radical scavenging enzyme MnSOD and mesothelioma cells are associated with higher nitrotyrosine staining (Kinnula et al., 2004). In breast cancer nitrotyrosine levels are significantly associated with lymph node metastasis and VEGF-C expression (Nakamura et al., 2006). This suggested that high nitrotyrosine levels might serve as a prognostic factor for long-term survival in breast cancer.

6.5 Model of T-Cell Tolerance Induced by MDSC and Mediated by Peroxynitrite

Recent studies have suggested a direct role for peroxynitrite in MDSC-mediated T-cell tolerance. During presentation of antigen MDSC closely interact with antigen-specific T cells. This provides an environment where peroxynitrite abundantly produced by MDSC can affect molecules on the surface of T cells. We proposed that peroxynitrite might modulate an immune response leading to anergy and suppression via modification of different amino acids in molecules involved in recognition of peptide MHC complexes: TCR, CD8, CD4. One example could be a nitration of the tyrosine residues. The concentrations of nitric oxide and superoxide are 10–100 and 0.1–1 nM, respectively, under physiologic conditions and can go up to micromolar concentrations under pathological conditions (Groves, 1999).

To address the potential role of MDSC and ROS in T-cell tolerance in cancer we used an experimental model with an adoptive transfer of transgenic $CD8^+$ T cells and tumor-derived MDSC into naïve congenic recipient (Kusmartsev, 2005). $CD8^+$ T cells from MDSC-treated mice were not able to produce IFN-γ and IL-2 in response to specific peptide and did not kill peptide-loaded target cells (Kusmartsev, 2005). At the same time, these cells responded well to stimulation with anti-CD3 antibody. In both tested experimental systems (OT-1 and 2C T cells) MDSC induced a marked decrease in the binding of specific peptide MHC (pMHC) complex to $CD8^+$ T cells (Nagaraj et al., 2007). Although this effect was not previously described in cancer, it is known that the changes in TCR binding avidity can play an important role in regulating antigen sensitivity (Fahmy et al., 2001). Tolerization of H-Y TCR transgenic mice by repeated pMHC administration resulted in a substantial decrease in MHC tetramer binding (Maile et al., 2005). Similar changes were observed in mice infected with the influenza virus (Drake et al., 2005). Among proposed mechanisms are downregulation of CD8 expression, changes in cellular localization of TCR, glycosylation of TCR (Cawthon and Alexander-Miller, 2002; Drake et al., 2005; Kao et al., 2005; Maile et al., 2005). However, in our experiments MDSC did not induce downregulation of CD8 or TCR.

The use of peroxynitrite scavenger in vivo completely eliminated MDSC-induced T-cell tolerance suggesting its direct role on MDSC-mediated $CD8^+$ T-cell tolerance (Nagaraj et al., 2007).

As we described above peroxynitrite can induce protein modifications via several mechanisms. Nitration of tyrosine residues is being long recognized as a marker of peroxynitrite activity.

We hypothesized that nitration of tyrosines in TCR or/and CD8 could alter binding of TCR/CD8 complex and pMHC that is obligatory for T-cell stimulation. Molecular modeling revealed a number of tyrosine residues in TCR and CD8 molecules that could be susceptible to nitration, and structural analysis showed that nitration of these tyrosine residues would result in the decreased flexibility and increased rigidity of TCR domains that might significantly alter the epitope-specific interactions between TCR and pMHC, observed in our study (Nagaraj et al., 2007).

The results described above suggest a potential mechanism of $CD8^+$ T-cell tolerance in cancer. Tumor-derived factors stimulate production and activation of MDSC that contain high levels of ROS and generate peroxynitrite. MDSC accumulate in lymph nodes of tumor-bearing host. Since ROS are short-lived and highly reactogenic, they are active only at very short distance. Interface of MDSC and $CD8^+$T cells interacting during antigen-TCR recognition phase provides such an environment. The amount of peroxynitrite produced by MDSC is sufficient to nitrate tyrosine residues exposed on the surface of contacting cells. Modified tyrosine on TCR and CD8 alter the conformational flexibility of TCR chains that lead to loss of the response to specific antigen (Fig. 1). We suggest that a similar scenario of tumor-induced peripheral immunological tolerance (anergy) may potentially operate

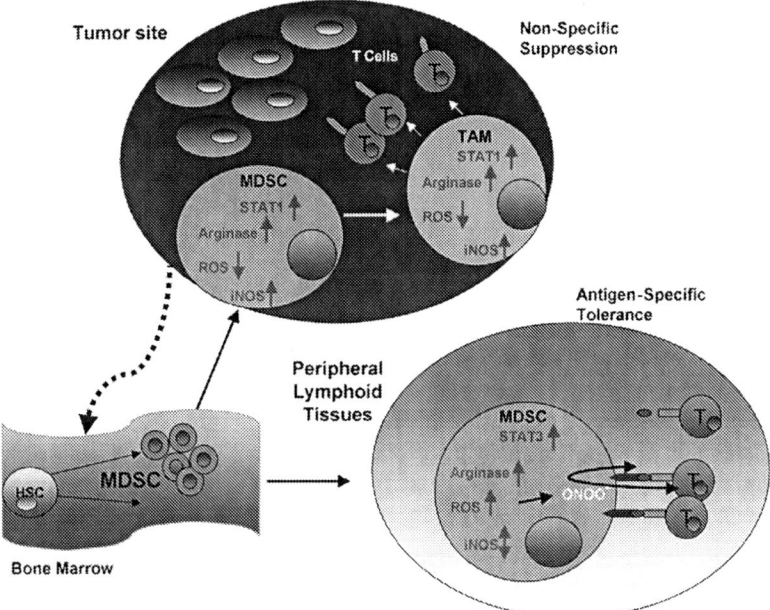

Fig. 1 Potential mechanism of ROS involvement in CD8$^+$ T-cell tolerance in cancer. MDSC are produced in bone marrow in response to tumor-derived factors. They contain high levels of ROS and generate peroxynitrite. MDSC accumulate in lymph nodes of tumor-bearing host. Since ROS are short-lived and highly reactogenic, they are active only at very short distance. Interface of MDSC and CD8$^+$T cells interacting during antigen-TCR recognition phase provides such an environment. The amount of peroxynitrite produced by MDSC is sufficient to nitrate tyrosine residues exposed on the surface of contacting cells. Modified tyrosine on TCR and CD8 alter the conformational flexibility of TCR chains that lead to loss of the response to specific antigen. MDSC also migrate into tumor site. There MDSC quickly differentiate into tumor-associated macrophages (TAM). Tumor microenvironment downregulates ROS production and induces iNOS expression possibly via upregulation of STAT1. TAM produce high levels of NO and immunosuppressive cytokines (like IL-10). These create immunosuppressive environment where T-cell function is inhibited in antigen non-specific manner

using other post-translational modifications of proteins from the TCR–pMHC complex. These modifications may include S-nitrosated cysteines, sulfated cytosines and methionines, nitrated cytosines, methionines and tryptophanes, dimers of tyrosines. This mechanism may explain the fact that T cells in peripheral organs of tumor-bearing mice and in peripheral blood of cancer patients retain their ability to respond to other stimuli including viruses, lectins, co-stimulatory molecules, IL-2 and stimulation with anti-CD3/CD28 antibody. This may also explain the fact that tumor-bearing hosts may not have a systemic immune deficiency unless treated with a high dose of chemotherapy or at terminal stages of the disease.

MDSC also migrate into tumor site. However, the fate of these cells there is different. MDSC quickly differentiate into tumor-associated macrophages (TAM). Tumor microenvironment downregulate ROS production and induce iNOS expression possibly via upregulation of STAT1 (Kusmartsev, 2005). TAM produce high levels of NO and immunosuppressive cytokines (like IL-10). These create immunosuppressive environment where T-cell function is inhibited in antigen non-specific manner (Fig. 1). The relative importance of systemic antigen-specific tolerance vs. local immune suppression needs to be elucidated.

7 Conclusion

The role of ROS in tumor development and progression is difficult to underestimate. Through alterations in nucleic acid structure and inhibition of repair enzymes, ROS contribute to the accumulation of genomic abnormalities that result in the malignant transformation of cells. Cellular apoptosis is prevented by ROS through blocked activation of caspases. ROS is intimately involved in inhibition of immune responses in cancer via various mechanisms involving direct effect on T-cell receptors, block of signal transduction and affecting gene transcription. Peroxynitrite emerged as one of the major factors involved in ROS-mediated immune suppression. Blocking peroxynitrite generation or using scavengers could represent an attractive opportunity to decrease or even eliminate MDSC-induced T-cell tolerance and enhance the effect of cancer immunotherapy. Other potential molecular targets could be discovered if all tumor-induced hypothetical post-translational modifications of proteins involved in T cell-antigen-presenting cell interactions are explored.

References

Agostinelli, E., and Seiler, N. (2006). Non-irradiation-derived reactive oxygen species (ROS) and cancer: therapeutic implications. *Amino Acids* 31:341–355.

Almand, B., Clark, J. I., Nikitina, E., English, N. R., Knight, S. C., Carbone, D. P., and Gabrilovich, D. I. (2001). Increased production of immature myeloid cells in cancer patients. A mechanism of immunosuppression in cancer. *J Immunol* 166:678–689.

Alvarez, B., and Radi, R. (2003). Peroxynitrite reactivity with amino acids and proteins. *Amino Acids* 25:295–311.

Amici, M., Lupidi, G., Angeletti, M., Fioretti, E., and Eleuteri, A. M. (2003). Peroxynitrite-induced oxidation and its effects on isolated proteasomal systems. *Free Radic Biol Med* 34:987–996.

Andreyev, A. Y., Kushnareva, Y. E., and Starkov, A. A. (2005). Mitochondrial metabolism of reactive oxygen species. *Biochemistry (Mosc)* 70:200–214.

Aoe, T., Okamoto, Y., and Saito, T. (1995). Activated macrophages induce structural abnormalities of the T cell receptor-CD3 complex. *J Exp Med* 181:1881–1886.

Assari, T. (2006). Chronic granulomatous disease; fundamental stages in our understanding of CGD. *Med Immunol* 5:4.

Babior, B. M. (1984). The respiratory burst of phagocytes. *J Clin Invest* 73:599–601.

Bagley, J., Singh, G., and Iacomini, J. (2007). Regulation of oxidative stress responses by ataxia-telangiectasia mutated is required for T cell proliferation. *J Immunol* 178:4757–4763.

Banfi, B., Clark, R. A., Steger, K., and Krause, K. H. (2003). Two novel proteins activate superoxide generation by the NADPH oxidase NOX1. *J Biol Chem* 278:3510–3513.

Banfi, B., Malgrange, B., Knisz, J., Steger, K., Dubois-Dauphin, M., and Krause, K. H. (2004). NOX3, a superoxide-generating NADPH oxidase of the inner ear. *J Biol Chem* 279: 16065–16072.

Bentz, B. G., Haines, G. K., 3rd and Radosevich, J. A. (2000). Increased protein nitrosylation in head and neck squamous cell carcinogenesis. *Head Neck* 22:64–70.

Bingisser, R. M., Tilbrook, P. A., Holt, P. G., and Kees, U. R. (1998). Macrophage-derived nitric oxide regulates T cell activation via reversible disruption of the Jak3/STAT5 signaling pathway. *J Immunol* 160:5729–5734.

Bolis, A., Corbetta, S., Cioce, A., and de Curtis, I. (2003). Differential distribution of Rac1 and Rac3 GTPases in the developing mouse brain: implications for a role of Rac3 in Purkinje cell differentiation. *Eur J Neurosci* 18:2417–2424.

Bonnefoy, M., Drai, J., and Kostka, T. (2002). [Antioxidants to slow aging, facts and perspectives]. *Presse Med* 31:1174–1184.

Boucher, J. L., Moali, C., and Tenu, J. P. (1999). Nitric oxide biosynthesis, nitric oxide synthase inhibitors and arginase competition for L-arginine utilization. *Cell Mol Life Sci* 55:1015–1028.

Boutard, V., Havouis, R., Fouqueray, B., Philippe, C., Moulinoux, J. P., and Baud, L. (1995). Transforming factor beta stimulates arginase activity in macrophages: implications for the regulation of macrophage cytotoxicity. *J Immunol* 155:2077–2084.

Bronte, V., Casic, T., Gri, G., Gallana, K., Borsellino, G., Marrigo, I., Battistini, L., Iafrate, M., Prayer-Galletti, U., Pagano, F., and Viola, A. (2005). Boosting antitumor responses of T lymphocytes infiltrating human prostate cancers. *J Exp Med* 201:1257–1268.

Bronte, V., Serafini, P., De Santo, C., Marigo, I., Tosello, V., Mazzoni, A., Segal, D. M., Staib, C., Lowel, M., Sutter, G., Colombo, M. P., and Zanovello, P. (2003). IL-4-induced arginase 1 suppresses alloreactive T cells in tumor-bearing mice. *J Immunol* 170:270–278.

Bronte, V., Wang, M., Overwijk, W., Surman, D., Pericle, F., Rosenberg, S. A., and Restifo, N. P. (1998). Apoptotic death of CD8+ T lymphocytes after immunization: induction of a suppressive population of Mac-1+/Gr-1+ cells. *J Immunol* 161:5313–5320.

Cawthon, A. G., and Alexander-Miller, M. A. (2002). Optimal colocalization of TCR and CD8 as a novel mechanism for the control of functional avidity. *J Immunol* 169:3492–3498.

Chen, J. J., and Yu, B. P. (1994). Alterations in mitochondrial membrane fluidity by lipid peroxidation products. *Free Radic Biol Med* 17:411–418.

Chiarugi, P., Pani, G., Giannoni, E., Taddei, L., Colavitti, R., Raugei, G., Symons, M., Borrello, S., Galeotti, T., and Ramponi, G. (2003). Reactive oxygen species as essential mediators of cell adhesion: the oxidative inhibition of a FAK tyrosine phosphatase is required for cell adhesion. *J Cell Biol* 161:933–944.

Chio, K. S., and Tappel, A. L. (1969). Synthesis and characterization of the fluorescent products derived from malonaldehyde and amino acids. *Biochemistry* 8:2821–2826.

Cobbs, C. S., Whisenhunt, T. R., Wesemann, D. R., Harkins, L. E., Van Meir, E. G., Samanta, M. (2003). Inactivation of wild-type p53 protein function by reactive oxygen and nitrogen species in malignant glioma cells. *Cancer Res* 63:8670–8673.

Dairou, J., Dupret, J. M., and Rodrigues-Lima, F. (2005). Impairment of the activity of the xenobiotic-metabolizing enzymes arylamine N-acetyltransferases 1 and 2 (NAT1/NAT2) by peroxynitrite in mouse skeletal muscle cells. *FEBS Lett* 579:4719–4723.

Datta, D., Vaidehi, N., Xu, X., and Goddard, W. A., 3rd (2002). Mechanism for antibody catalysis of the oxidation of water by singlet dioxygen. *Proc Natl Acad Sci U S A* 99:2636–2641.

Dean, R. T., Fu, S., Stocker, R., and Davies, M. J. (1997). Biochemistry and pathology of radical-mediated protein oxidation. *Biochem J* 324 (Pt 1):1–18.

Decoursey, T. E., and Ligeti, E. (2005). Regulation and termination of NADPH oxidase activity. *Cell Mol Life Sci* 62:2173–2193.

Denu, J. M., and Tanner, K. G. (1998). Specific and reversible inactivation of protein tyrosine phosphatases by hydrogen peroxide: evidence for a sulfenic acid intermediate and implications for redox regulation. *Biochemistry* 37:5633–5642.

Drake, D. R., 3rd, Ream, R. M., Lawrence, C. W., and Braciale, T. J. (2005). Transient loss of MHC class I tetramer binding after CD8+ T cell activation reflects altered T cell effector function. *J Immunol* 175:1507–1515.

Edgeworth, J., Gorman, M., Bennett, R., Freemont, P., and Hogg, N. (1991). Identification of p8,14 as a highly abundant heterodimeric calcium binding protein complex of myeloid cells. *J Biol Chem* 266:7706–7713.

Ekmekcioglu, S., Ellerhorst, J., Smid, C. M., Prieto, V. G., Munsell, M., Buzaid, A. C., and Grimm, E. A. (2000). Inducible nitric oxide synthase and nitrotyrosine in human metastatic melanoma tumors correlate with poor survival. *Clin Cancer Res* 6:4768–4775.

Eze, M. O. (1992). Membrane fluidity, reactive oxygen species, and cell-mediated immunity: implications in nutrition and disease. *Med Hypotheses* 37:220–224.

Fahmy, T. M., Bieler, J. G., Edidin, M., and Schneck, J. P. (2001). Increased TCR avidity after T cell activation: a mechanism for sensing low-density antigen. *Immunity* 14:135–143.

Ferraro, D., Corso, S., Fasano, E., Panieri, E., Santangelo, R., Borrello, S., Giordano, S., Pani, G., and Galeotti, T. (2006). Pro-metastatic signaling by c-Met through RAC-1 and reactive oxygen species (ROS). *Oncogene* 25:3689–3698.

Finan, P., Shimizu, Y., Gout, I., Hsuan, J., Truong, O., Butcher, C., Bennett, P., Waterfield, M. D., and Kellie, S. (1994). An SH3 domain and proline-rich sequence mediate an interaction between two components of the phagocyte NADPH oxidase complex. *J Biol Chem* 269: 13752–13755.

Forman, H. J., and Torres, M. (2002). Reactive oxygen species and cell signaling: respiratory burst in macrophage signaling. *Am J Respir Crit Care Med* 166:S4–S8.

Francke, U., Hsieh, C. L., Foellmer, B. E., Lomax, K. J., Malech, H. L., and Leto, T. L. (1990). Genes for two autosomal recessive forms of chronic granulomatous disease assigned to 1q25 (NCF2) and 7q11.23 (NCF1). *Am J Hum Genet* 47:483–492.

Fu, X., Kassim, S. Y., Parks, W. C., and Heinecke, J. W. (2003). Hypochlorous acid generated by myeloperoxidase modifies adjacent tryptophan and glycine residues in the catalytic domain of matrix metalloproteinase-7 (matrilysin): an oxidative mechanism for restraining proteolytic activity during inflammation. *J Biol Chem* 278:28403–28409.

Fu, Y., Watson, G., Jimenez, J., Wang, Y., and Lopez, D. (1990). Expansion of immunoregulatory macrophages by granulocyte-macrophage colony-stimulating factor derived from a murine mammary tumor.*Cancer Res* 50:227.

Gabrilovich, D. (2004). Mechanisms and functional significance of tumour-induced dendritic-cell defects. *Nat Rev Immunol* 4:941–952.

Gabrilovich, D., Bronte, V., Chen, S.-H., Colombo, M. P., Ochoa, A., Ostrand-Rosenberg, S., and Schreiber, H. (2007). The terminology issue for myeloid-derived suppressor cells. *Cancer Res* 67:425.

Gabrilovich, D. I., Velders, M., Sotomayor, E., and Kast, W. M. (2001). Mechanism of immune dysfunction in cancer mediated by immature Gr-1+ myeloid cells. *J Immunol* 166:5398–5406.

Gebhardt, C., Nemeth, J., Angel, P., and Hess, J. (2006). S100A8 and S100A9 in inflammation and cancer. *Biochem Pharmacol* 72:1622–1631.

Geiszt, M., Lekstrom, K., Witta, J., and Leto, T. L. (2003). Proteins homologous to p47phox and p67phox support superoxide production by NAD(P)H oxidase 1 in colon epithelial cells. *J Biol Chem* 278:20006–20012.

Gotoh, Y., and Cooper, J. A. (1998). Reactive oxygen species- and dimerization-induced activation of apoptosis signal-regulating kinase 1 in tumor necrosis factor-alpha signal transduction. *J Biol Chem* 273:17477–17482.

Groemping, Y., and Rittinger, K. (2005). Activation and assembly of the NADPH oxidase: a structural perspective. *Biochem J* 386:401–416.

Groves, J. T. (1999). Peroxynitrite: reactive, invasive and enigmatic. *Curr Opin Chem Biol* 3:226–235.

Grune, T., Reinheckel, T., and Davies, K. J. (1997). Degradation of oxidized proteins in mammalian cells. *FASEB J* 11:526–534.

Haataja, L., Groffen, J., and Heisterkamp, N. (1997). Characterization of RAC3, a novel member of the Rho family. *J Biol Chem* 272:20384–20388.

Hampton, M. B., Kettle, A. J., and Winterbourn, C. C. (1998). Inside the neutrophil phagosome: oxidants, myeloperoxidase, and bacterial killing. *Blood* 92:3007–3017.

Han, C. H., Freeman, J. L., Lee, T., Motalebi, S. A., and Lambeth, J. D. (1998). Regulation of the neutrophil respiratory burst oxidase. Identification of an activation domain in p67(phox). *J Biol Chem* 273:16663–16668.

Han, C. H., and Lee, M. H. (2000). Activation domain in P67phox regulates the steady state reduction of FAD in gp91phox. *J Vet Sci* 1:27–31.
Harari, O., and Liao, J. K. (2004). Inhibition of MHC II gene transcription by nitric oxide and antioxidants. *Curr Pharm Des* 10:893–898.
Huang, B., Pan, P. Y., Li, Q., Sato, A. I., Levy, D. E., Bromberg, J., Divino, C. M., and Chen, S. H. (2006). Gr-1+CD115+ immature myeloid suppressor cells mediate the development of tumor-induced T regulatory cells and T-cell anergy in tumor-bearing host. *Cancer Res* 66:1123–1131.
Husemann, J., Obstfeld, A., Febbraio, M., Kodama, T., and Silverstein, S. C. (2001). CD11b/CD18 mediates production of reactive oxygen species by mouse and human macrophages adherent to matrixes containing oxidized LDL. *Arterioscler Thromb Vasc Biol* 21:1301–1305.
Kamata, H., Manabe, T., Oka, S., Kamata, K., and Hirata, H. (2002). Hydrogen peroxide activates IkappaB kinases through phosphorylation of serine residues in the activation loops. *FEBS Lett* 519:231–237.
Kao, C., Daniels, M. A., and Jameson, S. C. (2005). Loss of CD8 and TCR binding to Class I MHC ligands following T cell activation. *Int Immunol* 17:1607–1617.
Kerkhoff, C., Klempt, M., and Sorg, C. (1998). Novel insights into structure and function of MRP8 (S100A8) and MRP14 (S100A9). *Biochim Biophys Acta* 1448:200–211.
Kinnula, V. L., Torkkeli, T., Kristo, P., Sormunen, R., Soini, Y., Paakko, P., Ollikainen, T., Kahlos, K., Hirvonen, A., and Knuutila, S. (2004). Ultrastructural and chromosomal studies on manganese superoxide dismutase in malignant mesothelioma. *Am J Respir Cell Mol Biol* 31: 147–153.
Kono, K., Salazar-Onfray, F., Petersson, M., Hansson, J., Masucci, G., Wasserman, K., Nakazawa, T., Anderson, P., and Kiessling, R. (1996). Hydrogen peroxide secreted by tumor-derived macrophages down-modulates signal-transducing zeta molecules and inhibits tumor-specific T cell-and natural killer cell-mediated cytotoxicity. *Eur J Immunol* 26:1308–1313.
Kusmartsev, S., and Gabrilovich, D. I. (2003). Inhibition of myeloid cell differentiation in cancer: the role of reactive oxygen species. *J Leukoc Biol* 74:186-196.
Kusmartsev, S., and Gabrilovich, D. (2005). STAT1 signaling regulates tumor-associated macrophage-mediated T cell deletion. *J Immunol* 174:4880–4891.
Kusmartsev, S., and Gabrilovich, D. I. (2006). Role of immature myeloid cells in mechanisms of immune evasion in cancer. *Cancer Immunol Immunother* 55:237–245.
Kusmartsev, S., Li, Y., and Chen, S. (2000). Gr-1+ myeloid cells derived from tumor-bearing mice inhibit primary T cell activation induced through CD3/CD28 costimulation. *J Immunol* 165:779.
Kusmartsev, S., Nagaraj, S., and Gabrilovich, D. I. (2005). Tumor-associated CD8+ T cell tolerance induced by bone marrow-derived immature myeloid cells. *J Immunol* 175:4583–4592.
Kusmartsev, S., Nefedova, Y., Yoder, D., and Gabrilovich, D. I. (2004). Antigen-specific inhibition of CD8+ T cell response by immature myeloid cells in cancer is mediated by reactive oxygen species. *J Immunol* 172:989–999.
Kyaw, M., Yoshizumi, M., Tsuchiya, K., Kirima, K., Suzaki, Y., Abe, S., Hasegawa, T., and Tamaki, T. (2002). Antioxidants inhibit endothelin-1 (1-31)-induced proliferation of vascular smooth muscle cells via the inhibition of mitogen-activated protein (MAP) kinase and activator protein-1 (AP-1). *Biochem Pharmacol* 64:1521–1531.
Laurent, A., Nicco, C., Chereau, C., Goulvestre, C., Alexandre, J., Alves, A., Levy, E., Goldwasser, F., Panis, Y., Soubrane, O., Weill, B., and Batteux, F. (2005). Controlling tumor growth by modulating endogenous production of reactive oxygen species. *Cancer Res* 65: 948–956.
Leonard, S. S., Harris, G. K., and Shi, X. (2004). Metal-induced oxidative stress and signal transduction. *Free Radic Biol Med* 37:1921–1942.
Leto, T. L., Adams, A. G., and de Mendez, I. (1994). Assembly of the phagocyte NADPH oxidase: binding of Src homology 3 domains to proline-rich targets. *Proc Natl Acad Sci U S A* 91:10650–10654.
Li, Q., Pan, P. Y., Gu, P., Xu, D., and Chen, S. H. (2004). Role of immature myeloid Gr-1+ cells in the development of antitumor immunity. *Cancer Res* 64:1130–1139.

Lockhart, D. C., Chan, A. K., Mak, S., Joo, H. G., Daust, H. A., Carritte, A., Douville, C. C., Goedegebuure, P. S., and Eberlein, T. J. (2001). Loss of T-cell receptor-CD3zeta and T-cell function in tumor-infiltrating lymphocytes but not in tumor-associated lymphocytes in ovarian carcinoma. *Surgery* 129:749–756.

Maile, R., Siler, C. A., Kerry, S. E., Midkiff, K. E., Collins, E. J., and Frelinger, J. A. (2005). Peripheral "CD8 tuning" dynamically modulates the size and responsiveness of an antigen-specific T cell pool in vivo. *J Immunol* 174:619–627.

Malmberg, K. J., Arulampalam, V., Ichihara, F., Petersson, M., Seki, K., Andersson, T., Lenkei, R., Masucci, G., Pettersson, S., and Kiessling, R. (2001). Inhibition of activated/memory (CD45RO(+)) T cells by oxidative stress associated with block of NF-kappaB activation. *J Immunol* 167:2595–2601.

Mantovani, G., Maccio, A., Madeddu, C., Mura, L., Gramignano, G., Lusso, M. R., Massa, E., Mocci, M., and Serpe, R. (2003). Antioxidant agents are effective in inducing lymphocyte progression through cell cycle in advanced cancer patients: assessment of the most important laboratory indexes of cachexia and oxidative stress. *J Mol Med* 81:664–673.

Marnett, L. J., Hurd, H. K., Hollstein, M. C., Levin, D. E., Esterbauer, H., and Ames, B. N. (1985). Naturally occurring carbonyl compounds are mutagens in Salmonella tester strain TA104. *Mutat Res* 148:25–34.

Mates, J. M., Perez-Gomez, C., and Nunez de Castro, I. (1999). Antioxidant enzymes and human diseases. *Clin Biochem* 32:595–603.

Melani, C., Chiodoni, C., Forni, G., and Colombo, M. P. (2003). Myeloid cell expansion elicited by the progression of spontaneous mammary carcinomas in c-erbB-2 transgenic BALB/c mice suppresses immune reactivity. *Blood* 102:2138–2145.

Modolell, M., Corraliza, I. M., Link, F., Soler, G., and Eichmann, K. (1995). Reciprocal regulation of the nitric oxide synthase/arginase balance in mouse bone marrow derived macrophages by Th1 and Th2 cytokines. *Eur J Immunol* 25:1101–1104.

Moll, J., Sansig, G., Fattori, E., and van der Putten, H. (1991). The murine rac1 gene: cDNA cloning, tissue distribution and regulated expression of rac1 mRNA by disassembly of actin microfilaments. *Oncogene* 6:863–866.

Mustelin, T., Vang, T., and Bottini, N. (2005). Protein tyrosine phosphatases and the immune response. *Nat Rev Immunol* 5:43–57.

Nagaraj, S., Gupta, K., Pisarev, V., Kinarsky, L., Sherman, S., Kang, L., Herber, D., Schneck, J., and Gabrilovich, D. (2007). Altered recognition of antigen is a novel mechanism of CD8+ T cell tolerance in cancer. *Nat Med* 13:828–835.

Nakamura, Y., Yasuoka, H., Tsujimoto, M., Yoshidome, K., Nakahara, M., Nakao, K., Nakamura, M., and Kakudo, K. (2006). Nitric oxide in breast cancer: induction of vascular endothelial growth factor-C and correlation with metastasis and poor prognosis. *Clin Cancer Res* 12:1201–1207.

Nefedova, Y., Huang, M., Kusmartsev, S., Bhattacharya, R., Cheng, P., Salup, R., Jove, R., and Gabrilovich, D. (2004). Hyperactivation of STAT3 is involved in abnormal differentiation of dendritic cells in cancer. *J Immunol* 172:464–474.

Newton, R. A., and Hogg, N. (1998). The human S100 protein MRP-14 is a novel activator of the beta 2 integrin Mac-1 on neutrophils. *J Immunol* 160:1427–1435.

Nicholls, S. J., and Hazen, S. L. (2005). Myeloperoxidase and cardiovascular disease. *Arterioscler Thromb Vasc Biol* 25:1102–1111.

Okada, F., Kobayashi, M., Tanaka, H., Kobayashi, T., Tazawa, H., Iuchi, Y., Onuma, K., Hosokawa, M., Dinauer, M. C., and Hunt, N. H. (2006). The role of nicotinamide adenine dinucleotide phosphate oxidase-derived reactive oxygen species in the acquisition of metastatic ability of tumor cells. *Am J Pathol* 169:294–302.

Otsuji, M., Kimura, Y., Aoe, T., Okamoto, Y., and Saito, T. (1996). Oxidative stress by tumor-derived macrophages suppresses the expression of CD3 zeta chain of T-cell receptor complex and antigen-specific T-cell responses. *Proc Natl Acad Sci USA* 93:13119–13124.

Pastore, A., Federici, G., Bertini, E., and Piemonte, F. (2003). Analysis of glutathione: implication in redox and detoxification. *Clin Chim Acta* 333:19–39.
Rabinovich, G. A., Gabrilovich, D., and Sotomayor, E. M. (2007). Immunosuppressive strategies that are mediated by tumor cells. *Annu Rev Immunol* 25:267–296.
Reeves, E. P., Lu, H., Jacobs, H. L., Messina, C. G., Bolsover, S., Gabella, G., Potma, E. O., Warley, A., Roes, J., and Segal, A. W. (2002). Killing activity of neutrophils is mediated through activation of proteases by K+ flux. *Nature* 416:291–297.
Ridley, A. J., Paterson, H. F., Johnston, C. L., Diekmann, D., and Hall, A. (1992). The small GTP-binding protein rac regulates growth factor-induced membrane ruffling. *Cell* 70:401–410.
Rivoltini, L., Carrabba, M., Huber, V., Castelli, C., Novellino, L., Dalerba, P., Mortarini, R., Arancia, G., Anichini, A., Fais, S., and Parmiani, G. (2002). Immunity to cancer: attack and escape in T lymphocyte-tumor cell interaction. *Immunol Rev* 188:97–113.
Rodaway, A. R., Teahan, C. G., Casimir, C. M., Segal, A. W., and Bentley, D. L. (1990). Characterization of the 47-kilodalton autosomal chronic granulomatous disease protein: tissue-specific expression and transcriptional control by retinoic acid. *Mol Cell Biol* 10:5388–5396.
Rodriguez, A. M., Carrico, P. M., Mazurkiewicz, J. E., and Melendez, J. A. (2000). Mitochondrial or cytosolic catalase reverses the MnSOD-dependent inhibition of proliferation by enhancing respiratory chain activity, net ATP production, and decreasing the steady state levels of $H(2)O(2)$. *Free Radic Biol Med* 29:801–813.
Ruiz de Morales, J., Velez, D., and Subiza, J. (1999). Ehrlich tumor stimulates extramedullar hematopoiesis in mice without secreting identifiable colony-stimulating factors and without engagement of host T cells. *Exp Hematol* 27:1757.
Ryter, S. W., and Tyrrell, R. M. (1998). Singlet molecular oxygen ((1)O2): a possible effector of eukaryotic gene expression. *Free Radic Biol Med* 24:1520–1534.
Salvadori, S., Martinelli, G., and Zier, K. (2000). Resection of solid tumors reverses T cell defects and restores protective immunity. *J Immunol* 164:2214.
Sathyamoorthy, M., de Mendez, I., Adams, A. G., and Leto, T. L. (1997). p40(phox) down-regulates NADPH oxidase activity through interactions with its SH3 domain. *J Biol Chem* 272:9141–9146.
Sauer, H., Wartenberg, M., and Hescheler, J. (2001). Reactive oxygen species as intracellular messengers during cell growth and differentiation. *Cell Physiol Biochem* 11:173–186.
Schmielau, J., and Finn, O. J. (2001). Activated granulocytes and granulocyte-derived hydrogen peroxide are the underlying mechanism of suppression of T-cell function in advanced cancer patients. *Cancer Res* 61:4756–4760.
Segal, A. W. (2005). How neutrophils kill microbes. *Annu Rev Immunol* 23:197–223.
Serafini, P., Borrello, I., and Bronte, V. (2006). Myeloid suppressor cells in cancer: recruitment, phenotype, properties, and mechanisms of immune suppression. *Semin Cancer Biol* 16:53–65.
Stadtman, E. R., and Oliver, C. N. (1991). Metal-catalyzed oxidation of proteins. Physiological consequences. *J Biol Chem* 266:2005–2008.
Starke-Reed, P. E., and Oliver, C. N. (1989). Protein oxidation and proteolysis during aging and oxidative stress. *Arch Biochem Biophys* 275:559–567.
St Clair, D., Zhao, Y., Chaiswing, L., and Oberley, T. (2005). Modulation of skin tumorigenesis by SOD. *Biomed Pharmacother* 59:209–214.
Szuster-Ciesielska, A., Hryciuk-Umer, E., Stepulak, A., Kupisz, K., and Kandefer-Szerszen, M. (2004). Reactive oxygen species production by blood neutrophils of patients with laryngeal carcinoma and antioxidative enzyme activity in their blood. *Acta Oncol* 43:252–258.
Szweda, L. I., Uchida, K., Tsai, L., and Stadtman, E. R. (1993). Inactivation of glucose-6-phosphate dehydrogenase by 4-hydroxy-2-nonenal. Selective modification of an active-site lysine. *J Biol Chem* 268:3342–3347.
Takada, Y., Mukhopadhyay, A., Kundu, G. C., Mahabeleshwar, G. H., Singh, S., and Aggarwal, B. B. (2003). Hydrogen peroxide activates NF-kappa B through tyrosine phosphorylation of I kappa B alpha and serine phosphorylation of p65: evidence for the involvement of I kappa B alpha kinase and Syk protein-tyrosine kinase. *J Biol Chem* 278:24233–24241.

Terabe, M., Matsui, S., Park, J. M., Mamura, M., Noben-Trauth, N., Donaldson, D. D., Chen, W., Wahl, S. M., Ledbetter, S., Pratt, B., Letterio, J. J., Paul, W. E., and Berzofsky, J. A. (2003). Transforming growth factor-beta production and myeloid cells are an effector mechanism through which CD1d-restricted T cells block cytotoxic T lymphocyte-mediated tumor immunosurveillance: abrogation prevents tumor recurrence. *J Exp Med* 198:1741–1752.

Tobiume, K., Saitoh, M., and Ichijo, H. (2002). Activation of apoptosis signal-regulating kinase 1 by the stress-induced activating phosphorylation of pre-formed oligomer. *J Cell Physiol* 191:95–104.

Valko, M., Izakovic, M., Mazur, M., Rhodes, C. J., and Telser, J. (2004). Role of oxygen radicals in DNA damage and cancer incidence. *Mol Cell Biochem* 266:37–56.

Valko, M., Rhodes, C. J., Moncol, J., Izakovic, M., and Mazur, M. (2006). Free radicals, metals and antioxidants in oxidative stress-induced cancer. *Chem Biol Interact* 160:1–40.

Vickers, S. M., MacMillan-Crow, L. A., Green, M., Ellis, C., and Thompson, J. A. (1999). Association of increased immunostaining for inducible nitric oxide synthase and nitrotyrosine with fibroblast growth factor transformation in pancreatic cancer. *Arch Surg* 134:245–251.

Waris, G., and Ahsan, H. (2006). Reactive oxygen species: role in the development of cancer and various chronic conditions. *J Carcinog* 5:14.

Weiss, S. J., Test, S. T., Eckmann, C. M., Roos, D., and Regiani, S. (1986). Brominating oxidants generated by human eosinophils. *Science* 234:200–203.

Werner, E., and Werb, Z. (2002). Integrins engage mitochondrial function for signal transduction by a mechanism dependent on Rho GTPases. *J Cell Biol* 158:357–368.

White, E., Shannon, J. S., and Patterson, R. E. (1997). Relationship between vitamin and calcium supplement use and colon cancer. *Cancer Epidemiol Biomarkers Prev* 6:769–774.

Winterbourn, C. C. (1993). Superoxide as an intracellular radical sink. *Free Radic Biol Med* 14:85–90.

Wu, G., and Morris, S. M. (1998). Arginine metabolism: nitric oxide and beyond. *Biochem J* 336:1–17.

Young, M., Newby, M., and Wepsic, T. (1987). Hematopoiesis and suppressor bone marrow cells in mice bearing large metastatic Lewis lung carcinoma tumors.*Cancer Res* 47:100–105.

Zea, A. H., Rodriguez, P. C., Atkins, M. B., Hernandez, C., Signoretti, S., Zabaleta, J., McDermott, D., Quiceno, D., Youmans, A., O'Neill, A., Mier, J., and Ochoa, A. C. (2005). Arginase-producing myeloid suppressor cells in renal cell carcinoma patients: a mechanism of tumor evasion. *Cancer Res* 65:3044–3048.

Zhu, Q. S., Xia, L., Mills, G. B., Lowell, C. A., Touw, I. P., and Corey, S. J. (2006). G-CSF induced reactive oxygen species involves Lyn-PI3-kinase-Akt and contributes to myeloid cell growth. *Blood* 107:1847–1856.

Tumor Stroma and the Antitumor Immune Response

Bin Zhang, Donald A. Rowley, and Hans Schreiber

In the saga of cancers[1] and the response to it, stroma plays a central and changing role from the beginning to the final stage of cancer development. The evolution of changes in immune and stromal components in tumors caused by cancer cells is as complicated and intertwined as the progressive genetic changes in the cancer cells themselves. Most of the complicated interactions of the numerous components in malignant tumors have been presented in the previous chapters. This being the case, we will limit our discussion to large established tumors where the objective is to stabilize or reverse tumor growth by adoptive transfer of T cells. In the process, we will necessarily consider briefly some of the biology covered in the previous chapters.

We start with five premises:

1. Experimentally in mice, the tumors we consider are at least 1 cm in diameter, 2 weeks old and in a stage of continuing growth. Most human malignant tumors are at least this size and contain comparable numbers of cells when they are first detectable clinically (Sweeney et al., 2003).[2]

H. Schreiber
The Department of Pathology and the Committee on Immunology, University of Chicago, 5841 S. Maryland, MC3083, Chicago, IL 60637, USA
E-mail: hszz@uchicago.edu

[1] The term cancer or cancer cell refers to genetically altered malignant cell; the word tumor refers to the mass produced by the cancer cells and stroma, the host-derived components consisting of a variety of non-malignant cells and extracellular matrix. Stromal cells include bone-marrow-derived cells such as macrophages, granulocytes and lymphocytes and non-bone-marrow-derived cells such as endothelium and fibroblasts.

[2] A 1 cm diameter solid tumor usually contains 10^9 cancer cells containing a variable number of cancer stem cells. Therefore, we use at least 1 cm in average diameter tumors (≥ 500 mm^3) in our studies. Some have argued that 0.5 g tumor in a mouse (25 g) is the equivalent of over 1 kg of tumor in a human (60 kg), but murine and human cell sizes are similar and therefore 1 cm diameter tumors in *mouse or man* contain about 10^9 cancer cells and have a similar chance of resistance to therapy due to variants. With these non-hemopoietic tumors, most currently used experimental immunotherapies are not curative and relapse occurs after initial shrinkage. These problems may occur even with smaller size tumors and when multiple different treatments are combined. We think it is very unrealistic to expect that any single treatment will eradicate all cancer cells.

2. Cancer cells are antigenic, i.e., cancers express epitopes which the individual bearing the tumor can respond to immunologically. These responses are obviously not effective at this stage of tumor growth because of tumor-induced immune suppression.
3. Growth of large tumors can be stabilized or reversed by cytolytic T cells; such T cells can be rescued and derived from the individual bearing the tumor after reversing the defects.
4. Complete eradication of large tumors by T cells requires killing of tumor stroma cross-presenting cancer cell antigen (Ag); otherwise, Ag loss variants (ALVs) escape with recurrence of cancers.
5. Because some cancers release too little cancer Ag, treatment such as radiation or chemotherapy is necessary for release of sufficient Ag for cross-presentation by stromal cells.

These five premises are derived from mouse models; the experiments have necessarily been reductionist in order to sort out the contributions of different components of host–cancer interactions. Work accomplished to date cannot verify all aspects of the assumptions made, but the inference is strong from other work and ours that the premises will indeed apply to many human cancers. Our emphasis is on premises 4 and 5 though clearly premises 2 and 3 are crucial and dealt with in the previous chapters and are an important component of our ongoing research.

1 The Model

The importance of our chapter depends on confirming premise 4 and showing details supporting premise 5. In our model, OT-1 transgenic mice ($H-2^b$) or $Rag1^{-/-}$ mice ($H-2^b$) were used as recipient hosts. Murine fibrosarcoma MC57-SIY-Hi ($H-2^b$), MC57-SIY-Lo ($H-2^b$) or PRO4L-SIY-Hi ($H-2^k$) cancer cells were injected s.c. into recipient mice (Spiotto et al., 2004). Solid tumors are established by 2 weeks after tumor challenge. The pre-activated 2C transgenic T cells that specifically recognize SIY epitope were adoptively transferred into the tumor-bearing mice. Chemotherapy or radiotherapy was combined with T-cell therapy for successful treatment of MC57-SIY-Lo tumors. High-affinity TCR tetramers were used for flow cytometry to visualize the presentation of the SIY epitope used as tumor-specific Ag and presented on cell surface K^b as peptide–MHC Class I complex (pMHC) (Zhang et al., 2007).

2 Eradication of Established Tumors Requires Stromal Killing

$CD8^+$ cytotoxic T lymphocytes (CTLs) kill cancer cells and stromal cells presenting cancer-specific peptides in the context of the appropriate MHC class I molecules; killing is predominantly by a perforin-mediated process. Obviously, procedures which kill stroma benefit the host. Killing of cancer cells is necessary, but often not sufficient because ALVs escape immune recognition, a phenomenon observed in both mouse (Liu et al., 2005; Sanchez-Perez et al., 2005; Zhou et al., 2004) and human tumors (Khong et al., 2004; Yamshchikov et al., 2005).

Tumor stroma serves no beneficial purpose to the host but is necessary for tumor growth. Targeting cancer cells as well as the tumor stroma is essential for eradicating ALVs by adoptively transferred CD8 T cells (Spiotto and Schreiber, 2005; Spiotto et al., 2004) (Fig. 1). Both the bone-marrow-derived and the non-bone-marrow-derived components of the stroma have to be targeted. When only the cancer cells (not the stroma) were targeted, variant cancer cells resistant to the T cells escaped and killed the host (Fig. 1). Tumor-specific T cells failed to eradicate the cancer and the tumor escaped destruction if the Ag could not be cross-presented because of inappropriate MHC class I molecules on the stromal cells even when cancer cells expressed high levels of Ag (Fig. 1). Tumor-specific T cells also failed to eradicate established tumors when cancer cells expressed insufficient levels of the tumor Ag to sensitize the stroma (Fig. 3). Targeting cancer cells that express high levels of Ag, and thus simultaneously targeting the stroma, eradicated the cancer (Fig. 1). When tumors expressed low levels of cancer Ag, the T cells eliminated all Ag-positive cancer cells in the tumor, but ALVs grew out progressively (Fig. 3). Similar results were found in two different tumor models and antigenic systems that have been analyzed (Spiotto and Schreiber, 2005; Spiotto et al., 2004). We also showed that targeting stromal cells alone arrested tumor growth at least in a short

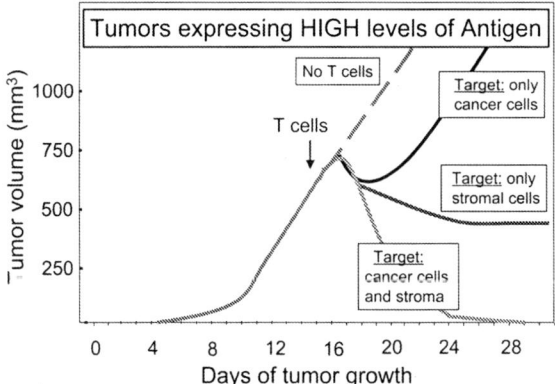

Fig. 1 The effects of adoptively transferred T cells on established tumors expressing high levels of antigen. Targeting cancer cells without targeting the stroma invariably leads to recurrence after a short reduction in tumor size. T cells fail to eradicate the tumor because Ag loss variants escape when tumor Ag could not be cross-presented by stromal cells from either MC57-L^d tumors or MC57-SIY-Hi tumors with inappropriate host MHC class I. Targeting stroma as well as cancer cells (MC57-SIY-Hi) eradicates the tumor and prevents the escape of these cancer variants. Targeting stroma alone (PRO4L-SIY-Hi) leads to the arrest of tumor growth but fails to cause complete regression. PRO4L-SIY-Hi ($H-2^k$) cancer cells failed to be lysed by CTL due to the inability to directly present $H-2^b$ restricted SIY-epitope. Only the stromal cells ($H-2^b$) can cross-present SIY Ag released from cancer cells

term (Fig. 1) (Spiotto et al., 2004). Cancer cells expressed high levels of Ag but could not be targeted directly because cancer cells expressed inappropriate MHC class I molecules. However, the stromal cells can cross-present Ag released from cancer cells; therefore, Ag-loaded stromal cells become tumor-specific targets for T cells. This suggested inhibition of tumor growth (not complete destruction) would probably occur in tumors where Ag released from cancer cells is loaded into the surrounding stroma although the cancer cells have lost MHC class I.

The mechanism for destruction of ALVs as bystanders has not been determined except that we already know that perforin and IFN-γ secretion by the T cells is required and bone-marrow-derived as well as non-bone-marrow-derived stroma must be targeted (Spiotto et al., 2004). Whether killing of the overwhelming majority of sensitive cancer cells will also kill a few residual cancer cells was studied three decades ago by outstanding investigators who reached seemingly discrepant conclusions (Klein and Klein 1972; Prehn, 1973; Weissman, 1973). It is clear today that the study that observed bystander killing also targeted stromal cells (Prehn, 1973). The endothelial cells may be the essential target in the non-bone-marrow-derived components of the stroma (Kim et al., 1993). $CD11b^+$ stromal cells are certainly targets in the bone-marrow-derived stroma (Spiotto et al., 2004). In addition, the tumor-specific cytolytic T cells may also target the other cell types in the tumor stroma, such as myofibroblasts (Seemayer et al., 1979).

When adoptively transferred T cells were directed against a non-mutant differentiation Ag expressed by cancer cells (Antony et al., 2005; Gattinoni et al., 2005a,b; Klebanoff et al., 2005), cancers usually recurred after a temporary yet potent antitumor effect despite additional treatment such as repeated high doses of IL-2. We speculate that the failure occurred because differentiation Ags and other self-Ags bind poorly or not at all to the presenting MHC class I molecules on stromal cells (Yu et al., 2004) and, without stromal killing, cancers grew out after initial regression. Alternatively or additionally, self-reactive T cells are at least partially tolerized in the host due to thymic or other mechanisms.

3 Induced Sensitization of Tumor Stroma Leads to Eradication of Established Tumor

Success of CTL therapy depends on destruction of stroma cross-presenting cancer Ag, which in turn depends on cancer cells releasing sufficient Ag to sensitize stromal cells. If this is the case, then CTL therapy should be effective against tumors expressing low levels of Ag if the tumor stroma is loaded with sufficient cancer cell Ag. Irradiation or chemotherapy causes apoptosis and necrosis of cancer cells; therefore, we tested whether these procedures could cause sufficient release of Ag from cancer cells expressing low levels of Ag to sensitize tumor stroma for killing by CTL. For this objective, we used engineered high-affinity TCR tetramers for detecting minute quantities of the relevant tumor-specific peptide–MHC complexes cross-presented by stromal cells, because wild-type TCR tetramers cannot detect this complex (Zhang et al., 2007). The single-chain m67 TCR consisted of the wild-type 2C Vβ-region, a 25 aa linker and the mutant m67 Vα-region fused to a C-terminal peptide that contained the recognition site for biotinylation by the BirA enzyme. Each biotin attaches to the single biotin-binding site on a streptavidin (SA) molecule that occurs as a homotetramer and is chemically coupled to the fluorescent dye phycoerythrin (PE), thereby generating mTCR-tetramer-SA-PE (Fig. 2a). The differences in peptide concentrations in the range of 0.1 nM–1 μM could be clearly distinguished. The probe is specific with minimal background binding to cells loaded with the irrelevant gp33 peptide at 1 μM concentration (Fig. 2b). Tumor stroma from established MC57-SIY-Lo tumors was not stained by the TCR tetramer (Fig. 2c). Also, the stromal cells from established MC57-SIY-Lo tumors were insensitive to lysis by specific T cells (Fig. 2d, left panel). We then tested whether radiation or chemotherapy caused sufficient tumor Ag release from MC57-SIY-Lo for uptake and presentation by $CD11b^+$ tumor stromal cells. High-affinity TCR tetramers detected SIY-K^b complexes on $CD11b^+$ stromal cells from the tumors of irradiated but not non-irradiated mice. In order to determine the time of maximal Ag cross-presentation by tumor-derived stroma in vivo, mice bearing 2-week-old tumors were sacrificed at 12 h, 1, 2, 3 and 4 days following local radiation. As shown in Fig. 2c, maximal cross-presentation by MC57-SIY-Lo tumor stroma was found 2 days following irradiation and then decreased with very little Ag remaining at day 4. Stromal cells derived from MC57-SIY-Lo tumors 2 days after irradiation were killed by CTL, but stromal cells isolated 4 days after irradiation were not killed, consistent with the kinetics of Ag uptake by stromal cells observed by staining (Fig. 2d, left and middle panels). MC57-SIY-Lo and MC57-gp33-Lo cancer cells were used as controls (Fig. 2c, right panel).

Consistent with the kinetics of stromal staining, locally irradiating 14-day tumors once with 10 Gy followed by T-cell transfer 2 days post-irradiation, but not 4 days post-irradiation, led to complete eradication of well-established MC57-SIY-Lo tumors (Table 1). The Lo Ag tumors, without irradiation, regularly escaped with outgrowth of cancer variants following T-cell therapy (Fig. 3). Thus, the time interval between radiation and adoptive T-cell therapy was crucial (Table 1). The effects observed post-irradiation also applied to a chemotherapeutic agent, gemcitabine. Maximal loading of the MC57-SIY-Lo tumor stroma also occurred at 2 days after drug treatment and complete rejection was achieved when T cells were transferred at 2 days but not at 4 days after gemcitabine treatment (Table 1).

Important work has been done on synergy between immunotherapy and irradiation (Ciernik et al., 1999; Ganss et al., 2002; Lugade et al., 2005; Reits et al., 2006), for review see Demaria et al. (2005), or chemotherapy (Casares et al., 2005; Lugade et al., 2005; Nowak et al., 2003), for review see Lake and Robinson (2005). However, unlike our study, these studies have dealt with the induction of immune responses or with direct effects of radiation or drugs on killing of cancer cells. For example, it was recently demonstrated that radiation, by activating the mTOR (mammalian target of rapamycin) pathway, could rapidly increase the intracellular pool of peptides derived from rapidly degraded proteins and increase MHC class I expression and peptide presentation by the cancer cells, thereby increasing the sensitivity

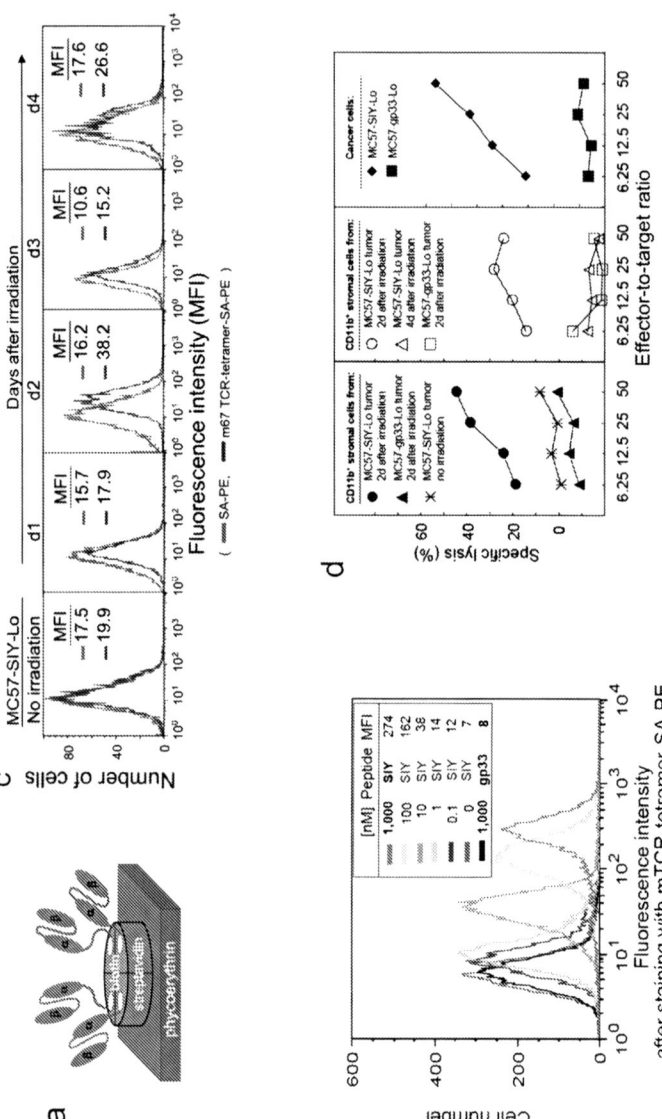

Fig. 2 Local radiation of tumors leads to the time-dependent stromal sensitization in established cancer. (a) Schematic structure of high-affinity T-cell receptor m67 tetramer. (b) The TCR tetramer can detect SIY-K^b complexes on T2K^b cells artificially loaded with peptide at the indicated concentrations as low as 0.1 nM. (c) Direct visualization of the transient appearance of tumor-specific peptide–MHC complexes on stromal cells using high-affinity T-cell receptor tetramers. Maximum loading of tumor stroma with cancer antigen occurred 2 days after local irradiation of tumors expressing low levels of antigen. (d) Two but not 4 days after local irradiation, CD11b$^+$ stromal cells purified from tumors expressing low levels of antigen were lysed by antigen-specific T cells in a ^{51}Cr release assay. Reproduced from *The Journal of Experimental Medicine*, 2007, 204:49–55. Copyright 2007 The Rockefeller University Press

Table 1 Stroma and timing are critical for complete elimination of established tumors

Treatment	Host	Tumor stroma	Rejection of tumors[a]	p value
Irradiation + T cells 2 days later	OT-1	H-2^b	12/12	–
Irradiation + T cells 4 days later	OT-1	H-2^b	0/9	0.001[b]
Irradiation + T cells 2 days later	C3H Rag2$^{-/-}$	H-2^k	0/9	0.001[b]
Gemcitabine + T cells 2 days later	OT-1	H-2^b	7/8	–
Gemcitabine + T cells 4 days later	OT-1	H-2^b	0/6	0.001[c]
Gemcitabine + T cells 2 days later	C3H Rag2$^{-/-}$	H-2^k	0/6	0.001[c]

Reproduced from *The Journal of Experimental Medicine*, 2007, 204:49–55. Copyright 2007 The Rockefeller University Press
[a] Data pooled from seven independent experiments
[b] Compared, respectively, to the group: OT-1 irradiation + T cell 2 days later
[c] Compared, respectively, to the group: OT-1 Gemcitabine + T cell 2 days later

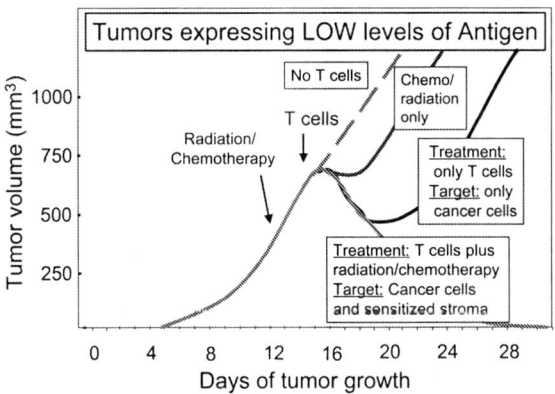

Fig. 3 Induced sensitization of tumor stroma that leads to eradication of established tumor expressing low levels of antigen by T cells. Established tumors expressing low levels of Ag (MC57-SIY-Lo) escape with outgrowth of cancer variants following adoptive transfer of T cells since antigenic cancer cells are only able to be targeted by T cells. This relapse is due to the escape of Ag loss variant cancer cells that cannot be killed by T cells. Radiation or chemotherapy alone cannot effectively control tumor growth. However, radiation or chemotherapy causes a "bolus" of tumor Ag release from cancer cells expressing low levels of Ag sufficient for uptake and presentation by tumor stromal cells. The Ag-load stroma thereby becomes a tumor-specific target for CTL. The complete tumor regression can be achieved by this radiation- or chemotherapy-induced stromal sensitization to T-cell killing. Reproduced from *The Journal of Experimental Medicine*, 2007, 204:49–55. Copyright 2007 The Rockefeller University Press

of cancer cells to direct killing by T cells (Reits et al., 2006). This enhancement of direct killing may be important and necessary, but may not be sufficient for treating large established solid tumors (Spiotto and Schreiber, 2005; Spiotto et al., 2004). The results of our experiments summarized in Fig. 3 show that radiation at doses insufficient to eradicate cancer synergizes with adoptively transferred T cells to eliminate cancers successfully. T cells need to target the stroma, and radiation was required to sensitize stroma with Ag released from cancer cells apoptosed by radiation. ALVs escaped when the stroma ($H\text{-}2^k$) was unable to cross-present the Ag (Zhang et al., 2007), and escape also occurred when T cells were transferred at a time when the stroma was no longer sensitized (Table 1). The direct effects of radiation or chemotherapy on tumor stroma may contribute to tumor rejection, as has been indicated by an earlier study using a chemotherapeutic agent (Ibe et al., 2001), but we have not examined this possibility.

4 Relevance to Studies Combining Active Immunization with Subsequent Radiation or Chemotherapy of Advanced Solid Tumors

Chemotherapy and/or radiation remain the treatment of choice for most advanced cancers. However, for advanced solid tumors in particular, these treatments alone or combined are rarely curative. Thus, it makes sense to combine them with other treatments, such as adoptive T-cell therapy shown above. An important alternative to adoptive immunotherapy with T cells is passive immunotherapy with antibodies or inhibitors against growth factors, growth factor receptors or other target molecules on the cancer cells or stromal cells (Kaminski et al., 1993; Lynch et al., 2004; Slamon et al., 2001; Yang et al., 2003). Some combinations of treatment have had remarkable success and have become parts of standard FDA-approved therapy. Similar to adoptive T-cell immunotherapy, passive antibody therapy does not require the tumor-bearing host to mount an active immune response and is therefore particularly suited to an immunosuppressed patient. By contrast, active immunization of the tumor-bearing host has to overcome tolerance and immunosuppression. Not only does the tumor-bearing state in itself cause some degree of immune suppression, but chemotherapy or radiation is also immunosuppressive. Combining active vaccination with chemotherapy may be counterintuitive (Gabrilovich, 2007), but we have found that adoptive T-cell therapy given shortly after radiation or chemotherapy is synergistic because the latter treatment resulted in loading up the stroma with Ag for destruction by the adoptively transferred T cells. Therefore, it is logical to propose that T-cell immunity once induced in the tumor-bearing host should be able to destroy stroma sensitized by Ag released from subsequent chemotherapy or radiation. Thus, the "unexpected" successes of several clinical trials (Antonia et al., 2006; Arlen et al., 2006; Gribben et al., 2005; Wheeler et al., 2004) using active vaccination followed by chemotherapy can be rationalized this way. Also, a solid T-cell memory-type response is relatively resistant to subsequent radiation or chemotherapy (Brent and Medawar, 1966). Thus, we would postulate that stromal sensitization occurs in both scenarios combining immunological treatments with chemotherapy or radiation. The reason for the beneficial effect of active vaccination on subsequent chemotherapy lasting for only weeks would be that the antitumor immune responses generated by active vaccination cannot be sustained for a long period in cancer patients (Gabrilovich, 2007), or that the cancer cells become so resistant to drugs or radiation that they no longer apoptose and release Ag into the stroma.

5 Targeting Stroma Only

To kill cancer cells expressing specific Ag, CTLs require recognizing the antigenic pMHC. However, the MHC class I molecule downregulation occurs frequently in many human cancers, and this abnormality might adversely affect the clinical course of cancer and the outcome of T-cell-based immunotherapy (Bubenik, 2003; Hicklin et al., 1999). Mutations in the MHC class I genes themselves, abnormalities in their regulation and/or defects in MHC class I-dependent antigen processing can underlie MHC class I downregulation. These mutations or regulatory changes modulate the susceptibility of cancer cells to lysis by CTL. Thus, direct killing of cancer cells is not always feasible. Immune selection of such cancer variants might explain the rapid progression and poor prognosis of cancers that exhibit MHC class I downregulation. In contrast, stroma cells are genetically much more stable, and targeting only tumor-derived stromal cells by T cells may contribute to effective control of tumor growth. Indeed, we found a long-term arrest or stabilization of progressive growth of established tumors by adoptive transfer of T cells in the absence of direct recognition of cancer cells (unpublished data). This tumor growth inhibition relying on stromal destruction is long lasting. We have found no tumors that escape from this inhibition and cancer cells reisolated from biopsy of these tumors retained the susceptibility to T cells (unpublished data).

CTL killing of stromal cells in normal organs and tissues cross-presenting Ag released from the cancer cells is unlikely for several reasons. For example, the destruction of cross-presenting tissue is completely Ag dose-dependent. This point is further supported by recent findings showing cross-presentation of intracellular peptides by transfer through gap junctions is limited to a few neighboring cells (Heath and Carbone, 2005; Neijssen et al., 2005). Although spread of apoptotic and necrotic material from killed cancer cells most likely extends beyond neighboring cells, tumor stroma is distinguished from other stromal cells of normal organs and tissues by its close proximity to cancer cells. Certainly we have not observed destruction of normal stroma away from the tumor. Tumor stroma consists of nonmalignant cells, but this stroma is by no means normal. Tumor stroma contains activated fibroblasts (Scott et al., 2003; Wesley et al., 1999), recently formed immature and leaky capillaries and many types of inflammatory cells comparable to those in a non-healing wound (Dvorak, 1986). Ag pick-up and presentation is therefore likely to be very different in such an active stroma. Thus, stroma may more readily acquire and cross-present highly expressed Ag to T cells.

Our studies emphasize the importance of the stromal elements for eradication of established cancers and preventing cancer escape not only by ALVs but also by drug or radio-resistant variants. Initiation of stromal cell destruction by T cells may trigger a sequence of events that leads to better tumor Ag cross-presentation and ultimately better direct cancer cell recognition and lysis. Other biological (Garin-Chesa et al., 1990), metabolic, inflammatory or angiostatic agents (Ferrara et al., 2007; Kelly, 2005; Nair et al., 2003; Niederman et al., 2002) that inhibit stromal elements may also potentiate immune elimination of cancer cells. Anti-angiogenic compounds, such as anti-VEGF monoclonal antibody, could theoretically stress cancer cells because of diminished perfusion and either make them more sensitive to T-cell-mediated killing or sensitize stroma by apoptosing some of the cancer cells. Inhibition of other stromal support cells, including fibroblasts, e.g., by inhibitors of fibroblast-activating protein, may have similar effects. Together, combining these or similar agents should be explored to determine the therapeutic potential.

6 Caveat and Concluding Remarks

Our model is artificial in terms of the neoantigen, transgenic T cells and immunodeficient hosts. However, the transfected Ag is a tumor-specific Ag, not a self-Ag. We believe that truly tumor-specific Ags exist and are the major Ags to be targeted on human tumors by autologous T cells whether the T cells are recovered directly from the host or cultured and "redirected" in vitro before use for therapy (Schreiber, 2003). The affinity of wild-type 2C TCR for the SIY-K^b complex is similar to most other "natural" CD8-dependent TCRs reactive with pMHC (Holler and Kranz, 2003; Holler and Kranz, 2003). Thus, there is nothing unnatural about the TCR/pMHC interaction we used. We think that self-Ag may be inferior targets for therapy because the affinity of TCR for pMHC is often low (Yu et al., 2004), or the T cells are partially tolerized because they are self-reactive.

We have explored conditions in which the target Ag is expressed at low levels, as may well occur in human cancers. We provide a solution to this problem by combining adoptive T-cell therapy with chemotherapy or radiation. Another concern is that chemotherapy or radiation will blunt a new immune response. However, we use pre-activated T cells; the additional priming by APC is thus no longer needed. An additional concern is that the tumor-bearing hosts we treat with adoptive T cells are T-cell-deficient. Since the pivotal and elegant experiments of Greenberg/Fefer/Cheever/Riddell (and Bob North's group) in the 1980s, the advantages and problems of lymphodepletion prior to adoptive T-cell therapy have become apparent. There has been major progress in the development of human T cells suitable for adoptive cellular immunotherapy, particularly when used in the lymphopenic patients (Dudley and Rosenberg, 2003; Gattinoni et al., 2006; Ho et al., 2003; Morgan et al., 2006). The OT-1 or OT-1 Rag1$^{-/-}$ mice we use may simulate such lymphodepleted patients, since the presence of the irrelevant OT-1 T cells does not prevent homeostasis-driven proliferation of transferred T cells expressing a different TCR (Spiotto and Schreiber, 2005; Spiotto et al., 2004). Thus, adoptive transfer of T cells into lymphodepleted hosts has its own therapeutic effect that can be exploited experimentally and clinically. Although our results still need to be confirmed in hosts bearing autochthonous tumors, there are good reasons to assume that the principles and concepts developed in our study will be applicable to established cancers in humans.

References

Antonia, S. J., Mirza, N., Fricke, I., Chiappori, A., Thompson, P., Williams, N., Bepler, G., Simon, G., Janssen, W., Lee, J. H., et al. (2006). Combination of p53 cancer vaccine with chemotherapy in patients with extensive stage small cell lung cancer. *Clin Cancer Res* 12: 878–887.

Antony, P. A., Piccirillo, C. A., Akpinarli, A., Finkelstein, S. E., Speiss, P. J., Surman, D. R., Palmer, D. C., Chan, C. C., Klebanoff, C. A., Overwijk, W. W., et al. (2005). CD8+ T cell immunity against a tumor/self-antigen is augmented by CD4+ T helper cells and hindered by naturally occurring T regulatory cells. *J Immunol* 174:2591–2601.

Arlen, P. M., Gulley, J. L., Parker, C., Skarupa, L., Pazdur, M., Panicali, D., Beetham, P., Tsang, K. Y., Grosenbach, D. W., Feldman, J., et al. (2006). A randomized phase II study of concurrent docetaxel plus vaccine versus vaccine alone in metastatic androgen-independent prostate cancer. *Clin Cancer Res* 12:1260–1269.

Brent, L., and Medawar, P. (1966). Quantitative studies on tissue transplantation immunity. 8. The effects of irradiation. *Proc Royal Soc Lond Ser B Biol Sci* 165:413–423.

Bubenik, J. (2003). Tumour MHC class I downregulation and immunotherapy (review). *Oncol Rep* 10:2005–2008.

Casares, N., Pequignot, M. O., Tesniere, A., Ghiringhelli, F., Roux, S., Chaput, N., Schmitt, E., Hamai, A., Hervas-Stubbs, S., Obeid, M., et al. (2005). Caspase-dependent immunogenicity of doxorubicin-induced tumor cell death. *J Exp Med* 202:1691–1701.

Ciernik, I. F., Romero, P., Berzofsky, J. A., and Carbone, D. P. (1999). Ionizing radiation enhances immunogenicity of cells expressing a tumor-specific T-cell epitope. *Int J Radiat Oncol Biol Phys* 45:735–741.

Demaria, S., Bhardwaj, N., McBride, W. H., and Formenti, S. C. (2005). Combining radiotherapy and immunotherapy: a revived partnership. *Int J Radiat Oncol Biol Phys* 63:655–666.

Dudley, M. E., and Rosenberg, S. A. (2003). Adoptive-cell-transfer therapy for the treatment of patients with cancer. *Nat Rev Cancer* 3:666–675.

Dvorak, H. F. (1986). Tumors: wounds that do not heal. Similarities between tumor stroma generation and wound healing. *N Engl J Med* 315:1650–1659.

Ferrara, N., Mass, R. D., Campa, C., and Kim, R. (2007). Targeting VEGF-A to treat cancer and age-related macular degeneration. *Annu Rev Med* 58:491–504.

Gabrilovich, D. I. (2007). Combination of chemotherapy and immunotherapy for cancer: a paradigm revisited. *Lancet Oncol* 8:2–3.

Ganss, R., Ryschich, E., Klar, E., Arnold, B., and Hammerling, G. J. (2002). Combination of T-cell therapy and trigger of inflammation induces remodeling of the vasculature and tumor eradication. Cancer Res 62:1462–1770.

Garin-Chesa, P., Old, L. J., and Rettig, W. J. (1990). Cell surface glycoprotein of reactive stromal fibroblasts as a potential antibody target in human epithelial cancers. *Proc Natl Acad Sci USA* 87:7235–7239.

Gattinoni, L., Finkelstein, S. E., Klebanoff, C. A., Antony, P. A., Palmer, D. C., Spiess, P. J., Hwang, L. N., Yu, Z., Wrzesinski, C., Heimann, D. M., et al. (2005a). Removal of homeostatic cytokine sinks by lymphodepletion enhances the efficacy of adoptively transferred tumor-specific CD8+ T cells. *J Exp Med* 202:907–912.

Gattinoni, L., Klebanoff, C. A., Palmer, D. C., Wrzesinski, C., Kerstann, K., Yu, Z., Finkelstein, S. E., Theoret, M. R., Rosenberg, S. A., and Restifo, N. P. (2005b). Acquisition of full effector function in vitro paradoxically impairs the in vivo antitumor efficacy of adoptively transferred CD8+ T cells. *J Clin Invest* 115:1616–1626.

Gattinoni, L., Powell, D. J., Jr, Rosenberg, S. A., Restifo, N. P. (2006). Adoptive immunotherapy for cancer: building on success. *Nat Rev Immunol* 6:383–393.

Gribben, J. G., Ryan, D. P., Boyajian, R., Urban, R. G., Hedley, M. L., Beach, K., Nealon, P., Matulonis, U., Campos, S., Gilligan, T. D., et al. (2005). Unexpected association between induction of immunity to the universal tumor antigen CYP1B1 and response to next therapy. *Clin Cancer Res* 11:4430–4436.

Heath, W. R., and Carbone, F. R. (2005). Coupling and cross-presentation. *Nature* 434:27–28.

Hicklin, D. J., Marincola, F. M., and Ferrone, S. (1999). HLA class I antigen downregulation in human cancers: T-cell immunotherapy revives an old story. *Mol Med Today* 5:178–186.

Ho, W. Y., Blattman, J. N., Dossett, M. L., Yee, C., and Greenberg, P. D. (2003). Adoptive immunotherapy: engineering T cell responses as biologic weapons for tumor mass destruction. *Cancer Cell* 3:431–437.

Holler, P. D., Chlewicki, L. K., and Kranz, D. M. (2003). TCRs with high affinity for foreign pMHC show self-reactivity. *Nat Immunol* 4:55–62.

Holler, P. D., and Kranz, D. M. (2003). Quantitative analysis of the contribution of TCR/pepMHC affinity and CD8 to T cell activation. *Immunity* 18:255–264.

Ibe, S., Qin, Z., Schuler, T., Preiss, S., and Blankenstein, T. (2001). Tumor rejection by disturbing tumor stroma cell interactions. *J Exp Med* 194:1549–1559.

Kaminski, M. S., Zasadny, K. R., Francis, I. R., Milik, A. W., Ross, C. W., Moon, S. D., Crawford, S. M., Burgess, J. M., Petry, N. A., Butchko, G. M., et al. (1993). Radioimmunotherapy of B-cell lymphoma with [^{131}I]anti-B1 (anti-CD20) antibody. *N Engl J Med* 329:459–465.

Kelly, T. (2005). Fibroblast activation protein-alpha and dipeptidyl peptidase IV (CD26): cell-surface proteases that activate cell signaling and are potential targets for cancer therapy. *Drug Resist Updat* 8:51–58.

Khong, H. T., Wang, Q. J., and Rosenberg, S. A. (2004). Identification of multiple antigens recognized by tumor-infiltrating lymphocytes from a single patient: tumor escape by antigen loss and loss of MHC expression. *J Immunother* 27:184–190.

Kim, K. J., Li, B., Winer, J., Armanini, M., Gillett, N., Phillips, H. S., and Ferrara, N. (1993). Inhibition of vascular endothelial growth factor-induced angiogenesis suppresses tumour growth in vivo. *Nature* 362:841–844.

Klebanoff, C. A., Gattinoni, L., Torabi-Parizi, P., Kerstann, K., Cardones, A. R., Finkelstein, S. E., Palmer, D. C., Antony, P. A., Hwang, S. T., Rosenberg, S. A., et al. (2005). Central memory self/tumor-reactive CD8+ T cells confer superior antitumor immunity compared with effector memory T cells. *Proc Natl Acad Sci USA* 102:9571–9576.

Klein, E., and Klein, G. (1972). Specificity of homograft rejection in vivo, assessed by inoculation of artificially mixed compatible and incompatible tumor cells. *Cell Immunol* 5:201–208.

Lake, R. A., and Robinson, B. W. (2005). Immunotherapy and chemotherapy—a practical partnership. *Nat Rev Cancer* 5:397–405.

Liu, K., Caldwell, S. A., and Abrams, S. I. (2005). Immune selection and emergence of aggressive tumor variants as negative consequences of Fas-mediated cytotoxicity and altered IFN-gamma-regulated gene expression. *Cancer Res* 65:4376–4388.

Lugade, A. A., Moran, J. P., Gerber, S. A., Rose, R. C., Frelinger, J. G., and Lord, E. M. (2005). Local radiation therapy of B16 melanoma tumors increases the generation of tumor antigen-specific effector cells that traffic to the tumor. *J Immunol* 174:7516–7523.

Lynch, T. J., Bell, D. W., Sordella, R., Gurubhagavatula, S., Okimoto, R. A., Brannigan, B. W., Harris, P. L., Haserlat, S. M., Supko, J. G., Haluska, F. G., et al. (2004). Activating mutations in the epidermal growth factor receptor underlying responsiveness of non-small-cell lung cancer to gefitinib. *N Engl J Med* 350:2129–2139.

Morgan, R. A., Dudley, M. E., Wunderlich, J. R., Hughes, M. S., Yang, J. C., Sherry, R. M., Royal, R. E., Topalian, S. L., Kammula, U. S., Restifo, N. P., et al. (2006). Cancer regression in patients after transfer of genetically engineered lymphocytes. *Science* 314:126–129.

Nair, S., Boczkowski, D., Moeller, B., Dewhirst, M., Vieweg, J., and Gilboa, E. (2003). Synergy between tumor immunotherapy and antiangiogenic therapy. *Blood* 102:964–971.

Neijssen, J., Herberts, C., Drijfhout, J. W., Reits, E., Janssen, L., and Neefjes, J. (2005). Cross-presentation by intercellular peptide transfer through gap junctions. *Nature* 434:83–88.

Niederman, T. M., Ghogawala, Z., Carter, B. S., Tompkins, H. S., Russell, M. M., and Mulligan, R. C. (2002). Antitumor activity of cytotoxic T lymphocytes engineered to target vascular endothelial growth factor receptors. *Proc Natl Acad Sci USA* 99:7009–7014.

Nowak, A. K., Lake, R. A., Marzo, A. L., Scott, B., Heath, W. R., Collins, E. J., Frelinger, J. A., and Robinson, B. W. (2003). Induction of tumor cell apoptosis in vivo increases tumor antigen cross-presentation, cross-priming rather than cross-tolerizing host tumor-specific CD8 T cells. *J Immunol* 170:4905–4913.

Prehn, R. T. (1973). Destruction of tumor as an "innocent bystander" in an immune response specifically directed against nontumor antigens. *Isr J Med Sci* 9:375–379.

Reits, E. A., Hodge, J. W., Herberts, C. A., Groothuis, T. A., Chakraborty, M., Wansley, E. K., Camphausen, K., Luiten, R. M., de Ru, A. H., Neijssen, J., et al. (2006). Radiation modulates the peptide repertoire, enhances MHC class I expression, and induces successful antitumor immunotherapy. *J Exp Med* 203:1259–1271.

Sanchez-Perez, L., Kottke, T., Diaz, R. M., Ahmed, A., Thompson, J., Chong, H., Melcher, A., Holmen, S., Daniels, G., and Vile, R. G. (2005). Potent selection of antigen loss variants of B16 melanoma following inflammatory killing of melanocytes in vivo. *Cancer Res* 65:2009–2017.

Schreiber, H. (ed.) (2003). *Tumor Immunology*, 5th edn. Philadelphia, PA: Lippincott-Williams & Wilkins, pp. 1557–1592.

Scott, A. M., Wiseman, G., Welt, S., Adjei, A., Lee, F. T., Hopkins, W., Divgi, C. R., Hanson, L. H., Mitchell, P., Gansen, D. N., et al. (2003). A phase I dose-escalation study of sibrotuzumab in patients with advanced or metastatic fibroblast activation protein-positive cancer. *Clin Cancer Res* 9:1639–1647.

Seemayer, T. A., Lagace, R., Schurch, W., and Tremblay, G. (1979). Myofibroblasts in the stroma of invasive and metastatic carcinoma: a possible host response to neoplasia. *Am J Surg Pathol* 3:525–533.

Slamon, D. J., Leyland-Jones, B., Shak, S., Fuchs, H., Paton, V., Bajamonde, A., Fleming, T., Eiermann, W., Wolter, J., Pegram, M., et al. (2001). Use of chemotherapy plus a monoclonal antibody against HER2 for metastatic breast cancer that overexpresses HER2. *N Engl J Med* 344:783–792.

Spiotto, M. T., Rowley, D. A., and Schreiber, H. (2004). Bystander elimination of antigen loss variants in established tumors. *Nat Med* 10:294–298.

Spiotto, M. T., and Schreiber, H. (2005). Rapid destruction of the tumor microenvironment by CTLs recognizing cancer-specific antigens cross-presented by stromal cells. *Cancer Immun* 5:8.

Sweeney, C. J., Miller, K. D., and Sledge, G. W., Jr (2003). Resistance in the anti-angiogenic era: nay-saying or a word of caution? *Trends Mol Med* 9:24–29.

Weissman, I. L. (1973). Tumor immunity in vivo: evidence that immune destruction of tumor leaves "bystander" cells intact. *J Natl Cancer Inst* 51:443–448.

Wesley, U. V., Albino, A. P., Tiwari, S., and Houghton, A. N. (1999). A role for dipeptidyl peptidase IV in suppressing the malignant phenotype of melanocytic cells. *J Exp Med* 190:311–322.

Wheeler, C. J., Das, A., Liu, G., Yu, J. S., and Black, K. L. (2004). Clinical responsiveness of glioblastoma multiforme to chemotherapy after vaccination. *Clin Cancer Res* 10:5316–5326.

Yamshchikov, G. V., Mullins, D. W., Chang, C. C., Ogino, T., Thompson, L., Presley, J., Galavotti, H., Aquila, W., Deacon, D., Ross, W., et al. (2005). Sequential immune escape and shifting of T cell responses in a long-term survivor of melanoma. *J Immunol* 174:6863–6871.

Yang, J. C., Haworth, L., Sherry, R. M., Hwu, P., Schwartzentruber, D. J., Topalian, S. L., Steinberg, S. M., Chen, H. X., and Rosenberg, S. A. (2003). A randomized trial of bevacizumab, an anti-vascular endothelial growth factor antibody, for metastatic renal cancer. *N Engl J Med* 349:427–434.

Yu, Z., Theoret, M. R., Touloukian, C. E., Surman, D. R., Garman, S. C., Feigenbaum, L., Baxter, T. K., Baker, B. M., and Restifo, N. P. (2004). Poor immunogenicity of a self/tumor antigen derives from peptide-MHC-I instability and is independent of tolerance. *J Clin Invest* 114:551–559.

Zhang, B., Bowerman, N. A., Salama, J. K., Schmidt, H., Spiotto, M. T., Schietinger, A., Yu, P., Fu, Y. X., Weichselbaum, R. R., Rowley, D. A., et al. (2007). Induced sensitization of tumor stroma leads to eradication of established cancer by T cells. *J Exp Med* 204:49–55.

Zhou, G., Lu, Z., McCadden, J. D., Levitsky, H. I., and Marson, A. L. (2004). Reciprocal changes in tumor antigenicity and antigen-specific T cell function during tumor progression. *J Exp Med* 200:1581–1592.

Subject Index

Acute myeloid leukemia, 54
AICD, in systemic T cells in cancer, gangliosides tumor secretion and, 75–78
AIRE, transcription factor, 10–11
All-*trans* retinoic acid, 159, 180
Alpha7 nicotinic acetylcholine receptor (alpha7nAChR), 209
AML, *see* Acute myeloid leukemia
Androgens, 18–20
Anti-CD25 antibody (7D4), 56–57
Antigen-presenting cells, 10–12, 15, 17, 31, 48, 89, 137, 139, 159, 197
    MDSC differentiation in, 180–181
    signaling pathways in, T-cell activation *vs.* tolerance and
    SOCS, 210–212
    STAT3, 205–210
    tyrosine kinase receptors, 201–205
    and tolerance to tumor antigens, 198–200
Antigen recognized by Treg cells 1, 55
Anti-HA CD4$^+$ T cells, 51
Antitumor T cells, cancer-induced signaling defects in, 69
    AICD in systemic T cells, enhanced sensitivity to, 75–78
    cAMP-dependent modulation of proximal TCR signal transduction, Csk activity and, 86–90
    defective proliferation, 74–75
    function, systemic defects in, 71–74
    p56^{lck} activity regulation, 83–86
    signal transduction, in cytolytic T cells, 80–82
    soluble reactive oxygen and nitrogen metabolites and, 78
    TIL signal transduction inhibition by tumor, 78–80
APCs, *see* Antigen-presenting cells

Arginase (ARG), 160
    ARG I/II, 221
    arginine and cancer, 222–224
    dependent suppression, 174
    MDSC and production of, 224–226
L-Arginine
    arginase and cancer, 222–224
    metabolism, 174–175
    metabolism, in M1 and M2 macrophages, 142–143
    T-cell function and, 221–222
β-Arrestin, 86
ARTC1, *see* Antigen recognized by Treg cells 1
L-Ascorbic acid, 262
ATRA, *see* All-*trans* retinoic acid

B and T-lymphocyte attenuator, 30
B-cell lymphomas, 9, 12
Bcl-2 protein expression, 241
B16F10, melanoma cell line, 44
B7 homolog 1 (B7-H1)
    mediated suppression
        CTL lysis, resistance to, 35
        T-cell anergy and exhaustion, generation and maintenance, 35–37
        T-cell deletion/apoptosis, 33–35
    upregulation in cancer, chronic viral infections and, 32–33
B7-H1-PD-1 pathway, 32, 37
B7 ligand family, 31
B16 melanoma model, 57
5-Bromo-2′-deoxyuridine (BrdU), 51
BTLA, *see* B and T-lymphocyte attenuator

Cancer
    arginine, arginase and, 222–224
    B7-H1 upregulation in, chronic viral infections and, 32–33

Cancer (cont.)
  dysfunctional immune response in, 219–221
  immune-suppressive mechanisms and, 1
  characteristics of, 3
  chemotherapy and, 4
  specific vs. non-specific, 2–3
  strategies for, 3–4
  immunobiology of dendritic cells in, see Dendritic cells
  induced signaling defects, in antitumor T cells, 69
    AICD in systemic T cells, enhanced sensitivity to, 75–78
    cAMP-dependent modulation of proximal TCR signal transduction, Csk activity and, 86–90
    defective proliferation, 74–75
    function, systemic defects in, 71–74
    $p56^{lck}$ activity regulation, 83–86
    signal transduction, in cytolytic T cells, 80–82
    soluble reactive oxygen and nitrogen metabolites and, 78
    TIL signal transduction inhibition by tumor, 78–80
  MDSC in, see Myeloid-derived suppressor cells
  ROC in T-cell defects in, see Reactive oxygen species
  Treg in, 41
    expansion in tumor bearers, mechanism, 50–54
    functional inactivation, 58–59
    in human cancer, 45–46
    in murine tumors, 43–45
    neutralization, by depletion, 56–58
    suppression in anti-tumor immunity, targets of, 46–50
    tumor-associated, antigen specificity of, 54–55
Carbohydrate-recognition domain, 235, 244
Catalase, 262
Cationic amino acid transport system (CAT), 224–225
Cbp, see Csk binding protein
CCAAT/enhancer-binding protein β, 224
CCL22, macrophage-derived chemokine, 51, 140
CCL17, thymus- and activation-regulated chemokine, 140
CCR4 receptor, 51
CD11b$^+$/Gr1$^+$ cells, 158–160, 175

Cdc42 and Rac1 genes, 114
CD4$^+$CD25$^-$ T cells, 53–54
CD4$^+$ CD25$^+$ Treg homeostasis, regulation, 176–178
CD4 cells, 9, 11–13, 19
CD11c^{int} myeloid DC, 118
CD8+ cytolytic T-cell, 33–34, 47, 49, 106, 157, 282, 284
  signal transduction in, 80–82
CD23, Fc receptor for IgE, 139
CD163, hemoglobin scavenger receptor, 139
CD22, inhibitory signaling receptor, 84
CD28-like receptors, 30–31
CD4$^+$ T-cells, 41, 49, 56, 133, 141, 198
CD8$^+$ T cells, 9–10, 15, 79, 108, 133, 158, 173, 179
"Cellular (phagocytic) theory of immunity," 131
CGD, see Chronic granulomatous disease
Chemotherapy, 4
CHO cells, 236
CHOP gene, 224
Chronic granulomatous disease, 263
Chronic myelogenous leukemia, 203
c-kit and imatinib mesylate, 203–205
"Classically activated" (M1) and "alternatively activated" (M2) macrophages, 138–141
  arginine metabolism in, 142–143
  and M1 and M2 paradigm, 141–142
  M1/M2 nomenclature, 144–145
  transcriptional signatures of, 143–144
CML, see Chronic myelogenous leukemia
Colony stimulating factor-1, 135, 163
Complete Freund Adjuvant (CFA), 199
Core-2-β-1,6-N-acetylglucosaminyltransferase (C2GnT), 240
COS-7 L cells, 224
COX-2, see Cyclooxygenase-2
CRD, see Carbohydrate-recognition domain
CSF-1, see Colony stimulating factor-1
Csk activity, cAMP-dependent modulation of TCRs signal transduction and, 86–90
Csk binding protein, 87
CTL, see CD8+ cytolytic T-cell
CTLA-4, see Cytotoxic T-lymphocyte antigen-4
CXCL14 protein, 108–109
Cyclooxygenase-2, 54, 111, 134, 169, 181
Cyclophosphamide (Cytoxan), 17

Cytokines, 53, 139
  galectin-3 and, 247
  proinflammatory and anti-inflammatory, galectin-1 and, 239
Cytotoxic T-lymphocyte antigen-4, 30, 42, 44, 48, 50, 58, 119

DC, see Dendritic cells
Dectin-1 (β-glucan receptor), 140
Delayed-type hypersensitivity, 220
Dendritic cells, 12, 32, 47, 51, 53, 76, 158–159, 199
  immunobiology, in cancer, 101–102
    abnormalities of, 104–105
    alterations, 103–109
    dysfunction mechanisms, 109–114
    tumor escape mechanisms, 114–120
Dinileukin diftitox, CD25-depleting agent, 46, 57
DTH, see Delayed-type hypersensitivity

EL-4 cells, 79
Electromobility shift assay (EMSA), 206
Env-specific T cells, 8
Epidermal growth factor (EGF), 136
EP2 receptor, 111
Epstein–Barr virus proteins, 69
Extracellular regulated kinase (ERK), 114

Flavin adenine dinucleotide (FAD), 259, 263
Foxp3 expression, 11, 42, 51, 53, 57

β-Galactosides, 235
Galectin-1, 76–77
  adaptive immune responses regulation and, 237–241
  biochemistry and cell biology, 236
  cell adhesion, migration and angiogenesis and, 243
  monocytes, macrophages, dendritic cells and neutrophils, 242–243
  tumor-immune privilege, establishment and maintenance, 241–242
  in tumor microenvironment, 237
Galectin-3
  biochemistry and cellular biology, 244–245
  expression, in tumors, 245
  immune cell activation modulation and, 247–248
  T-cell responses regulation and, 245–247
  in tumor-immune escape, 249–250
Ganglioside receptor, 76
GAS6, see Growth arrest-specific protein 6
Gastrointestinal sarcoma tumors, 203

GCN2 kinase, 224
GIST, see Gastrointestinal sarcoma tumors
Glucocorticoid-induced tumor necrosis factor receptor (GITR), 42, 45, 51, 58
Glutathione (GSH), 262
Glutathione-S-transferase, 262
Glycan-binding proteins, see Galectins
GM-CSF, see Granulocyte macrophage-colony stimulating factor
gp91phox, 263
gp100 tumor antigen, 33
Graft-versus-host disease, 32, 176
Granulocyte macrophage-colony stimulating factor, 45, 59, 117, 159, 165–166, 178, 225, 269
Gr1$^+$ cells, 158, 161, 173
GRIFIN (galectin-related inter fiber protein), 235
Growth arrest-specific protein 6, 202
GST, see Glutathione-S-transferase
GVHD, see Graft-versus-host disease

hCG, see Human chorionic gonadotropin
Head and neck squamous cell carcinoma, 106–108, 242
Helicobacter pylori, 36
Hepatitis B/C virus (HBV/HCV), 33, 36
Hepatocellular carcinoma, 106
Her-2-expressing tumor, 50
Herpes virus entry mediator, 30
Her-2 receptor, 55
HLA-DR expression, 118, 178
HLA-DR$^+$ immature cells (DR+IC), 107
HNSCC, see Head and neck squamous cell carcinoma
Hodgkin's disease (HD), 46, 220
Human cancer, Treg in, 45–46
Human chorionic gonadotropin, 112
Human immunodeficiency virus (HIV), 33, 36
Human leukocyte antigen G (HLA-G), 112
Human MDSCs, 178–179
"Humoral" theory, of immunity, 131
HVEM, see Herpes virus entry mediator
Hydrogen peroxide (H_2O_2), 260
3-Hydroxyanthranilic acid, 115
Hydroxyl radical (OH·), 260
Hypochlorous acid (HOCl), 260–261
Hypoxiainducible factor (HIF) 1α and 2α, 134

IDO, see Indoleamine-2,3-dioxygenase
IFN-γ, see Interferon-γ

IgG-TGF-β complex, 173
IL-2, see Interleukin-2
IL-6, 164
IL-10, 166, 202, 208
IL-13, 167
IL-10+CCR7+CD45RO+CD8+ regulatory T cells, 115
IL-13/IL4Rα/STAT6 pathway, 183–184
IL-2 plus anti-CD3ε-activated killer cells, 71
ILT4, see Immunoglobulin-like transcript 4
Immature myeloid dendritic cells (IMDC), 51, 53
Immune suppression mechanisms, MDSCs and, 172
   indirect, CD4$^+$ CD25$^+$ Treg homeostasis regulation, 176–178
   L-arginine metabolism, 174–175
   reactive oxygen species (ROC), 175–176
   TGF-β, 173
Immunoglobulin (Ig) superfamily, 30
Immunoglobulin-like transcript 4, 112
Immunoreceptor tyrosine-based activation motif, 30, 84
Immunoreceptor tyrosine-based inhibition motif, 30, 32, 84–85
Immunoreceptor tyrosinebased switch motif, 30, 32
Indoleamine dioxygenase, 2
Indoleamine-2,3-dioxygenase, 115, 163
   activation, 47–48
   macrophages and suppression of tumor immunity and, 146–147
Inducible costimulator (ICOS), 30
Inducible costimulator ligand (ICOS-L), 119–120
Inducible nitric oxide synthase, 134, 140, 142
"Infectious tolerance," 52
iNOS, see Inducible nitric oxide synthase
αvβ3 Integrin, 167
Interferon-γ, 139, 141, 167–169, 267
Interleukin-2, 41, 47, 51, 57, 107
Interleukin-3 (IL-3), 225
ITAM, see Immunoreceptor tyrosine-based activation motif
ITIM, see Immunoreceptor tyrosine-based inhibition motif
ITSM, see Immunoreceptor tyrosinebased switch motif

Janus activated kinase (JAK) family, 114, 170, 205, 210
Janus kinase 2 (JAK2), 145, 171, 210
Jurkat cells, 76–77, 246

Kinase inhibitory region (KIR), 210
Kynurenine pathway, 115

LAGE1, 55
Langerhans cells, 105
Latency-associated protein (LAP), 47
LCMV, see Lymphocytic choriomeningitis virus
Leishmania, 141
Lewis lung carcinoma, 166, 181
Lipopolysaccharide (LPS), 138–139
Listeria monocytogenes, 79
LMB-2, fusion protein, 57
Low molecular weight proteins (LMP)-2 and LMP-7, 167–168
Lung squamous cell carcinoma, 111
Ly49 molecules, 83
Lymphocytic choriomeningitis virus, 33, 36

Macrophage colony stimulating factor, 163, 200, 201
Macrophage inhibitory factor, 136
Macrophages, tumor development and, see Tumor-associated macrophages
MAPK, see Mitogen-activated protein kinase
Matrix metalloprotease-9, 136, 164
MCA38 tumor cells, 72, 74, 79–80, 220
MC57-gp33-Lo cancer cells, 285
M-CSF, see Macrophage colony stimulating factor
MC57-SIY-Lo tumors, 285
M15, CTL clone, 33
MDSC, see Myeloid-derived suppressor cells
624mel, human melanoma cell line, 33
MHC– peptide complex and TCRs, interaction, 29–30
MIF, see Macrophage inhibitory factor
Mitogen-activated protein kinase, 113–114, 266
Mixed leukocyte reaction (MLR), 166
M1/M2 nomenclature, 144–145
MMP-9, see Matrix metalloprotease-9
mTECs, see Thymic medullary epithelial cells
MUC1 glycoprotein, 112
Murine tumors, Treg in, 43–45
Myeloid-derived suppressor cells, 48, 54, 78, 85, 142, 219, 267
   arginase production and, 224–226
   biology and function, 158–160
   dependent immune suppression, mechanism, 172–178
   differentiation in fully mature APCs, 180–182

expansion, recruitment and activation regulation, tumor-derived factors and, 161–169
functions, transcription factors regulating, 170–172
history and nomenclature, 157–158
human, 178–179
IL-13/IL4Rα/STAT6 pathway, 183–184
and ROS, 268–269
suppressive pathways, pharmacological inhibition, 182–183
TAMS and, 147–149
and tumor progression, 160–161
in vivo depletion of, 179–180

β 1,6-N–acetylglucosaminyltransferase V, 247
Naïve T cells, 52–56, 70, 101, 241
Natural killer (NK) cells, 47–48, 133, 139, 157
Natural suppressor (NS) cells, 157–158
NF-κB, transcription factor, 138, 207, 265
Nicotinamide adenine dinucleotide phosphate (NADPH) oxidase
biological roles, 264–266
components of, 263–264
Nitric oxide synthase (NOS2), 160, 173, 221, 223
and T-Cell Function, 226–227
Non-CD8$^+$ T cells, 72
Non-Hodgkin lymphoma, 46
Nonobese diabetic (NOD) mice, 36
Norepinephrine, 103
Nor-NOHA, arginase inhibitor, 225
North's suppressor T cells, 44
NOS-dependent suppression, 174–175
NOX2, see gp91phox

Ontak, see Dinileukin diftitox
OT-1 cells, 36
Ovalbumin (OVA), 176
OX40, 59

Pathogen-associated molecular patterns (PAMPs), 12
PBL T cells, 76
PC61 antibody, 50
PD-1, B7 receptor family member, 32, 36
Peripheral blood mononuclear cells (PBMC), 248
Peroxynitrite (ONOO$^-$), 261, 270–272
PGEs, see Prostaglandins
Phosphodiesterases (PDE), 88–89

Phosphoinositide 3-kinase (PI3K), 114
inhibitors, 202
PI3K/AKT pathway, 202
Plasmacytoid DC (pDC), 115, 117–119
p56^{lck} activity, 79, 81, 82, 87
regulation, by inhibitory phosphatases, 83–86
p38 MAPK, 114
Pmel-17/gp100, 8
p47phox, 263–264
Program death 1 (PD-1), 30
Programmed death-ligand 2 (B7-DC; CD273), 109, 166
Prostaglandins, 111–112, 169
Prostate-specific antigen (PSA), 112
Protein tyrosine phosphatases (PTPs), 266
P815 tumor model, 33–34

Rac proteins, 264
RCC, see Renal cell carcinoma
Reactive nitrogen intermediates, 175
Reactive oxygen species, 175–176
T-cell defects in cancer and, 259
cellular defenses, 262
myeloid cell-mediated immune suppression in cancer and, 266–274
NADPH, as source in myeloid cells, 263–264
NADPH oxidase, biological roles, 265–266
types and biological function, 260–261
Regulatory T cells, in cancer, 41, 115
expansion in tumor bearers, mechanism, 50–54
functional inactivation, 58–59
in human cancer, 45–46
in murine tumors, 43–45
neutralization, by depletion, 56–58
suppression in anti-tumor immunity, targets of, 46–50
tumor-associated, antigen specificity of, 54–55
RELB1 proteins, 207
Renal cell carcinoma, 33, 226
RNI, see Reactive nitrogen intermediates
ROC, see Reactive oxygen species

Shp-1 binding domain (SH2), 83, 85, 87
Shp-1, nonmembrane protein tyrosine phosphatase, 83–86

Signal transducer and activator of transcription (STAT) family, 113
STAT1, 170–171, 208
STAT3, 171–172, 205–210
  activation, 112–113
  inhibition, 181–182
STAT6, 145–146, 172
Singlet oxygen, 260
Smad signaling pathway, 113
S100 proteins, 264
Src homology 2-containing inositol-5' phosphatase (SHIP), 146
STAT1, 170–171, 208
STAT3, 171–172, 205–210
  activation, 112–113
  inhibition, 181–182
STAT6, 145–146, 172
ST6Gal-I sialyltransferase, 240
Superoxide dismutases (SOD), 260, 262
Suppressors of cytokine signaling 1/3 (SOCS1/3), 210–213
Systemic lupus erythematosus (SLE), 211–212

TAMs, see Tumor-associated macrophages
T-cell
  adhesion
    and extravasation, galectin-1 in, 239–240
    and migration, galectin-3 and, 247
  anergy and exhaustion, generation and maintenance, 35–37
  co-signaling pathways, 29–31
  deletion/apoptosis, 33–35
  function
    arginine and, 221–222
    nitric oxide synthase (NOS2) and, 226–227
  tolerance mechanisms, tumor-associated, 7
    Env-specific T cells, 8
    hormones and tumor immunity and, 18–21
    peripheral self-antigen- and tumor-associated antigen-induced, 10–18
    tumor-HA vs self-HA, tolerance, 12
  See also Specific types
T-cell receptors (TCRs), 9, 13, 54–55, 73, 226
  MHC– peptide complex, interaction, 29–30, 81
  signaling complex, 72, 77, 80

signal transduction
  cAMP-dependent modulation of, Csk activity and, 86–90
  galectin-1 and, 237
TCRαβ, antigen receptor, 80
TDFs, see Tumor-derived factors
TEMs, see Tie2-expressing monocytes
Th17 cells, 53
T-helper lymphocytes, 49
Thymic medullary epithelial cells, 10–11
Tie2-expressing monocytes, 161
TIL, see Tumor-infiltrating lymphocytes
TKRs, see Tyrosine kinase receptors
TLR, see Toll-like receptors
TNF-α, 53, 221
Tocopherol, 262
Toll-like receptors, 59, 166, 198
Transforming growth factor (TGF)-β, 161, 173, 202, 221
  receptor II, 51
  signaling, 49
Transporters associated with antigen processing (TAP)-1 and TAP-2, 167
Treg, see Regulatory T cells
Tumor-associated macrophages, 171
  anti-tumor immunity inhibition and, 137
  cancer therapies and, 145–146
  hypoxia and, 133–134
  indolamine 2,3-dioxygenase (IDO) and, 146–147
  and MDSCs, 147–149
  NF-κB, transcription factor, 138
  RNA signatures of, 144
  tumor cell invasiveness and metastasis, 134–136
Tumor-derived factors, MDSCs expansion, recruitment and activation and, 161–162
  CSF-1, 163
  GM-CSF, 165–166
  IFN-γ, 167–169
  IL-6, 164
  IL-10, 166
  IL-13, 167
  PGEs, 169
  VEGF, 164–165
Tumor-infiltrating lymphocytes, 74–75, 85–86, 220, 271
  signal transduction inhibition, by tumor, 78–80
Tumor microenvironment, galectin-1 in, 237

Subject Index

Tumor necrosis factor (TNF) receptor/ligand superfamily, 30
Tumor stroma, antitumor immune response and, 281
 chemotherapy and radiation, 288–289
 eradication of established tumors and, 282–284
 induced sensitization of, 284–288
Tyro3 family, 201–203
Tyrosinase, 8

Tyrosine kinase receptors
 c-kit and imatinib mesylate, 203–205
 Tyro 3 family receptors, 201–203

Vascular endothelial growth factor (VEGF), 109, 111, 133–134, 161, 164–165, 171, 200, 225
Virus envelope protein (env), 8
v-*myc* and v-*raf* oncogenes, 166

ZAP70, 81–82

Printed in the United States
102801LV00002B/202-216/A